G.R. Newkome, C.N. Moorefield, F. Vögtle

Dendritic Molecules

Concepts · Syntheses · Perspectives

Further Titles on Interest

Lehn, J.-M.
Supramolecular Chemistry. Concepts and Perspectives
Hardcover: ISBN 3-527-29312-4
Softcover: ISBN 3-527-29311-6

Fuhrhop, J./Penzlin, G.
Organic Synthesis. Concepts, Methods, Starting Materials
Hardcover: ISBN 3-527-29086-9
Softcover: ISBN 3-527-29074-5

Nicolaou, K.C./Sorensen, E.J.
Classics in Total Synthesis
Hardcover: ISBN 3-527-29284-5
Softcover: ISBN 3-527-29231-4

Nógrádi, M.
Stereoselective Synthesis. A Practical Approach
Second, Thoroughly Revised and Updated Edition
Hardcover: ISBN 3-527-29242-X
Softcover: ISBN 3-527-29243-8

© VCH Verlagsgesellschaft mbH, D-69451 Weinheim (Federal Republic of Germany), 1996

Distribution

VCH, Postfach 10 11 61, D-69451 Weinheim (Federal Republic of Germany)

Switzerland: VCH, Postfach, CH-4020 Basel (Switzerland)

United Kingdom and Ireland: VCH (UK) Ltd., 8 Wellington Court, Cambridge CB1 1HZ (United Kingdom)

USA and Canada: VCH, 333 7th Avenue, New York, NY 10001 (USA)

Japan: VCH, Eikow Building, 10-9 Hongo 1-chome, Bunkyo-ku, Tokyo 113 (Japan)

ISBN 3-527-29325-6

G.R. Newkome, C.N. Moorefield, F. Vögtle

Dendritic Molecules

Concepts · Syntheses · Perspectives

Weinheim · New York · Basel · Cambridge · Tokyo

Prof. Dr. G.R. Newkome
Department of Chemistry
University of South Florida
4202 East Fowler Ave., ADM 200
Tampa, FL 33620-5950
USA

Prof. Dr. C. N. Moorefield
Department of Chemistry
University of South Florida
4202 East Fowler Ave., ADM 200
Tampa, FL 33620-5950
USA

Prof. Dr. F. Vögtle
Institut für Organische Chemie
und Biochemie der Universität
Gerhard-Domagk-Str. 1
53121 Bonn
Germany

Published jointly by
VCH Verlagsgesellschaft mbH, Weinheim (Federal Republic of Germany)
VCH Publishers, Inc., New York, NY (USA)

Editorial Director: Dr. Gudrun Walter
Production Manager: Dipl.-Ing. (FH) Bernd Riedel

Library of Congress Card No. applied for.

A catalogue record for this book is available from the British Library.

Die Deutsche Bibliothek – CIP-Einheitsaufnahme

Newkome, George R.:
Dendritic molecules : concepts, syntheses, perspectives / G. R.
Newkome ; C. N. Moorefield ; F. Vögtle. – Weinheim ; New
York ; Basel ; Cambridge ; Tokyo : VCH, 1996
   ISBN 3-527-29325-6
NE: Moorefield, Charles N.:: Vögtle, Fritz:

Composition: Mitterweger Werksatz GmbH, D-68723 Plankstadt
Printing: Strauss Offsetdruck GmbH, D-69509 Mörlenbach
Bookbinding: Osswald + Co. GmbH KG, D-67400 Neustadt/Weinstr.
Printed in the Federal Republic of Germany.

Prologue and Acknowledgments

A mutual interest in heterocyclic and macrocyclic chemistries was the thread that initially wove together the work of the Vögtle and Newkome research groups. Then in 1978, G. Newkome spent his first sabbatical leave with F. Vögtle at the University of Bonn in Germany. Now nearly 20 years later, those same interests and several more recent ones continue to bind our research groups.

The study of pyridinophanes resulted in joint publications on the complexation of water and the encapsulation of halocarbons within pyridino-crown hosts. Additional professional exchanges occurred over the years in which both groups were pursuing mutual synthetic interests in heterocyclic chemistry, stereochemistry, and cyclophanes. Central to those interests was a better understanding of molecular inclusion and recognition phenomena, now more uniquely defined as an aspect of supramolecular chemistry.

In the early 1970s, when F. Vögtle began to extend his work on podands, polyaminoligands, many-armed "octopus-type" molecules, and neutral ligands with numerous donor centers, as in macropolycyclic azacrowns, the Bonn group considered the possibility of "cascade" architecture and construction. This technology – first documented by F. Vögtle et al. in 1978 – ultimately led to the creation of the macromolecules later, in 1985, termed "dendrimers" or "arborols." Common interests associated with a better understanding of molecular encapsulation and macromolecular construction were expanded to include the investigation of "cascade" or "dendritic" materials.

In 1993 during a visit to Bonn, we decided to document the early years of this dynamic topic, and *"Dendritic Molecules: Concepts, Syntheses, Perspectives"* was conceived. The expansion of cascade construction into the macromolecular regime has in essence proven that there is no longer a molecular (weight or size) ceiling for chemists and others interested in materials science. We hope that this book will spark the reader's imagination to create new routes to dendritic or hyperbranched materials. In view of the increasing interest in macromolecular construction via dendrimers, there is little doubt in our minds that the surface of this topic has just been scratched.

We are grateful to the many coworkers and colleagues who have assisted us in compiling this book. First, we wish to thank Professor Wayne Mattice (the Alex Schulman Professor at the Maurice Morton Institute of Polymer Science at the University of Akron) for his first chapter, which melds the history and theory underlying dendritic polymer chemistry; it is also interesting to note that it was in 1978 that he did a sabbatical at Stanford with Professor Flory, who proposed structure-imperfect branched polymers in the 1940s.

The monumental task of newly styling and crafting all the illustrations, drawings, and figures was superbly handled in Bonn by Drs. Felat Bozkurt, Thomas Dünnwald, Stephan Ibach, Jörg Issberner, Ralf Jäger, Susanne Laufenberg, Stephan Ottens-Hildebrandt, Volker Prautzsch, Thomas Schmidt, Dirk Udelhofen, and in particular Stefan Ohren, who optimized the drawings, for without their efforts and expertise, completion of this book in a timely manner would not have been possible.

For support in acquiring the literature references and ensuring their completeness, helping with the graphs and figures in Chapter 2, constantly inquiring as to the status of "the book", and contributing substantially to nomenclature in Chapter 3, we gratefully acknowledge Dr. Gregory R. Baker (USF). Recognition goes to Mr. Peter Sutter (Dye Division, Ciba-Geigy Corp., Basel, Switzerland) and Dr. Claus Weis (USF, formerly with Ciba-Geigy in Basel) for the acquisition of patent information. Ms. Mubetcel Ayra and Professor Robert L. Potter (USF) are both acknowledged by C.N.M. for their encouragement, support, and enlightening theoretical discussions. Acknowledgment and gratitude also go to our wives, for their patience so neccessary for us to complete this book; to Rhonda Hendrix, who afforded quiet time to think at work; to Dr. Norbert Bikales at the National Science Foundation, who in those early years (1985) had faith in our initial thoughts about this budding area of macromolecular chemistry; to the financial backing from the Petroleum Research Fund, administered by the American Chemical Society, the Army Research Office, and NATO for supporting stimulating stays in Strasbourg; and lastly to our research groups at Bonn and USF, comprised of very supportive and productive students, postdoctoral associates, and visiting scholars.

January, 1995   George R. Newkome (Tampa, Florida, USA)

Charles N. Moorefield (Tampa, Florida, USA)

Fritz Vögtle (Bonn, Germany)

Contents

6   Synthetic Methodologies: One-Step (Hyperbranched) Procedures

7   Chiral Dendritic Macromolecules

8   Dendrimers Containing Metal Sites

1 Masses, Sizes, and Shapes of Macromolecules from Multifunctional Monomers*

1.1 Introduction

Macromolecules, as the name implies, have large molecular masses, from which it follows that they also have large sizes, or extensions in space. Furthermore, the presence of a large number of atoms within each macromolecule permits the possession of an enormous variety of conformations with different shapes and sizes. Each of these topics will be discussed here, first in the context of macromolecules in general, and then specifically for those cases where the macromolecule is highly branched.

The discussion of the distribution of molecular weights focuses on the polydispersity index, $\overline{M}_w/\overline{M}_n$, and the requirements for gelation (or the avoidance of gelation) when multifunctional monomers are incorporated into the macromolecule. Even moments of the radius of gyration are employed in the discussion of sizes. Shapes are evaluated using the principal moments of the radius of gyration tensor, with special emphasis on their use for the detection of asphericity. The collapse of chains to a dense globular state is described in the last portion of this chapter, along with the asphericity of model dendrimers.

1.2 Molecular Mass and Chain Connectivity in Macromolecules

As implied by the name, one of the fundamental distinctions between macromolecules and small molecules lies in their mass, or molecular weight, M. The lower limit for the molecular weight of a macromolecule is ill defined, but it is often considered to lie in the range $M = 10^3 - 10^4$. The upper limit for the molecular weight of a macromolecule is unbounded, and for all practical purposes it can be considered to be infinite. Of course, an infinite molecular weight is unattainable with a finite amount of material. However, molecular masses in covalently bound polymeric networks are conveniently expressed in units of kg molecule^{-1}, rather than units of kg mol^{-1}, and they are certainly an acceptable approximation to "infinite" in the context of this work. These polymeric networks contain numerous branch points, either as a consequence of the use of multifunctional monomers during the polymerization or as a consequence of crosslinking reactions after polymerization has occurred. In such systems, the molecular mass is of less fundamental interest than the mass of the network chains between crosslinks.

As a consequence of the strong tendency for chemical reactions to proceed in a random manner, the simplest and most common procedures used for the polymerization of ordinary monomers do not produce a set of molecules of exactly the same mass, but instead produce a population of macromolecules with a distribution of molecular weights. Therefore, descriptions of the molecular weights of macromolecules must seek a characterization of the distribution of the masses that are present. Usually the width of the distribution is characterized by the ratios of different average molecular weights.

1.2.1 Distribution with Linear Chains

The most probable distribution of molecular weights is expected in condensation polymerization of bifunctional monomers when the principle of the equal reactivity of all

---

* This introductory chapter was written by Professor Wayne L. Mattice, the Alex Schulman Professor at the Maurice Morton Institute of Polymer Science, The University of Akron, Akron, Ohio, USA.

functional groups is obeyed.[1] The reactants can be of the type A–B or they can be a mixture of A–A and B–B. The reactivity of each terminal functional group is assumed to be independent of the length of that part of the molecule denoted by "–". When only linear products are formed, the mole fraction and weight fraction, respectively, of the x-mer are given by

$$X_x = (1-p)^2 \, p^{x-1} \tag{1.1}$$

$$W_x = x(1-p)^2 \, p^{x-1} \tag{1.2}$$

where p denotes the fraction of the functional groups that have reacted. These distributions yield number- and weight-average degrees of polymerization of

$$\bar{x}_n = \frac{1}{1-p} \tag{1.3}$$

$$\bar{x}_w = \frac{1+p}{1-p} \tag{1.4}$$

The ratio of the weight- and number-average molecular weights is given by

$$\frac{\overline{M}_w}{\overline{M}_n} = 1 + p \tag{1.5}$$

which is indistinguishable from 2, if the average degree of polymerization is large (that is, if p is close to 1). The ratio $\overline{M}_w/\overline{M}_n$ is sometimes called the *polydispersity index,* and is used as a measure of the breadth of the distribution of molecular weights.

Other polymerization mechanisms can produce distributions of molecular weights that are different from the most probable distribution. Much narrower distributions are obtained if a specified number of chains is initiated at the same time, and these chains grow exclusively by the addition of monomer to the reactive end. Then

$$\frac{\overline{M}_w}{\overline{M}_n} = 1 + \frac{\xi}{(\xi + 1)^2} \tag{1.6}$$

where ξ denotes the ratio of the monomers that have reacted to the number of growing chains.[2] For large chains prepared by this mechanism, $\overline{M}_w/\overline{M}_n$ may become experimentally indistinguishable from unity.

A polymerization that provides a transition into a discussion of gelation is the condensation of an excess of A–B with a small amount of an *f*-functional monomer, $R–A_f$, that contains f equivalent functional groups of Type A, but no functional groups of Type B.[3] Linear chains are obtained when f is 1 or 2, but multichain condensation polymers are produced when $f > 2$. At high conversion the polydispersity index depends only on f.

$$\frac{\overline{M}_w}{\overline{M}_n} \cong 1 + \frac{1}{f} \tag{1.7}$$

The distribution becomes narrower as the functionality of the multifunctional monomer increases. Formation of an infinite network is prevented because all of the molecules derived from each $R–A_f$ contain only the functional group A at the reactive ends and it is assumed that A reacts only with B, and not with another A. A much different result would be obtained if the system contained B–B, or if the multifunctional monomer contained both A and B. This topic is discussed in Sect. 1.2.2.

1.2.2 Branched Structures and Gelation with Multifunctional Monomers

The formation of multichain condensation polymers using an excess of A–B with a small amount of the multifunctional monomer $R–A_f$, under conditions where no A can react with another A was described above in Sect. 1.2.1. Polyfunctional condensation polymerization using multifunctional monomers in other contexts can produce infinite networks.[4, 5] If the multifunctional $R-A_f$ is polymerized with a mixture of both A–A and B–B, the system is different from the simpler polymerization of $R-A_f$ with A–B in two important ways. At various stages in the polymerization of the molecules derived from $R-A_f$ can now have some arms that terminate with functional group B, and other arms that terminate with functional group A. When this combination of both functional groups is present simultaneously, intramolecular reaction, leading to the formation of rings of various sizes, becomes possible. This intramolecular cyclization occurs, but another process has a more dramatic consequence. The second process is the formation of macromolecules that contain more than one f-functional site derived from $R-A_f$, as the consequence of intermolecular reaction. Eventually, this second process can produce a single gigantic molecule with many branch points that extends throughout the macroscopic volume available to the polymerizing system; in essence the system forms a gel.

The point in the reaction at which gelation occurs has been deduced by Flory.[4] The gel point is developed in terms of the branching coefficient, α, which is the probability that a given functional group on the multifunctional monomers leads, via a chain that can contain any number of bifunctional units, to another multifunctional monomer. The critical value of the branching coefficient, denoted by α_c, at which gelation occurs is

$$\alpha_c = \frac{1}{f-1} \qquad (1.8)$$

which is just that value of α at which the expected number of chains that succeed any of the A in the multifunctional $R-A_f$, due to branching of one or more of them, exceeds one. The ratio $\overline{M}_w/\overline{M}_n$ diverges when the system gels, because some of the material will not be covalently incorporated into the network, but will instead be present as extractable monomer, oligomers, and polymers.

Gelation will also occur in the polycondensation of $R-A_f$ and A–A if the chemistry of the functional groups permits the formation of a covalent bond between one A and another A. An interesting variant of this process is the gelation that occurs in a thermoreversible manner when the groups denoted by A can aggregate into clusters by noncovalent interactions.[6, 7] This thermal gelation has been simulated in systems containing symmetric triblock copolymers, $A_{NA}-B_{NB}-A_{NA}$, in a medium that is a good solvent for block B, but a poor solvent for block A.[8] Aggregation of $x+f$ molecules of the triblock copolymer into a micelle, with x molecules placing both of their A blocks in the micellar core, and the remaining f molecules placing only one of their A blocks in the core, produces a micelle that has the characteristics of the f-functional monomer described by Flory, in the sense that there are f dangling chains with 'sticky' ends comprised of their A blocks. Simulations of this system have detected gelation, and the gel point can be determined from the simulation.[8] The critical gel point determined by simulation is in good agreement with the result predicted by the application of Flory's theory to this reversibly associating system.

1.2.3 Branching without Network Formation

Another type of behavior is obtained in the polymerization of a multifunctional monomer constructed so that it has one functional group of the type A, and $f-1$ functional groups of the type B, with the only type of reaction being between A and B. This multifunctional monomer can be written as $A-R-B_{f-1}$. The reaction produces highly branched structures, but without gelation.[9] For this system, the branching probability is

$$\alpha = \frac{p}{f-1} \qquad (1.9)$$

where p denotes here the fraction of the A groups that have reacted. Since p cannot be larger than 1, α cannot exceed the critical value for gelation, specified by Eq. 1.8.

If the polyfunctional condensation of $A-R-B_{f-1}$ proceeds under the conditions of equal reactivity, the polydispersity increases as the reaction proceeds (as α increases), and approaches infinity as $p \rightarrow 1$.[9] The uncontrolled polycondensation of $A-R-B_{f-1}$

does not produce a unique molecular species, but instead produces highly branched molecules with an extremely broad range of molecular weights.

A remarkable narrowing of the distribution becomes possible if $R-B_{f-1}$ initially reacts with $f-1$ molecules of $A-R-C_{f-1}$ under conditions where the only reaction is of A with B, C is subsequently converted to B, and then the cycle is repeated again and again. If all reactions are quantitative, the branched molecules obtained after each cycle all have the same covalent structure, and $\overline{M}_w/\overline{M}_n = 1$. The broadening of the molecular weight distribution that is expected if the reaction is less than quantitative during some, or all cycles has been considered recently by Mansfield.[10] The influence of a less than quantitative reaction on the polydispersity is least if it occurs in the final cycle, and is greatest if it occurs in the first cycle. Therefore, molecules with $\overline{M}_w/\overline{M}_n$ very close to unity must be kept perfect for as many cycles as possible for each generation. With any real system, a point must be reached at which future generations cannot remain perfect, because the mass of the molecule grows faster than the available volume as the number of cycles increases, implying the eventual attainment of a stage beyond which subsequent reaction can no longer be quantitative.

1.3   Overall Extension in Space: Molecular Sizes

The overall molecular shape, or conformation, is one of the most important properties of a macromolecule. Indeed, their large extension in space is one of the most important ways in which macromolecules differ from small molecules. With large linear chains, the overall extension is frequently discussed in terms of the mean square end-to-end distance, $\langle r^2 \rangle$, or the mean square radius of gyration, $\langle s^2 \rangle$.[11–20] For very long flexible homopolymeric chains, these two properties are approximately related as $\langle r^2 \rangle = 6\langle s^2 \rangle$. This approximation becomes exact when a very long flexible homopolymer is unperturbed by long-range interactions, as it is in the Θ state defined by Flory.[11, 21] When $\langle r^2 \rangle$ and $\langle s^2 \rangle$ are equally pertinent to a discussion of the size of a chain, there is often a preference for the mean square end-to-end distance because theoretical expressions for $\langle r^2 \rangle$ are usually simpler than their counterparts for $\langle s^2 \rangle$. However, the mean square end-to-end distance becomes ambiguous when the macromolecule is branched, and it is undefined when the macromolecule is cyclic. In both cases, however, the mean square radius of gyration remains well-defined. Therefore, $\langle s^2 \rangle$ will be considered in preference to $\langle r^2 \rangle$.

1.3.1   Radius of Gyration

The most fundamental characterization of the molecular size of a macromolecule is its radius of gyration, s. The squared radius of gyration of a rigid collection of $n + 1$ particles indexed by i running from 0 to n, and with particle i weighted as m_i, is

$$s^2 = \frac{\sum_{i=0}^{n} m_i \, (r_i - g) \cdot (r_i - g)}{\sum_{i=0}^{n} m_i} \tag{1.10}$$

where g denotes the vector to the center of mass, expressed in the same coordinate system used for the vector to atom i, r_i.

$$g = \frac{\sum_{i=0}^{n} m_i r_i}{\sum_{i=0}^{n} m_i} \tag{1.11}$$

The weighting factors m_i are useful in the study of copolymers, and especially so when the copolymer is studied by scattering measurements in which the effective scattering power differs strongly for the various components of the macromolecule. Here, however, we shall assume that these differences are not important, and will adopt the simpler definition of s^2 that arises when all of the m_i are identical:

$$s^2 = \frac{1}{n+1} \sum_{i=0}^{n} (r_i - g) \cdot (r_i - g) \tag{1.12}$$

1.3.1.1 Linear Chains

One of the simplest models for a flexible macromolecule is the freely jointed chain comprised of n bonds of the length l. For this model,

$$\langle s^2 \rangle_0 = \left(\frac{n+2}{n+1} \right) \frac{nl^2}{6} \tag{1.13}$$

The zero as a subscript denotes Flory's Θ state, where the chain is unperturbed by long-range interactions. This same model has $\langle r^2 \rangle_0 = nl^2$ at all n, which with Equation (1.13) yields the limiting behavior $\langle r^2 \rangle_0 \to 6\langle s^2 \rangle_0$ as $n \to \infty$. This very simple model correctly leads to the conclusion that flexible linear chains do not prefer the compact globular state. The globular state is characterized by $v = 1/3$ in

$$\langle s^2 \rangle \approx n^{2v} \tag{1.14}$$

but the freely jointed unperturbed chain has $v = 1/2$, implying that the freely jointed chain is more expanded than the globule. Values of v associated with five conformations are presented in Table 1.1.

Table 1.1 Values of v for various conformations ($\langle s^2 \rangle \approx n^{2v}$).

Conformation	v
Globule	1/3
Random coil (Θ solvent)	1/2
Random coil (good solvent)	$\approx 3/5$[a]
Rigid rod	1
Dendrimers (good solvent)	0.15–0.21[b]

[a] The best estimate[22, 23] for a linear random coil in a good solvent is $v = 0.588 \pm 0.001$.
[b] Estimated from the radii of gyration obtained in simulations.[24] The value for v is not constant, but instead tends to decrease with increasing generations. The largest dendrimer considered is generation nine.

Extremely detailed models for unperturbed flexible linear homopolymers can be constructed using rotational isomeric state theory.[14, 20] In the limit as $n \to \infty$, these detailed models yield

$$\lim_{n \to \infty} \frac{\langle s^2 \rangle_0}{nl^2} = \frac{C}{6} \tag{1.15}$$

$$\lim_{n \to \infty} \frac{\langle r^2 \rangle_0}{\langle s^2 \rangle_0} = 6 \tag{1.16}$$

Equation (1.16) specifies the same limiting behavior as was obtained with the simple freely jointed chain. However, Equation 1.15 specifies the same limiting $\langle s^2 \rangle_0$ as the freely jointed chain only in the special case where the characteristic ratio, C, is equal to unity. In general, the restrictions on bond angle and torsion angles in real chains produce $C > 1$, meaning that real flexible unperturbed polymers tend to have larger dimensions than those suggested by the random flight chain with bonds of the same number and lengths. For example, $C = 6.7$ for poly(ethylene) at $140\,°C$.[25] The restriction on the bond angle alone raises the value of C from 1 for the freely jointed chain to 2.20 in the freely rotating chain with a bond angle of $112°$. The additional restrictions on the torsion angles due to first- and second-order interactions bring the result to $C = 6.7$ for poly(ethylene).[26] The first-order interaction causes the *gauche* states to be of higher energy than the *anti* periplanar states. The most important second-order interaction strongly penalizes two successive *gauche* states of opposite sign, which produce the repulsive interaction known as the 'pentane effect'.

1.3.1.2 Macrocycles

The average extension of the $n + 1$ chain atoms in space can be reduced by changing their connectivity. A conceptually simple means by which this reduction can be obtained is by the introduction of a covalent bond that links the two ends of the chain. This macro-cyclization leads to a reduction in $\langle s^2 \rangle$ that, in the case of large unperturbed chains, is a factor of 1/2.[27]

$$\lim_{n \to \infty} \frac{\langle s^2 \rangle_{0, \text{ cyclic}}}{\langle s^2 \rangle_{0, \text{ linear}}} = 1/2 \tag{1.17}$$

The value of $\langle s^2 \rangle_0$ for the cyclized chain remains far above the result expected for the same collection of $n + 1$ atoms in the globular state for large unperturbed macrocycles, with $v = 1/2$.

1.3.1.3 Branched Macromolecules

A larger reduction in the mean square unperturbed dimensions can be achieved by rearrangement of the $n + 1$ atoms into an f-functional star-branched polymer. In this architecture, the macromolecule contains f branches ($f > 2$), each with n/f bonds, that emanate from a common atom. The same terminology is frequently used for branched macromolecules with very large f if the branches emanate from a collection of atoms that are constrained to remain close together, so that the origin of all of the branches is clustered in a volume much smaller than $\langle s^2 \rangle^{3/2}$. The influence of the star-branched architecture on the mean square dimensions is traditionally designated by a factor g that is defined as[28]

$$g = \frac{\langle s^2 \rangle_{0, \text{ branched}}}{\langle s^2 \rangle_{0, \text{ linear}}} \tag{1.18}$$

It is understood that the branched and linear molecules contain the same number of bonds. If n_j denotes the number of bonds in branch j, the application of random flight statistics leads to a very simple expression for g.[28]

$$g = \frac{1}{n^3} \sum_j (3nn_j^2 - 2n_j^3) \tag{1.19}$$

For the special case where all of the branches contain the same number of bonds, with that number being n/f, Equation (1.19) becomes

$$g = \frac{3f-2}{f^2} \tag{1.20}$$

Equation (1.20) suggests that the mean square dimensions of the macrocycle are approximated by the star-branched polymer with five arms, for which $g = 0.52$. It also suggests that a macromolecule with a closer approach to the condensed state could be obtained by using $f > 5$.

As f increases without limit, Equation (1.20) yields a value of g that approaches zero, implying the attainment of a globular state of arbitrarily high density. This implication is not realized with real systems due to a breakdown in the validity of the assumptions behind Equation (1.19), as f increases. Deviation between the prediction from Equation (1.20) and the experimental measurements become significant at $f \approx 6$ and become more important as f increases. As the density of segments near the branch point increases, one must abandon the assumption of equivalent random flight statistics for the linear chain and for the star-branched polymer. As f increases, the real ratio of the two mean square radii of gyration in Equation (1.18) becomes larger than the prediction in Equation (1.20).

$$\left(\frac{\langle s^2 \rangle_{0, \text{ branched}}}{\langle s^2 \rangle_{0, \text{ linear}}} \right)_{\text{experimental}} > \frac{3f-2}{f^2} \text{at large } f \tag{1.21}$$

Star-branched polymers with f larger than 10^2 have been prepared.[29, 30] Their dimensions are much larger than the prediction from Equation (1.20). These highly branched structures have been called 'fuzzy spheres',[30] because the comparison of the thermodynamic radii deduced from equilibrium measurements with the hydrodynamic radii from transport measurements suggests that the macromolecules behave as spheres with a hydrodynamically penetrable surface layer.

1.3.2 Breadth of the Distribution for the Squared Radius of Gyration

The value of $\langle s^2 \rangle_0$ gives the average over the distribution function for the squared radius of gyration, $P(s^2)$.

$$\langle s^{2p} \rangle_0 = \frac{\int_0^\infty s^{2p} P(s^2) \mathrm{d}s}{\int_0^\infty P(s^2) \mathrm{d}s} \tag{1.22}$$

By itself, $\langle s^2 \rangle_0$ provides no information about the shape of this distribution function. In particular, it does not reveal whether the distribution is broad, implying that the macromolecule can populate conformations with very different extensions, or whether it is narrow, as would be the case if a single conformation is populated in preference to all others. Information about the shape of the distribution function for the squared dimensions can be assessed by evaluation of the higher even moments. The breadth of the distribution function is deduced from $\langle s^4 \rangle / \langle s^2 \rangle^2$. If only a single conformation is accessible, this dimensionless ratio has the value of unity. The freely jointed chain provides another useful benchmark for interpreting larger values of $\langle s^4 \rangle_0 / \langle s^2 \rangle_0^2$. At any value of n, this model has[31]

$$\frac{\langle s^4 \rangle_0}{\langle s^2 \rangle_0^2} = \frac{19n^3 + 45n^2 + 32n - 6}{15n(n+1)(n+2)} \tag{1.23}$$

which is equal to 1 for $n = 1$, and is larger than 1 for $n \geq 2$. It increases with n to a limiting value of $19/15 = 1.267 \ldots$[32]

The freely jointed chain has a distribution function for s^2 that is narrower than the distribution function for r^2,

$$\frac{\langle r^4 \rangle_0}{\langle r^2 \rangle_0^2} = \frac{5n - 2}{3n} \tag{1.24}$$

with the limit as $n \rightarrow \infty$ being $\langle r^4 \rangle_0 / \langle r^2 \rangle_0^2 \rightarrow 5/3$, which is larger than the limit $\langle s^4 \rangle_0 / \langle s^2 \rangle_0^2 \rightarrow 19/15$.

The dimensionless ratios of the form $\langle s^{2p} \rangle_0 / \langle s^2 \rangle_0^p$ are easily evaluated from a theoretical model for the distribution function, $P(s^2)$, using Equation (1.22). For small p, they can also be calculated for unperturbed rotational isomeric state chains by efficient generator matrix methods.[33]

1.4 Shape Analysis from the Radius of Gyration Tensor

Analysis based on the averaged principal moments of the radius of gyration tensor provides information about the shape of the accessible conformations. This information is different from that contained in the $\langle s^{2p} \rangle_0 / \langle s^2 \rangle_0^p$. Consider a rather strange molecule that can adopt any of three shapes (a sphere, a disk, and a cylinder), all three of which have exactly the same radius of gyration. The ratio $\langle s^{2p} \rangle_0 / \langle s^2 \rangle_0^p$ provides no insight into which of these three conformations might be preferred, because $\langle s^{2p} \rangle_0 / \langle s^2 \rangle_0^p = 1$ for any combination of preferences, due to the postulate that the sphere, disk, and cylinder had the same radius of gyration. Distinctions in the shape become possible using the tensorial representation of the radius of gyration.

For a rigid array of $n + 1$ particles, the radius of gyration tensor, **S**, can be expressed as a symmetric 3×3 matrix

$$\mathbf{S} = \begin{bmatrix} X^2 & XY & XZ \\ XY & Y^2 & YZ \\ XZ & YZ & Z^2 \end{bmatrix} \tag{1.25}$$

or as a column in which the nine elements of Equation (1.25) are listed in 'reading order'.

$$\mathbf{S}^{\text{col}} = \begin{bmatrix} X^2 \\ XY \\ XZ \\ XY \\ Y^2 \\ YZ \\ XZ \\ YZ \\ Z^2 \end{bmatrix}$$

(1.26)

When the $n + 1$ particles are all of the same mass, the latter representation is generated from the vectors s_i (each from the center of mass to atom i, $s_i \equiv r_{0i} - g$) as

$$\mathbf{S}^{\text{col}} = \frac{1}{n+1}\sum_{i=0}^{n} s_i^{x2} = \frac{1}{n+1}\sum_{i=0}^{n} r_{0i}^{x2} - g^{x2}$$

(1.27)

where x2 as a superscript denotes the self direct product. The squared radius of gyration that was the subject of Equation (1.12) is the trace of $\mathbf{S}$, $s^2 = X^2 + Y^2 + Z^2$. This trace is, of course, independent of the orientation of the coordinate system used for the expression of $\mathbf{S}$. However, in the absence of spherical symmetry, the sizes of the off-diagonal elements, as well as each of X^2, Y^2, and Z^2 individually (but not their sum) depends on the orientation of the coordinate system used to express $\mathbf{S}$.

1.4.1 Principal Moments

Preparatory to the shape analysis, the coordinate system is rotated by a similarity transform so that the 3×3 representation of $\mathbf{S}$ for each individual conformation is rendered in diagonal form.

$$\mathbf{S}_{\text{diag}} = \mathbf{TST}^{\text{T}} = \begin{bmatrix} L_1^2 & 0 & 0 \\ 0 & L_2^2 & 0 \\ 0 & 0 & L_3^2 \end{bmatrix} = \text{diag}\,(L_1^2, L_2^2, L_3^2)$$

(1.28)

where $\mathbf{T}$ is the transformation matrix, $\mathbf{T}^{\text{T}}$ is its transpose, and $L_1^2 + L_2^2 + L_3^2 = X^2 + Y^2 + Z^2 = s^2$ because the trace is an invariant. Let us further stipulate that the subscripts on the principal moments, L_i^2, be assigned so that $L_1^2 \geq L_2^2 \geq L_3^2$.

The shape of a conformation can be characterized by various manipulations of the principal moments. The asymmetry of any one of the conformations is characterized by the dimensionless ratios $1 \geq L_2^2/L_1^2 \geq L_3^2/L_1^2 \geq 0$.[34, 35] Spherical symmetry requires $L_2^2/L_1^2 = L_3^2/L_1^2 = 1$. Averaging of the corresponding principal moments over all conformations permits discussion of the asymmetry of the population of conformations in terms of $\langle L_2^2\rangle/\langle L_1^2\rangle$ and $\langle L_3^2\rangle/\langle L_1^2\rangle$. Examples of these dimensionless ratios are presented in Table 1.2 for four types of macromolecules, unperturbed by long-range interactions. In none of the four cases do the conformations have spherical symmetry.

Deviations from spherical symmetry are largest for the linear chain, and smallest for the tetrafunctional star. The macrocycle occupies a position intermediate between the tri- and tetra-functional star.

Table 1.2 $\langle L_2^2\rangle_0/\langle L_1^2\rangle_0$ and $\langle L_3^2\rangle_0/\langle L_1^2\rangle_0$ for large unperturbed macromolecules.

Architecture	$\langle L_2^2\rangle_0/\langle L_1^2\rangle_0$	$\langle L_3^2\rangle_0/\langle L_1^2\rangle_0$
Linear chain[34, 35]	0.23	0.08
Trifunctional star[a] [36, 37]	0.33	0.12
Macrocycle[36, 37]	0.36–0.37	0.15–0.16
Tetrafunctional star[a] [36, 37]	0.39–0.41	0.15–0.16

[a] All arms contain the same number of bonds.

1.4.2   Asphericity, Acyclindricity, and Relative Shape Anisotropy

Three additional measurements that are useful in detecting other types of symmetry can be derived from the traceless form of the tensor, defined as

$$S_{\text{diag, traceless}} = \text{diag}\,(L_1^2, L_2^2, L_3^2)\,\frac{-s^2}{3}\text{diag}\,(1,1,1) \tag{1.29}$$

and comparing this traceless diagonal tensor with an analogous traceless tensor used in the treatment of the polarizability,[38]

$$S_{\text{diag, traceless}} = b\,\text{diag}\,(2/3,-1/3,-1/3) + c\,\text{diag}\,(0,1/2-1/2) \tag{1.30}$$

which defines the asphericity b and the acylindricity c as[39]

$$b = L_1^2 - \frac{1}{2}(L_2^2 + L_3^2) \tag{1.31}$$

$$c = L_2^2 - L_3^2 \tag{1.32}$$

The value of b will be zero if the collection of $n + 1$ points has tetrahedral or higher symmetry; otherwise $b > 0$. For long linear unperturbed chains, $\langle b\rangle_0/\langle s^2\rangle_0 = 0.66$.[39] If the shape is cylindrically symmetric, $c = 0$, and $c > 0$ otherwise. For long linear unperturbed chains, $\langle c\rangle_0/\langle s^2\rangle_0 = 0.11$.[39]

Another parameter, called the relative shape anisotropy, is denoted by $\varkappa^2$ and defined as

$$\varkappa^2 = \frac{b^2 + (3/4)c^2}{s^4} \tag{1.33}$$

It has the property that $\varkappa^2 = 0$ when the structure has tetrahedral or higher symmetry, $\varkappa^2 = 1$ when the points describe a linear array, and $0 < \varkappa^2 < 1$ otherwise. For long linear unperturbed chains, $\langle \varkappa^2\rangle_0 = 0.41$.[39]

For a perfectly spherical globule, $\langle L_2^2\rangle_0/\langle L_1^2\rangle_0 = \langle L_3^2\rangle_0/\langle L_1^2\rangle_0 = 1$, which will automatically produce $\langle b\rangle_0 = \langle c\rangle_0 = \langle \varkappa^2\rangle_0 = 0$.

1.5   Approach to the Globular State

For a relatively small linear poly(ethylene) chain of molecular weight 10,000, Equation (1.15) provides an estimate for $\langle s^2\rangle_0^{1/2}$ of 43 Å. This value is several times larger than the value (≈ 10 Å) expected for the same collection of atoms if they were arranged into the most compact globule consistent with the density of liquid n-alkanes.

$$s_{\text{min}}^2 = \frac{3}{5}\left(\frac{3\bar{v}M}{4\pi L}\right)^{2/3)} \tag{1.34}$$

Here $\bar{v}$ is the partial specific volume and L is Avogadro's number. A similar conclusion is obtained for virtually all unperturbed linear chains; $\langle s^2\rangle_0$ is much larger than the result expected for a compact globule.

The average extension of the chain in space can be modified by its interaction with the environment. This modification will occur in dilute solution, except in the special case where the chain is dissolved in a Θ solvent. The typical dilute solution is prepared using a 'good' solvent, in which case the polymer–solvent interaction produces a positive excluded volume that causes expansion of a flexible chain. The introduction of the consequences of long-range interactions into the conformational description of a linear chain greatly complicates the theoretical description.[15–18] The focus is on an expansion factor that can be defined as

$$\alpha_s^2 = \frac{\langle s^2\rangle}{\langle s^2\rangle_0} \tag{1.35}$$

where $\langle s^2 \rangle$ without the zero as a subscript denotes the mean square radius of gyration in the presence of the long-range interactions. In good solvents, both experiment and theory show that $\alpha_s^2 > 1$, with α_s^2 increasing (meaning the chain expands) as n for the quality of the solvent increases. Therefore, typical flexible polymer chains in the usual solvents adopt a swollen conformation that is very different from a globule.

1.5.1   Collapse of a Linear Chain

Collapse of the chain toward dimensions approaching the small size characteristic of a globule requires use of a solvent that is poorer than a Θ solvent. These solutions are handled with difficulty, because the reduction in the dimensions of the chain requires that the quality of the solvent must be so poor that suppression of aggregation due to intermolecular attractions, and maintenance of stable solutions with measurable concentrations of the homopolymer, becomes a formidable challenge. The transition of a linear homopolymer toward the globular state is studied with least difficulty when the chain is of very high molecular weight, so that the individual chains are, on the average, separated by large distances when the concentration (expressed as mass/volume) is in the range where measurements of the mean dimensions become feasible. Studies of these systems (notably high molecular weight poly(styrene) in poor solvents) show that the collapse to the globular state is a sharp transition.[40–42] Behavior typical of a globule ($\nu \approx 1/3$) can be achieved in terms of the scaling of $\langle s^2 \rangle$ with n, although $\langle s^2 \rangle$ itself remains substantially larger than $s^2_{\min}$, the value expected for completion of the coil $\rightarrow$ globule transition.

The smallest value of α_s obtained experimentally by this procedure is about 0.7. It is very difficult to obtain stable solutions of the globules formed by the intramolecular collapse of linear homopolymers. The collapse of a linear chain to a structure closely approximating a globule can more easily be achieved in molecular dynamics simulations of a single chain.[43] Values of α_s as small as 0.33 have been achieved for a poly(vinyl chloride) chain with a degree of polymerization of 300. The complete collapse, to a structure with radius of gyration given by Equation (1.33), would require $\alpha_s = 0.28$ for this chain. Therefore, the radius of gyration achieved in the simulation was within 20 % of the value expected for the idealized space-filling sphere. The values of $\langle L_2^2 \rangle / \langle L_1^2 \rangle$ and $\langle L_3^2 \rangle / \langle L_1^2 \rangle$ for this collapsed poly(vinyl chloride) chain were 0.78 and 0.56, respectively. Therefore, the collapsed structure does not have spherical symmetry, but it is much closer to spherical symmetry than any of the structures considered in Table 1.2. Qualitatively similar behavior is seen in the simulation of the collapse of poly(1,4-*trans*-butadiene).[44]

The collapse produced in both of these simulations arises from the intramolecular short-range attractive two-body interactions experienced by a polymer in a poor solvent. Another interesting type of collapse was recently proposed,[45] then identified in a simulation on a diamond lattice,[46] and subsequently verified by experiment with copolymers of acrylamide and *N*-isopropylacrylamide.[47] This collapse mechanism operates in grafted layers of chains when the short-range binary interactions are repulsive, but higher order interactions in '*N*-clusters' are attractive. The value of N is found experimentally to be three for copolymers of acrylamide and *N*-isopropylacrylamide,[46] as was assumed in the simulation.[47]

1.5.2   Shapes from the Simulations of Dendrimers

Some conclusions about the shape of highly branched macromolecules based on successive perfect generations of the controlled condensation of A–R–B$_2$ are accessible from simulations on a diamond lattice.[24] This type of simulation has the disadvantage that it is not atomistic, and therefore does not accurately describe in detail any particular macromolecule. But the advantages, namely the ability to convincingly demonstrate the attainment of equilibrated structures and the likelihood that the overall trends are reflective of this class of molecules in general, are compelling, at least at the current state-of-the-art.

In the simulations, the exponent v in Equation (1.14) is not constant, but instead decreases as the number of generations increases. It is always much smaller than the values of v for more common conformations, as shown by comparison of the entries in Table 1.1.

The shape deduced from the analysis of the radius of gyration tensor also changes as the generation number increases, as shown in Table 1.3. The entries in this table were calculated from the results for the moments of inertia that were reported by Mansfield and Klushin.[24] Even as soon as the second generation, these model dendrimers (**1**) are closer to being spherically symmetric than are any of the conformations listed in Table 1.2.

Table 1.3 Ratios of the averaged principal moments of the radius of gyration tensor, asphericity, acylindricity, and anisotropic shape factor for model dendrimers (**1**, Fig. 1.1)[a].

Generation	$\dfrac{\langle L_2^2 \rangle}{\langle L_1^2 \rangle}$	$\dfrac{\langle L_3^2 \rangle}{\langle L_1^2 \rangle}$	$\dfrac{\langle b \rangle}{\langle s^2 \rangle}$	$\dfrac{\langle c \rangle}{\langle s^2 \rangle}$	$\dfrac{\langle \varkappa^2 \rangle}{\langle s^2 \rangle^2}$
1	0.42	0.14	0.46	0.18	0.24
2	0.49	0.20	0.39	0.17	0.17
3	0.58	0.28	0.31	0.16	0.12
4	0.64	0.38	0.24	0.13	0.07
5	0.68	0.44	0.21	0.11	0.05
6	0.77	0.55	0.15	0.09	0.03
Ideal Sphere	1	1	0	0	0

[a] Numerical results for 1–6 generations of the A–R–B$_2$ series were calculated from the moments of inertia obtained in the simulations.[24]
[b] Limiting results if very large dendrimers adopt the symmetry of idealized spheres, either of uniform density, or with a density that is a function of r.

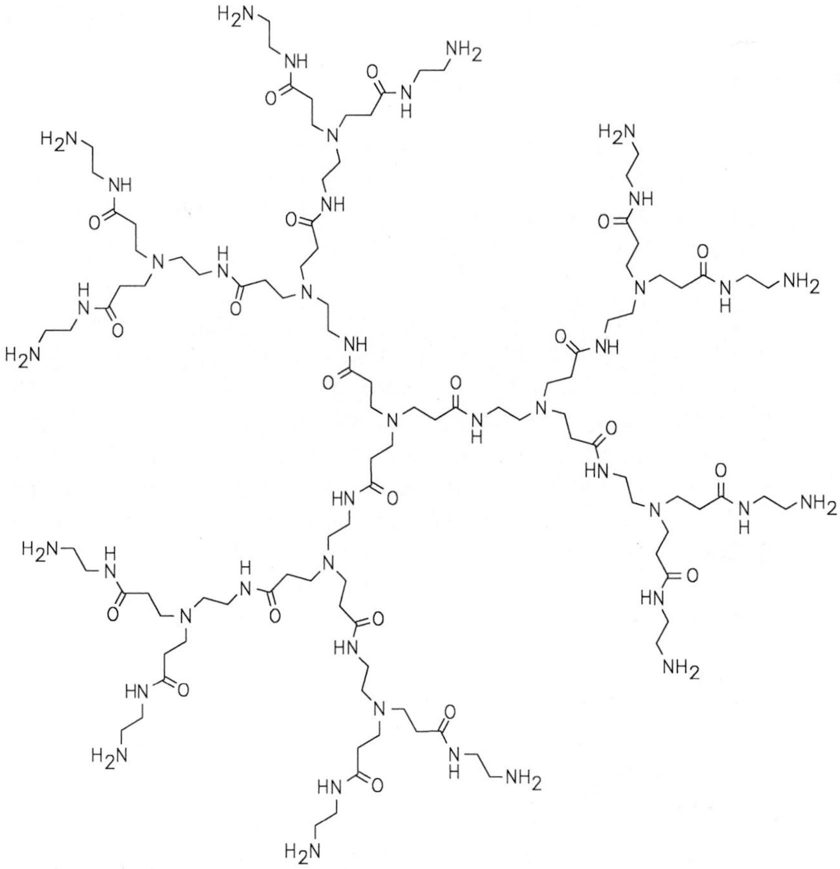

1

Figure 1.1. Example of a structure for which $\dfrac{L_2^2}{L_1^2} = 1,\ \dfrac{L_3^2}{L_1^2} = 0,\ \dfrac{b}{s^2} = \dfrac{1}{4},\ \dfrac{c}{s^2} = \dfrac{1}{2},\ \dfrac{\varkappa^2}{s^2} = \dfrac{1}{4}.$

However, perfect adherence to spherical symmetry is not attained until a much higher generation number (beyond the range covered in the simulations), if indeed it is attained at all.

The behavior of the asphericity with generation number is depicted in Figure 1.2. The best linear extrapolation of the six points suggests that the asphericity would fall to zero by generation nine. However, there is no reason to expect that the linear extrapolation is valid. Instead one anticipates that the asphericity should exhibit positive curvature with each increasing generation, as is evident from close inspection of Figure 1.1. Therefore the proper interpretation is that generations 1–9 are all aspherical, with the asphericity decreasing with increasing generations. Perhaps the asphericity would fall to zero at some generation above the ninth.

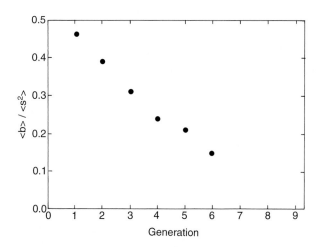

Figure 1.2. Asphericity of model dendrimers, as calculated from the moments of inertia described by Mansfield and Klushin.[24]

1.6   References

[1]   P. J. Flory, *J. Am. Chem. Soc.* **1936**, *58*, 1877.
[2]   P. J. Flory, *J. Am. Chem. Soc.* **1940**, *62*, 1561.
[3]   J. R. Schaefgen, P. J. Flory, *J. Am. Chem. Soc.* **1948**, *70*, 2709.
[4]   P. J. Flory, *J. Am. Chem. Soc.* **1941**, *63*, 3083.
[5]   P. J. Flory, *Chem. Rev.* **1946**, *39*, 137.
[6]   W. H. Stockmayer, *Macromolecules* **1991**, *24*, 6368.
[7]   F. Tanaka, W. H. Stockmayer, *Macromolecules* **1994**, *27*, 3943.
[8]   M. Nguyen-Misra, W. L. Mattice, *Macromolecules* **1995**, *28*, 1444.
[9]   P. J. Flory, *J. Am. Chem. Soc.* **1952**, *74*, 2718.
[10] M. L. Mansfield, *Macromolecules* **1993**, *26*, 3811.
[11] P. J. Flory, *Principles of Polymer Chemistry*, Cornell University Press, Ithaca, New York, **1953**.
[12] M. V. Volkenstein, *Configurational Statistics of Polymer Chains*, Wiley, New York, **1963**.
[13] T. M. Birshtein, O. B. Ptitsyn, *Conformations of Macromolecules*, Wiley, New York, **1966**.
[14] P. J. Flory, *Statistical Mechanics of Chain Molecules*, Wiley, New York, **1969**; reprinted with the same title by Hanser, Munich, **1989**.
[15] H. Yamakawa, *Modern Theory of Polymer Solutions*, Harper & Row, New York, **1971**.
[16] P. G. de Gennes, *Concepts in Polymer Physics*, Cornell University Press, Ithaca, New York, **1979**.
[17] M. Doi, S. F. Edwards, *The Theory of Polymer Dynamics*, Clarendon Press, Oxford, **1986**.
[18] K. R. Freed, *Renormalization Group Theory of Macromolecules*, Wiley, New York, **1987**.
[19] J. des Cloizeaux, G. Jannink, *Polymers in Solution. Their Modelling and Structure*, Clarendon Press, Oxford, **1990**.
[20] W. L. Mattice, U. W. Suter, *Conformational Theory of Large Molecules. The Rotational Isomeric State Model in Macromolecular Systems*, Wiley, New York, **1994**.
[21] P. J. Flory, *J. Phys. Chem.* **1949**, *53*, 197.
[22] J. C. Le Guillou, J. Zinn-Justin, *Phys. Rev. Lett.* **1977**, *39*, 95.
[23] Y. Oono, *Adv. Chem. Phys.* **1985**, *61*, 301.

[24] M. L. Mansfield, L. I. Klushin, *Macromolecules* **1993**, *26*, 4268.

[25] R. Chiang, *J. Phys. Chem.* **1965**, *69*, 1645.

[26] A. Abe, R. L. Jernigan, P. J. Flory, *J. Am. Chem. Soc.* **1966**, *88*, 631.

[27] H. A. Kramers, *J. Chem. Phys.* **1946**, *4*, 415.

[28] B. H. Zimm, W. H. Stockmayer, *J. Chem. Phys.* **1949**, *17*, 1302.

[29] J. Roovers, L. L. Zhou, P. M. Toporowski, M. van der Zwan, H. Iatrou, N. Hadjichristidis, *Macromolecules* **1993**, *26*, 4324.

[30] L. Willner, O. Jucknischke, D. Richter, J. Roovers, L.-L. Zhou, P. M. Toporowski, L. J. Fetters, J. S. Huang, M. Y. Lin, N. Hadjichristidis, *Macromolecules* **1994**, *27*, 3821.

[31] K. Sienicki, W. L. Mattice, *J. Chem. Phys.* **1989**, *90*, 1956.

[32] M. Fixman, *J. Chem. Phys.* **1962**, *36*, 306.

[33] P. J. Flory, *Macromolecules* **1974**, *7*, 381.

[34] K. Solc, W. H. Stockmayer, *J. Chem. Phys.* **1971**, *54*, 2756.

[35] K. Solc, *J. Chem. Phys.* **1971**, *55*, 355.

[36] K. Solc, *J. Chem. Phys.* **1973**, *6*, 378.

[37] W. L. Mattice, *Macromolecules* **1980**, *13*, 506.

[38] R. P. Smith, E. M. Mortensen, *J. Chem. Phys.* **1960**, *32*, 502.

[39] D. N. Theodorou, U. W. Suter, *Macromolecules* **1985**, *18*, 1206.

[40] I. H. Park, Q.-W. Wang, B. Chu, *Macromolecules* **1987**, *20*, 1965.

[41] I. H. Park, Q.-W. Wang, B. Chu, *Macromolecules* **1987**, *20*, 2883.

[42] B. Chu, R. Xu, J. Zhuo, *Macromolecules* **1988**, *21*, 273

[43] G. Tanaka, W. L. Mattice, *Macromolecules* **1995**, *28*, 1049.

[44] Y. Zhan, W. L. Mattice, *Macromolecules* **1994**, *27*, 7056.

[45] P. G. de Gennes, *C. R. Acad. Sci. Paris II* **1991**, *313*, 1117.

[46] W. L. Mattice, S. Misra, D. H. Napper, *Europhys. Lett.* **1994**, *28*, 603.

[47] K. Turner, P. W. Zhu, D. H. Napper, personal communication.

2 From Theory to Practice: Historical Perspectives

2.1 Introduction

In Chapter 1, Professor Mattice laid the foundation for a general understanding of macroassemblies by considering molecular masses, sizes and shapes. Mathematical analyses of these physical properties were extended from an examination of classically prepared, long chain, linear polymers to highly branched polymers and finally to iteratively constructed macromolecules commonly known today as dendrimers or cascade (macro)-molecules. With respect to their macromolecular assembly and utility, a notable concept was established by investigation of the averaged principal moments of the radius of gyration tensors, which allowed analysis of the asphericity, acylindricity, and shape anisotropy for the particular macromolecular models examined.

The concept, manifested in the discussion, is precise and deliberate control over macromolecular geometry. This far-reaching idea can be gleaned from the general trend toward pseudospherical or globular symmetry that was observed for dendrimers synthesized via repetitive condensations with $1 \rightarrow 2$ branching, $A-R-B_2$ monomers. Extension of this type of analysis to cascades constructed with symmetrical, four-directional cores and $1 \rightarrow 3$ branching $A-R-B_3$ monomers would lead presumably to even more precise spherical assemblies. Control over macromolecular configuration is therefore a key axiom in dendritic chemistry. Ramifications extend far beyond command of the overall molecular shape to include choice over such parameters as internal and external rigidity, lipophilicity and hydrophilicity, degrees of void volume and excluded volume, density gradients, complimentary functionalities, and environmental cooperativity. The iterative synthetic method is, therefore, the quintessential macromolecular construction technique.

Development of the dendritic method for macroassembly did not simply arise from the first examples of deliberately prepared branched molecules. Rather, it was a logical progression of synthetic approaches derived from the efforts of countless researchers to realize new materials with novel properties and uses.

2.2 Branched Architectures

From a historical perspective, progress towards the deliberate construction of macromolecules possessing branched architecture can be considered to have occurred during three general eras. The first period occurred roughly from the late 1860's to the early 1940's, when branched structures were considered as being responsible for insoluble and intractable materials formed in polymerization reactions. Synthetic control, mechanical separations, and physical characterization were primitive at best as judged by current standards; isolation and proof of structure were simply not feasible.

The early 1940's to the late 1970's defines the second period, in which branched structures were considered primarily from a theoretical vantage point with initial attempts at preparation via classical, or single-pot, polymerization of functionally differentiated monomers. As noted by Flory[1] in the 1950's: "The breadth of the distribution coupled with the impossibility of selectivity fractionating 'branching' and 'molecular weight' separately make this approach impractical. Attempts to investigate 'branching' by such means consequently have been notably fruitless."

The late 1970's and early 1980's recorded initial successful progress toward macromolecular assembly based on the iterative method that became the cornerstone of dendritic chemistry and thus defined the start of the third period of development. At this stage, the concept of control over macroassembly construction was better developed. Advances in physical isolation and purification, as well as the introduction of diverse spectroscopic procedures, had reached the level of sophistication necessary for the support of this blossoming field. In essence, the expansion of molecular construction limits of single, stepwise prepared, chemical structures beyond the approximate 2000 amu historical threshold was effected.

2.2.1   Early Observations

Initial reports leading to speculation about non-linear or branched polymeric connectivity were provided by researchers such as Zincke,[2] who isolated an insoluble hydrocarbon-based material upon the treatment of benzyl chloride with copper. Later, in 1885, reaction of benzyl chloride with aluminium chloride led Friedel and Crafts[3] to report similar observations. When benzyl chloride was subjected to the action of a zinc–copper couple analogous materials were obtained.[4]

A significant aspect, at approximately this time, relating to the basic understanding of polymers, linear and branched, was the idea that virtually any substance could exist in a "colloidal state",[5, 6] analogous to the gaseous, liquid, and solid states. The term "colloid", coined by Graham[7] in 1861 and synonymous with "gluelike", was introduced to describe polymers possessing a negligible diffusion ratio in solution and which did not pass through semi-impermeable membranes. Materials that could be obtained in the crystalline form were described as "crystalloids". Extended use of "colloid" terminology[5, 6] was unfortunate from a perspective of covalent molecular assembly because a simple physical change of state will not, in general, disassemble a polymer into its starting components. For many years, the notion that polymers obtained their properties via non-covalent small molecule intermolecular interactions supplanted the view of a polymeric molecule consisting of units covalently connected.

Regardless of the misleading colloid concept, the study of polymers remained unabated. Hlasiwetz and Habermann[8] in 1871 concluded that proteins and carbohydrates were comprised of polymeric units with differing degrees of condensation. Certain members of these substances were noted as "soluble and unorganized" while others were described as "insoluble and organized". Much later, Flory[9] would liken these categories to crystalline and non-crystalline polymers.

Thirty-five years after Gladstone and Tribe's observation[4] concerning the reaction of benzyl chloride with the zinc–copper couple, reports concerning the preparation of unclassifiable substances continued. Hunter and Woollet[10] described the synthesis of an amorphous material by polymerization of triiodophenolic salts and salicylic acid. It is interesting to note that, in 1922, Ingold and Nickolls[11] reported the preparation of the branched, small molecule "methanetetraacetic acid". This is perhaps the earliest example of a deliberately constructed, symmetrically substituted, polyfunctional, dendritic paradigm, or archetype. As if to forecast the potential utility of branched architecture with respect to supramolecular host–guest chemistry, the authors noted that crystals of this tetraacid "contained 2–3 % nitrogen, which was only slowly removed by boiling with 50 % sulfuric acid or with 40 % potassium hydroxide."

Staudinger[12] (Nobel Prize in Chemistry, 1953) at about the same time, suggested that materials such as natural rubber were composed of long-chain, high molecular weight molecules and should not be considered as small molecule aggregates in the colloidal state. Carothers' studies[12, 13] on condensation polymerizations aided in the formation of Staudinger's theory of a macromolecular composition. Intractable properties of substances like Bakelite-C were attributed to a three-dimensional network-like structure,[13] while Jacobson[14] suggested the presence of a three-dimensional structure to account for insoluble materials that were present after polymerization processes.

2.2.2 Prelude to Practice

A second and distinct era in the development of branched macromolecular architecture encompasses the time between 1940 to 1978, or approximately the next four decades. Kuhn[15] published the first report of the use of statistical methods for analysis of a polymer problem in 1930. Equations were derived for molecular weight distributions of degraded cellulose. Thereafter, mathematical analyses of polymer properties and interactions flourished. Perhaps no single person has affected linear and non-linear polymer chemistry as profoundly as P. J. Flory. His contributions were rewarded by receipt of the Nobel Prize for Chemistry in 1974.

With respect to understanding the historical development of polymer chemistry in general, it is interesting to note a prevailing attitude[16] that material preparation was acceptable mainly from a perspective of isolation of discrete molecules with definite structure. As Flory[1] noted in his seminal treatise on polymers: "To be eligible for acceptance in the chemical kingdom, a newly created substance or a material of natural origin had to be separated in such a state that it could be characterized by a molecular formula." This was largely the result of the progress in synthetic organic chemistry.

Evolution of dendritic chemistry, and material science in general, might well have occurred more rapidly had the field of polydisperse polymer chemistry been more widely recognized and accepted earlier. For example, during the 1860's, Lourenco[17, 18] and Kraut[19] observed and reported the synthesis of materials consisting of difficult to separate molecular mixtures from the condensation of ethylene glycol with an ethylene dihalide and thermal condensation of acetylsalicylic acid, respectively. Their papers noted the basic nature of condensation polymer connectivity and formulation. It was not until many years later (ca. 1910) that the general scientific community began to significantly recognize and accept polydisperse macromolecular materials. Flory even suggests[20] that "evidence of retrogression could be cited."

During 1941 and 1942, Flory[21–24] disseminated theoretical and experimental evidence for the appearance of branched-chain, three-dimensional* macromolecules. These papers discussed a feature of polymerization reactions called "gelation". Descriptive terminology used by Flory to categorize differing polymeric fractions included the terms "gel" and "sol" referring to polymers that were insoluble or soluble, respectively. Flory showed statistically that branched polymeric products began to appear after polymerization had progressed a definite extent. Molecular size distributions, the number average degree of polymerization, as well as derivations relating to tri- and tetra-functional branching units (monomers) were also addressed. Flory's concept of a branched macromolecule is illustrated in Figure 2.1 by a two-dimensional drawing of a 2,3,2,3,1,0 macromolecule (**1**). The prefix numbering denotes the number of branch points occurring at each successive generation.

Stockmayer[25] subsequently developed equations relating to branched-chain polymer size distributions and "gel" formation, whereby branch connectors were of unspecified length and branch functionality was undefined. An equation was derived for the determination of the extent of reaction where a three-dimensional, network ("gel") forms; this relation was similar to Flory's, although it was derived using another procedure. Stockmayer likened gel formation to that of a phase transition and noted the need to consider: (a) intramolecular reactions, and (b) unequal reactivity of differing functional groups. This work substantially corroborated Flory's earlier studies.

In 1949, Flory[26] examined branched polymer "scaling" properties, e. g., the number of chain monomers relative to the mean squared end-to-end chain distance. Pursuing synthetic experiments with branched polymers, he reported[1] the preparation of a highly branched polymer without "insoluble gel formation". Flory employed a $1 \rightarrow 2$ branching, AB_2 monomer. Figure 2.2 depicts a branched structure **2**, as envisioned by Flory. With

---

* The term "three-dimensional", as used in this context, loosely refers to substantial molecular dimension in the direction of each of the x, y, and z coordinates that define the geometry of a particular macroassembly.

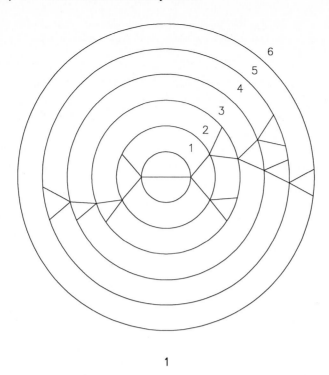

Figure 2.1. Flory's depiction of a branched 2,3,2,3,1,0 macromolecule.

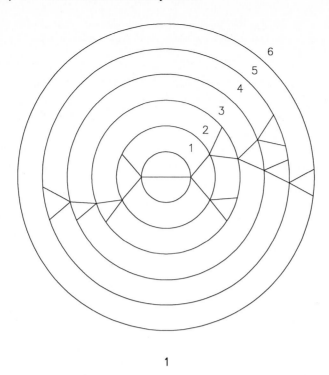

2

Figure 2.2. Branched polymer architecture as demonstrated by Flory by the assembly of AB$_2$-type monomers.

respect to separating monodisperse (or nearly so) fractions of these polymers he noted the difficulties, see above, inherent in the fractionation of high molecular weight polydisperse materials.

The same year (1952) Goldberg[27] presented a theory for the reaction of a multivalent antigen with a bivalent or univalent antibody. Like Stockmayer[25] and Flory,[1, 26] he considered a critical extent of reaction and suggested that it was "the point at which the system changes from one chiefly composed of small aggregates into one composed chiefly of relatively few exceedingly large aggregates" (visually observed as the precipitation point in some systems). Accordingly, Goldberg[28] extended the theory (1953) to encompass multivalent antigen and multivalent antibody reactions. Many other researchers contributed to the theoretical analysis of macromolecules. For example, Rouse[29] based a theory of the linear viscoelastic properties of dilute solutions of coiling polymers on the coordinated motions of different polymer units, whereas Zimm[30] considered the "problem of motions of a chain molecule diffusing in a viscous fluid under the influence of external forces or currents".

2.2.3 Incipient Macromolecular Progression

2.2.3.1 Initial Concept and Practice

"For the construction of *large* molecular cavities and pseudocavities that are capable of binding ionic guests and molecules (as [in] a complex or inclusion compounds) in a host–guest interaction, synthetic pathways allowing a frequent repetition of similar steps would be advantageous." The modern era of cascade or dendrimer chemistry came to life when Vögtle published this introductory sentence in his 1978 paper entitled "Cascade and Nonskid-Chain-like Syntheses of Molecular Cavity Topologies".[31] Ramifications of this statement by Vögtle et al. in 1978 extended far beyond the use of iterative methodology for the construction of host–guest assemblies. Repetition of similar and complimentary synthetic steps has since been used for the preparation of many new and exciting materials.

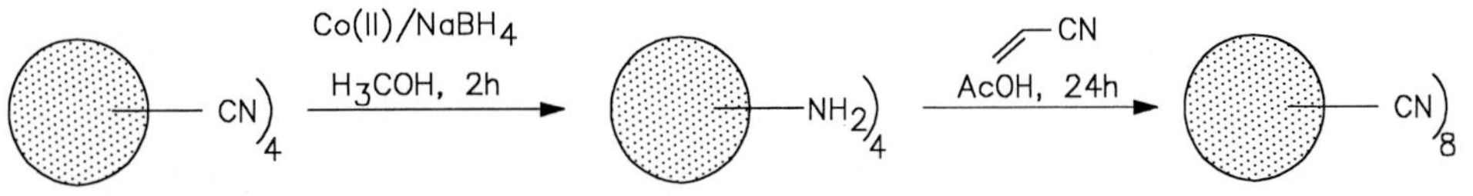

Scheme 2.1. The concept of "cascade" or "repeating" syntheses.

The essence of the "cascade"* synthesis is depicted in Scheme 2.1. Two procedures, alkylation and reduction, comprised the [a → b → a → b →] sequence. Thus, treatment of a diamine with acrylonitrile afforded tetranitrile **3**. Cobalt-mediated nitrile reduction gave tetraamine **4**. Further amine alkylation provided the second generation octanitrile **5**. The "nonskid-chain-like"[2] synthesis is shown in Scheme 2.2. Construction of polycyclic **6** was accomplished by repetitive alkylation, reduction, acylation, and reduction (a → b → c → d → a → b → ...) sequences. Again, repetitive and multiple reaction sequences were employed for the generation of new molecular assemblies. Most notable about these syntheses is that for the first time, 'generational' molecules were prepared and characterized at each stage of the construction process.

2.2.3.2 Implied Dendritic Construction

As it is perhaps the case with many scientific and technological advances, earlier accounts of synthetic efforts towards novel (macro) molecular assemblies logically implied the utility of the iterative method. Consider Figure 2.3, whereby Lehn[32] (Lehn, Pedersen, and Cram; Nobel Prize for Chemistry, 1987) pictorially delineates stepwise synthetic strategies for the construction of macrocyclic organic complexing agents. Multiple component attachment (in this case two) is used for ring and cage preparation of the "molecular cavities". Divergent as well as convergent methods are envisioned. Similar strategies for ligand construction can be found, as described in the work of Cram et al.[33] entitled "Chiral Recognition in Complexation of Guests by Designed Host Molecules". These two accounts were pivotal to the seminal cascade methodology and were noted by Vögtle.[31]

Finally, to underscore and stress the impact of seemingly unrelated works, an examination of critical reviews by pioneering scientists, such as Lehn,[32–35] Ringsdorf,[36] and Lindsey,[37] reveals unparalleled insight into potential utility of cascade-related macro-assemblies including, of course, dendrimers.

---

* "'Cascade syntheses' meant reaction sequences that could be carried out repeatedly, whereby a functional group is made to react in such a way as to appear twice in the subsequent molecule. However, 'nonskid syntheses' meant stepwise construction of polycyclic ring compounds by repeatedly occurring reaction sequences; thus, a ring system is connected by a new bridge, which possesses functional groups for the annexation of further bridges."

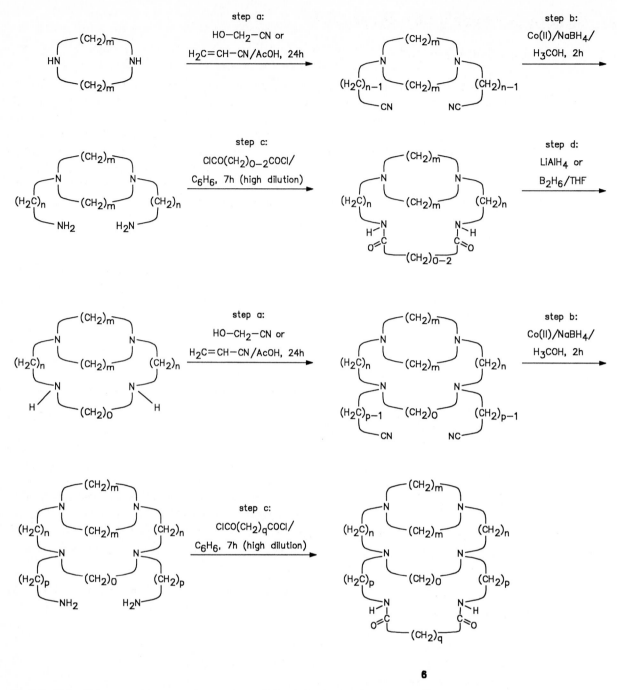

Scheme 2.2. Repetitive a → b → c → d sequences formed the basis of this so-called "nonskid-chain-like" synthesis.

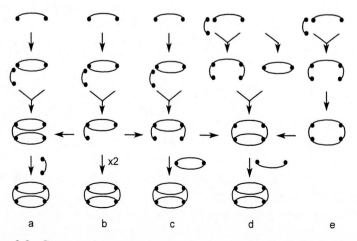

Figure 2.3. Construction pathways for "molecular cavities". Redrawn with permission of Jean-Marie Lehn.

2.2.3.3   Alternative Architectures

After the initial disclosure of "a viable" iterative synthetic method for the construction of polyfunctional macromolecules, a small number of articles explored the use of repetitive chemistry for the preparation of dendritic materials. Figure 2.4 illustrates the early branched architectures that were constructed, in most cases, by employing protection–deprotection schemes. Detailed descriptions of the synthetic procedure can be found in Chapters 4 (structures **7–9**) and 5 (structures **10** and **11**).

Denkewalter, Kolc, and Lukasavage[38] patented a method for the synthesis of poly-lysine-based dendrimers (**7**). Interesting features of these dendritic polymers include a 1 → 2 asymmetric branching pattern and the incorporation of multiple chiral centers at each tier. Aharoni, Crosby, and Walsh[39] studied these lysine dendrimers and reported that each member of the series was monodisperse; higher generations behaved as non-draining spheres (i.e., they trapped solvent within the void regions). Shortly thereafter, Kricheldorf, Zang, and Schwarz[40] demonstrated a renewed interest in branched topologies prepared via single-pot procedures, and they are currently referred to as 'hyper-branched' polymers (see Chapter 6).

Approximately six years after Denkewalter's patent, two new architectures were published in 1985. Newkome et al.[41] relied on triester amidation with 'tris' for the construction of polyols (**8**) possessing maximal 1 → 3 C-based branching. Repetition of the sequence was precluded by surface steric inhibition towards nucleophilic substitution. Tomalia et al.[42a] reported the preparation of an entire series of dendrimers (**9**, up to the 7th tier) possessing trigonal, 1 → 2 N-based, branching centers. To date, availability has

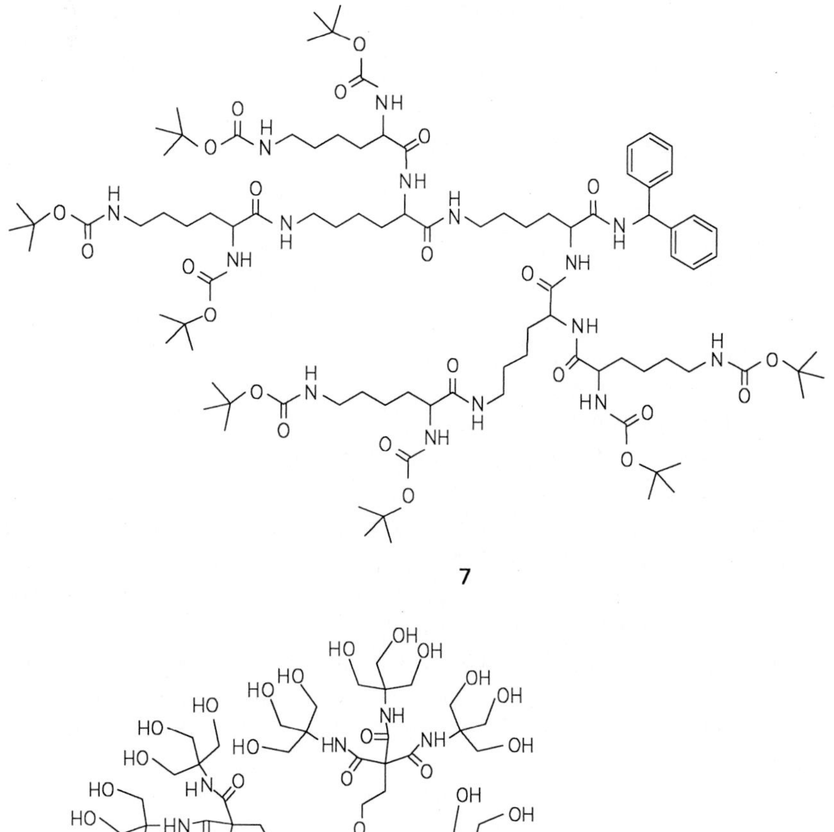

7

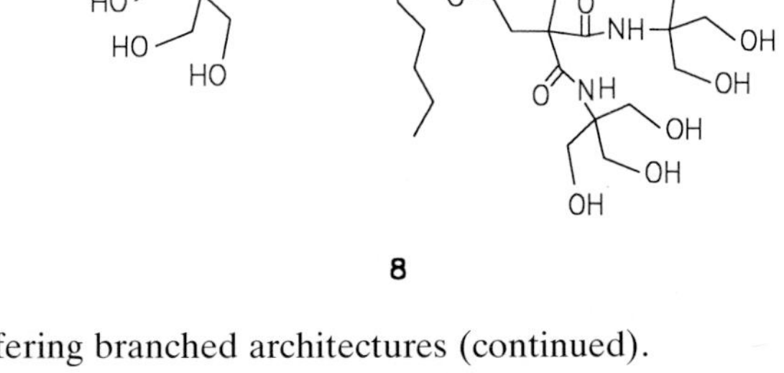

8

Figure 2.4. Illustration of differing branched architectures (continued).

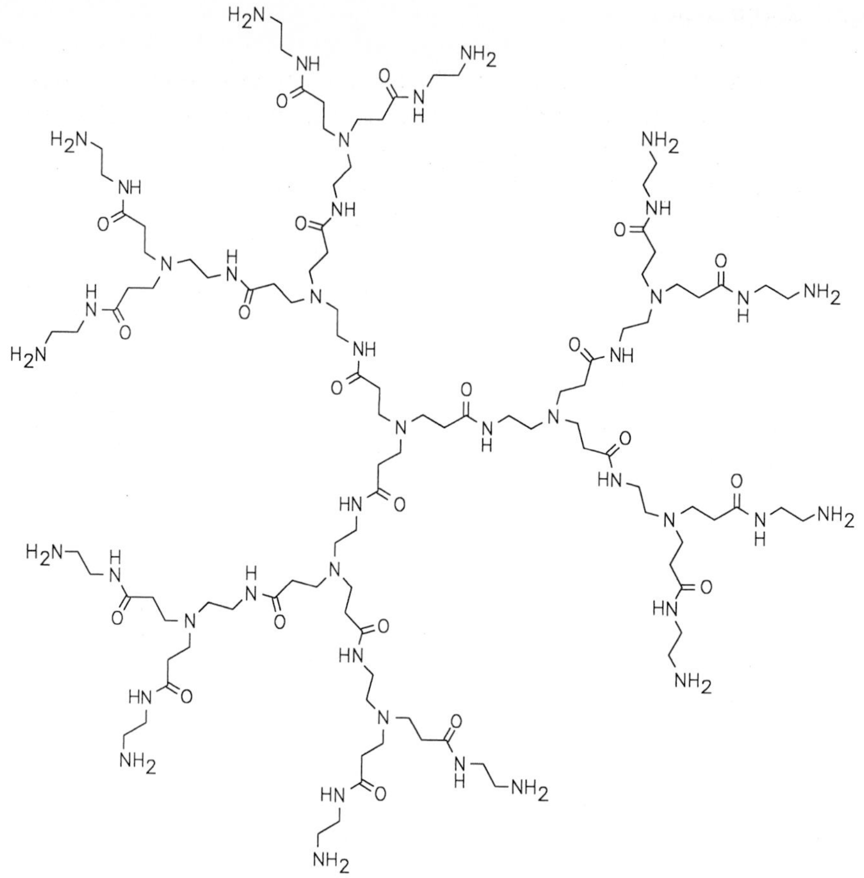

9

Figure 2.4. Illustration of differing branched architectures (continued).

made this series of dendrimers one of the most widely investigated families. It is interesting to note that, during 1985, Bidd and Whiting[42b] described the iterative-based synthesis of pure, linear alkyl hydrocarbons possessing 102, 150, 198, 246, and 390 carbon atoms. Preparative methodology involved the use of repetitive Wittig coupling and cyclic acetal hydrolysis reactions.

The first convergent preparation of dendrimers resulted in poly(aryl ether) architecture (**10**) as reported by Fréchet and Hawker[43] in 1990. Innovative use of their pivotal phenoxide-based, benzylic bromide displacement sequence has led to many creative and novel macromolecular assemblies. During that same year (1990), Miller and Neenan[44] published their efforts with respect to the convergent preparation of the first series of aromatic-based, all-hydrocarbon dendrimers (**11**). High rigidity was inherent in this series; other notable pioneers in the area of convergently constructed rigid, branched topologies, included Moore and Xu.[45]

After Flory[46] and Stockmayer[47] published books describing the statistical mechanics of chain molecules and molecular fluids, respectively, which further laid the foundation of modern polymer analyses, theoretical examination of branched macromolecules continued to advance.

Notable theoretical investigations of branched polymers at this time included the treatise by Maciejewski[48] which examined "trapping topologically by shell molecules". This author alludes to numerous dendritic properties such as the concepts of *dense packing* whereby building block connectivity is precluded due to steric hindrance and *void volume entrapment* of solvent (or guest) molecules much like the non-draining spheres described by Denkewalter et al.[38] Speculation about the preparation of macromolecules with infrastructures consisting of all quaternary carbons (e.g., a 'diamond' polymer) as well as assemblies possessing cylindrical topologies was further included. Buchard, Kajiwara, and Nerger[49] examined dynamic and static light scattering of "regularly branched chain molecules" and formulated a soft-sphere microgel model. Sponge[50] investigated particle aggregation whereby pair-wise bonding formed "tree-structures".

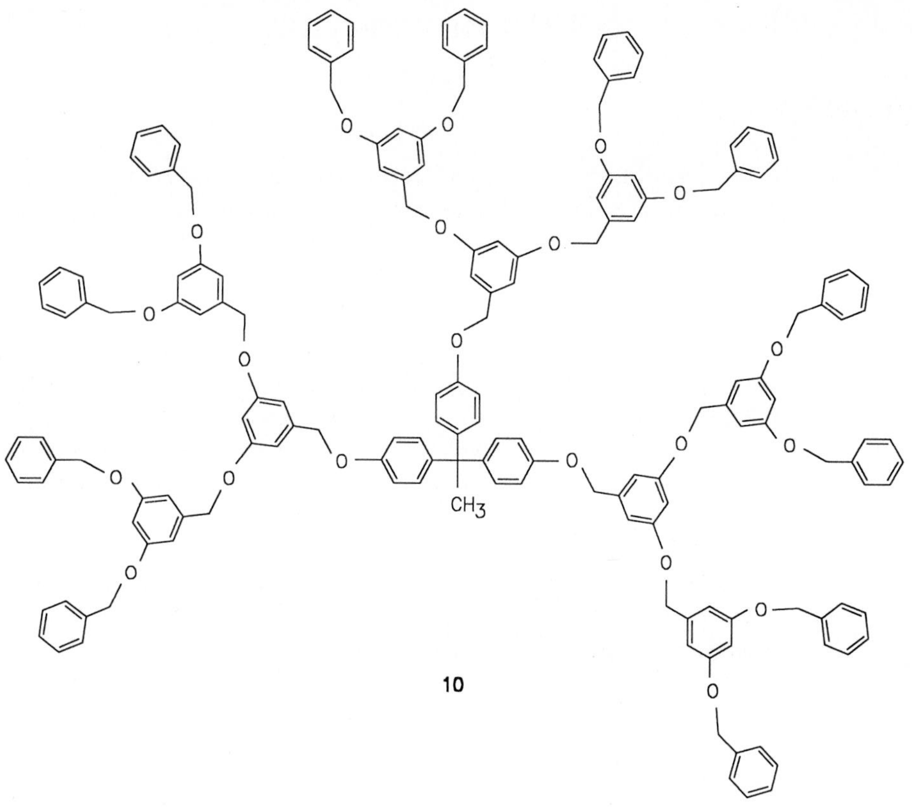

10

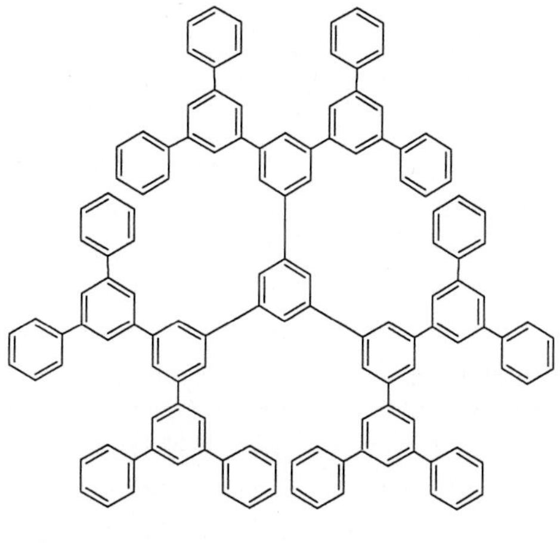

11

Figure 2.4. Illustration of differing branched architectures.

Consideration was given to multibranching processes: finite aggregation distribution (a *sol*), infinite aggregate (a *gel*) existence criteria, and relations for the determination of mole- and weight-average molecular masses of the finite aggregates.

De Gennes (Nobel prize in Physics, 1991) and Hervet[51] published a statistical treatment of starburst dendrimers. They concluded from a mathematical growth model that steric hindrance limiting continued tier addition was dictated by the length of the spacer units that connect the branching centers. This reinforced the dense packing concept.

These initial theoretical investigations, as well as the early alternative architectures that were synthesized, serve as an exordium for the expanding and synergistic discourse of dendritic material science. In-depth discussion pertaining to more recent synthetic and theoretical reports can be found beginning at Chapter 4.

2.3 The Fractal Geometry of Macromolecules

2.3.1 Introduction

Understanding (molecular) geometry leads to a better understanding of (molecular) physical properties, and visa versa. The intrinsic geometrical beauty of the hexagonal symmetry of benzene is inextricably related to its physical properties. For example, the chemical and magnetic equivalence of the constituent carbon and hydrogen nuclei, for unsubstituted benzene, are directly related to the atomic juxtaposition. Hence, just classical Euclidean geometrical shapes such as the regular hexagon, circle, or cone are indispensable to a discussion of small molecules, 'fractal' geometry allows insight into the structures and properties of macromolecules, such as dendrimers.

This section attempts to examine macromolecular geometry, and in particular dendritic surface characteristics, from the perspectives of self-similarity and surface irregularity, or complexity, which are fundamental properties of basic fractal objects. It is further suggested that analyses of dendritic surface fractality can lead to a greater understanding of molecule/solvent/dendrimer interactions based on analogous examinations of other materials (e.g., porous silica and chemically reactive surfaces such as found in heterogeneous catalysts).[52]

2.3.2 Fractal Geometry

'Fractal' geometry was introduced and pioneered in the mid-70's by the brilliant mathematician Professor Benoit B. Mandelbrot.[53–55] His development of this new mathematical language has led to a greater understanding of seemingly highly disordered objects and shapes. In short, fractal geometry provides a rational description of complicated structures.

Intuitively, objects such as lines, squares, and cubes possess dimensionalities of 1.0, 2.0, and 3.0, respectively. It is also rational to expect that many natural, as well as man-made, objects possess non-integral, or fractional, dimensionalities due to complicated patterns. Classical examples of common fractal patterns and forms include naturally-occurring objects such as coastlines, clouds, mountains, and snowflakes.[53–55]

A rigorous treatment of fractal mathematics is beyond the scope of this treatise; therefore, only essential concepts relating to a descriptive understanding of fractal shapes and analyses will be discussed. More complete and comprehensive reviews of this growing area are available.[52, 56–61]

2.3.2.1 Self-Similarity or Scale-Invariance

For a limited discussion of fractal geometry, some simple descriptive definitions should suffice. *Self-similarity* is a characteristic of basic fractal objects. As described by Mandelbrot,[58] "When each piece of a shape is geometrically similar to the whole, both the shape and the cascade that generate it are called *self-similar*." Another term that is synonymous with self-similarity is *scale-invariance*, which also describes shapes that remain constant regardless of the scale of observation. Thus, the *self-similar* or *scale-invariant* macromolecular assembly possesses the same topology, or pattern of atomic connectivity,[62] in small as well as large segments. Self-similar objects are thus said to be invariant under dilation.

2.3.2.2 Fractal Dimension (*D*)

Fractal dimension, D, is another crucial property that is used to describe fractal objects and shapes. It is a measure of the amount of irregularity, or complexity, possessed by an object. For lines, $1 \leq D < 2$ and for surfaces, $2 \leq D < 3$. The greater the value of D, the more complex the object. As described by Avnir,[63] "D is obtained from a resolution analysis: the rate of appearance of new features of the irregularity as a function of the size of the probing yardstick (or degree of magnification) is measured. An object is fractal if the rate is given by the power law, $n \propto r^{-D}$, where n is the number of yardsticks of the size r needed to measure the total length of the wiggly line."

"More generally, a power-law scaling relation characterizes one or more of the properties of an object or of a process carried out near the object:

$$property \propto scale^{\beta} \tag{2.1}$$

Examples for <<property>> are the surface area, the rate of a heterogeneous reaction or shape of adsorption isotherm. The scales, or yardsticks, would be pore diameter, cross-sectional area of an adsorbate, particle size, or layer thickness. The exponent β is an empirical parameter which indicates how sensitive is the property to changes in scale, and depending on the case, it can be either negative (e.g., in length measurement) or positive (e.g., in measurements of mass distribution)."[64]

Wegner and Tyler[65] define "the fractal dimension of an object as a measure of its degree of irregularity considered at all scales, and it can be a fractional amount greater than the classical geometrical dimension of the object. The fractal dimension is related to how fast the estimated measurement of the object increases as the measurement device becomes smaller. A higher fractal dimension means the fractal is more irregular, and the estimated measurement increases more rapidly. For objects of classical geometry, the dimension of the object and its fractal dimension are the same. A fractal is an object that has a fractal dimension that is strictly greater than its classical dimensions." Thus, "if the estimated property of an object becomes arbitrarily large as the measuring stick, or scale, becomes smaller and smaller, then the object is called a fractal object."

2.3.3 Applied Fractal Geometry

Fractal analysis provides an indication of complexity and a convenient method of categorization. In chemistry, it has been applied to the interpretation and understanding of macromolecular surface phenomena. Fractal geometry thus offers new insights and perspectives relating to nanoscale chemistry. It has further been applied to modeling of growth in plants and biological objects such as arterial and bronchial organs.

2.3.3.1 Fractals in Chemistry

During 1979, de Gennes[66] emphasized the importance of fractal geometry for the study of macromolecules with the publication of *Scaling Concepts in Polymer Physics*. His work delineated the potential to obtain information relating to polymeric properties via examination of scaling relationships. Later in 1983, de Gennes and Harvet[51] reported that the <<property>> of dendritic radius (R) was proportional to the corresponding molecular weight (M) raised to a fractional exponent (i.e., $R \approx M^{0.2}$ for low M; $R \approx M^{0.33}$ for M at or near the growth limit). No specific reference to dendritic fractality was mentioned; however, a fractional exponential, or scaling, relationship between dendritic properties was established. Dendritic surfaces were thus described as "showing some interesting cusps, which may become a natural locus for stereochemically active sites."

Pfeifer, Wely, and Wippermann[67] examined surface portions of a lysozyme protein molecule. A fractal dimension of $D = 2.17$ was determined and related to the rate of substrate trapping at the enzyme active site. The root mean square substrate displacement via diffusion at the surface increases as time $(t)^{1/D}$; whereas surface capture of substrates is greater when compared with a smooth ($D = 2$) surface. From their results, the authors suggested that the observed value of $D = 2.17$ corresponds to an optimum overall capture rate.

Lewis and Rees[68] determined values of D equal to 2.44, 2.44, and 2.43 for the proteins lysozyme, ribonuclease A, and superoxide dismutase, respectively. Protein regions associated with tight complexes (i. e., interfaces and antibody-combining regions) were shown to be more irregular than transient complex areas, such as active sites.

Muthukumar[69] described a theory of a fractal polymer possessing solution viscosity. Solutions containing dilute, semidilute, and high concentrations of fractal polymers were examined; intra- and interfractal hydrodynamic interactions as well as excluded volume effects were included in the treatment.

Klein, Cravey, and Hite[70] delineated the fractality of benzenoid hydrocarbons constructed from regularly repeating hexagonal patterns. It was suggested that these fractal hydrocarbons might serve as "zero-order" models for carbonaceous materials such as coals, lignites, chars, and soot. Generalized protection–deprotection schemes were described for the construction of various fractal ring systems without specifically proposing the complimentary functional groups to be employed.

Avnir et al.[63] employed improved computerized image analysis of boundary lines of objects that possess irregular surfaces. Standard fractal line analyses were demonstrated to be insensitive leading possibly to erroneous conclusions. Objects analyzed included the protein α-cobratoxin ($D = 1.11$), Pt-black catalyst ($D = 1.20$), the Koch curve ($D = 1.34$), and an arbitrary object – a *rabbit* ($D = 1.06$). The authors point out that caution should be taken in the interpretation of experimental values of $D < 1.2$ as demonstrated by obtaining low D values for any "low irregularity line, even for lines that are neither fractals nor self-similar" (e. g., as evidenced by the implication of the fractal nature of a *rabbit* outline). As Avnir suggests, prudence is necessary due to a general 'smoothing effect' that log–log plots impart on data as well as the points which are plotted are sometimes indicative of limited or localized regions only.

Aharoni et al.[71] prepared hyperbranched polymers possessing rigid $1 \rightarrow 2$ branching centers and stiff "rod-like" spacers, or connectors. Scanning electron microscopy of dried material revealed a fractal morphology. Porosimetry experiments supported the fractal supposition. These authors proposed that single-step polymerizations form fractal polymers. As the polymerization proceeds, a contiguous network is formed, described as an "infinite cluster of polymeric fractals". Gelation occurs at this point.

Abad-Zapatero and Lin[72] examined globular protein surfaces and suggested that the exponent of the Box–Cox transformation[73] is a function of the fractal dimension D and the shape parameter S. D was approximated at 2.2 for two lysozymes and 2.4 for superoxide dismutase, which agrees well with previously reported values of 2.19[74] and 2.43,[68] respectively.

Avnir and Farin[64, 75–77] have published numerous articles examining the fractal dimensions of reactive and adsorptive surfaces. The authors reported[75] that, in many situations, the reaction rate, v, with respect to substrates diffusing from the surrounding solution and a surface, scale with reactant radius, R, where $R^{D_R-3} \propto v$, and D_R is the 'reaction dimension'. D_R, it was suggested, could be thought of as the fractal dimension of the surface reaction sites. It was demonstrated that reaction efficiency could be enhanced by controling the geometrical parameters of the reacting material, i. e. decreasing particle size for $D_R < D, m > 1$ and increasing particle size for $D_R > D, m < 1$ reactions, where $m = (D_R-3)/(D-3)$. Also, D_R is not necessarily equal to D, the surface fractal dimension; this could perhaps be explained using a multifractal concept.[78]

Catalytic activity of a variety of dispersed metals, such as Pt, Pd, Ir, Ag, Rh, Fe, and Ni, as well as bimetallic catalysts dispersed on supports like SiO_2 and Al_2O_3 was also examined.[76] A wide range of values of D_R was found. For example $D_R = 0.2$ for ethylene oxidation with Ag on SiO_2 suggesting little activity dependence on particle size and a low concentration of reactive surface sites. At the other extreme, D_R was determined to be 5.8 for NH_3 synthesis on Fe/MgO suggesting severe sensitivity to structure. It should be noted at this point that D_R is derived via consideration of an 'effective' surface site geometry that has implications for the reaction under investigation but does not necessarily equate with a 'true' geometry. Therefore, D_R can vary over a greater range than that observed for D.

Fractal geometry was employed to study surface geometry effects on adsorption conformations of polymers.[77] Using poly(styrene) for the case study, it was shown, that for highly porous objects, solution conformation changed very little after solvent adsorption.

Avnir and Farin[64] extended and reviewed the fractal geometry studies of molecule–surface interactions by evaluating surface accessibility to physisorption of small molecules, surface accessiblity of proteins, surface–geometry effects on the adsorption of polymers, and surface accessibility with respect to adsorbate energy transfer.

Mansfield[79] performed Monte Carlo calculations on model dendrimers and determined that as a result of the unique architecture of the branches, even when similar chemically, they are well segregated. Further, he concluded that dendrimers are fractal (D ranges from 2.4 to 2.8) and self-similar only over a rather narrow scale of lengths.

2.3.3.2 Fractals in Biology

Fractal geometry is not only ubiquitous in dendritic topologies it is abundantly obvious in naturally occurring objects; this is acutely evident in physiological objects. West and Goldberger[80] assert that "the mathematical concept of fractal scaling brings an elegant new logic to the irregular structure, growth, and function of complex biological forms." Their review includes an excellent discussion of the fractal properties of motifs created by such objects as the human 'bronchial tree' that is essentially the lung superstructure and the human heart, which can be considered as being comprised of a 'fractal hierarchy' (i.e., it possesses layered fractal objects).

Rabouille, Cortassa, and Aon[81] dried protein, glycoprotein, or polysaccharide containing brine solutions that resulted in dendritic-like fractal patterns. The fractal dimension, $D = 1.79$, was determined for the pattern afforded by an ovomucin–ovalbumin mixture (0.1 M NaCl). Similar D values were obtained for dried solutions of fetuin, ovalbumin, albumin, and starch; the authors subsequently suggest that 'fractal patterning' is characteristic of biological polymers.

Plants also exhibit fractal patterns. Figure 2.5 illustrates the Leeuwenberg model of tree architecture as depicted by Professor P. B. Tomlinson[82] in a treatise on tree motifs and a primitive 27-arborol, prepared by Newkome et al.[41] Features of the Leeuwenberg model include 1 → 3 branching centers and a crowded 'canopy', or periphery, much like the morphology possessed by a dendrimer constructed via building blocks that incorporate 1 → 3 C-based branching. It is interesting to note that this model is but one of many architectures delineated in Tomlinson's fascinating article.

During 1990, Lindenmayer systems, or L-systems, were developed as a means to graphically model the fractal geometry of plants. Iteration of 'compressed' mathematical descriptions of plant architecture generates the models that are delineated in the text entitled "The Algorithmic Beauty of Plants".[83] Figures 2.6 and 2.7 illustrate the use of L-system algorithms for the creation of "tree-like" patterns.[84] Algorithmic iteration resembles divergent, generational, dendritic growth resulting in dense packed surfaces and internal void regions.

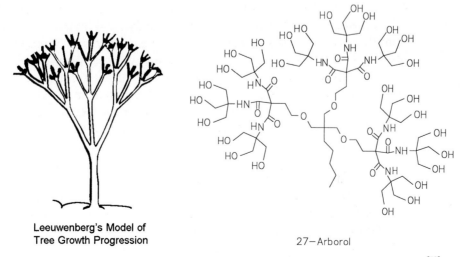

Leeuwenberg's Model of
Tree Growth Progression

27–Arborol

Figure 2.5. Leeuwenberg's model of tree architecture (redrawn with permission of *American Scientist*[82]) as compared to a small dendrimer.

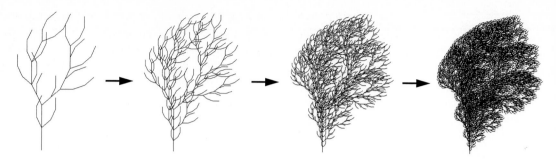

Figure 2.6. A bush-like geometry created by a "L-system" algorithm.

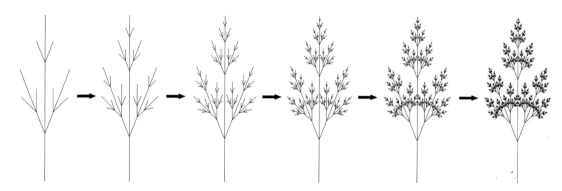

Figure 2.7. An "L-system" algorithm was employed for the generation of this tree-like architecture with $1 \rightarrow 3$ branching.

2.3.3.3   Solvent Accessible Surface Area (A_{SAS})

In 1990, it had been proposed[85] that the repetitive, generational, branching topology inherent in dendritic structures could be characterized as being fractal. Support for this conjecture was provided by a computational[86] examination of the *solvent accessible surface area* (A_{SAS}) of dendrimers at different generations.

Figure 2.8 illustrates the SAS concept as well as the method used for its determination. SASs are essentially computed by generating a three-dimensional, graphical representation of the dendrimer and computationally 'rolling' probes (p) of various radii (r) over the surface. Intuitively, as well as physically, the larger the probe radius the less chance for contact the probe has within the internal void region of the dendrimer. For probes with a small radius, and in particular at the limit $r = 0$ Å, the total internal surface area can be determined. Typically, the solvent accessible surface area ($\sqrt{A_{SAS}}$) is plotted versus probe radius or diameter. Thus, a measure of dendritic porosity can be derived.

Figure 2.9 illustrates the porosity of a series of polyamido, acid terminated dendrimers[87, 88] (generations 1 through 4) as determined by a plot of $(A_{SAS})^{1/2}$ vs. probe radius. For an ideal, completely space filled sphere, $A_{SAS} = 4\pi(R + P)^2$, where R is the radius of the sphere and P is the probe radius. Thus, a perfect sphere displays a linear plot of $A_{SAS}^{1/2}$ vs. P with slope of $2\pi^{1/2}$ and an ordinate intercept proportional to R at $P = 0$. Extrapolation of the linear portion of the curves to $P = 0$ (linear regression analysis) gives dashed lines that would be obtained by determination of A_{SAS} for a smooth surfaced sphere (fractal dimension $D = 2$). Plotted solid curves, at generations 3 and 4 (108 and 324 carboxylic acid moieties, respectively) reveal a greater *internal* surface area than *external* surface area (with radius R at $P = 0$).

These curves thus suggest an open and vacuous structure characterized by dynamic channels and pockets. The solvent accessible surface area, as measured for the first generation 12-acid, indicates a non-porous structure possessing little or no void regions as evidenced by experimental values falling below the ideal surface area. Data for the second generation 36-acid reveal a structure intermediate between "dendritic" and simply "branched".

These data are corroborated[85] by A_{SAS} measurements for the polyamidoamine (PAMAM) dendrimers (see Figure 2.4; **9**) as well as a series of compact polyethereal

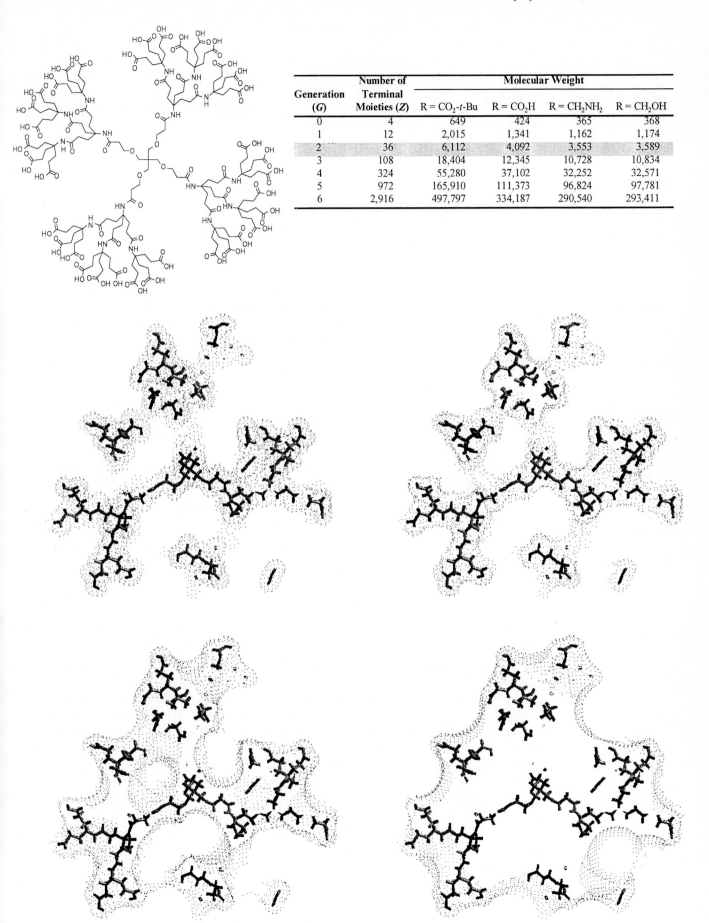

Generation (G)	Number of Terminal Moieties (Z)	Molecular Weight			
		R = CO₂-*t*-Bu	R = CO₂H	R = CH₂NH₂	R = CH₂OH
0	4	649	424	365	368
1	12	2,015	1,341	1,162	1,174
2	36	6,112	4,092	3,553	3,589
3	108	18,404	12,345	10,728	10,834
4	324	55,280	37,102	32,252	32,571
5	972	165,910	111,373	96,824	97,781
6	2,916	497,797	334,187	290,540	293,411

Figure 2.8. Solvent accessible surface areas calculated for a second generation amide-based dendrimer using probes with increasing radii.

Porosity of Acid Terminated Polyamido Cascades

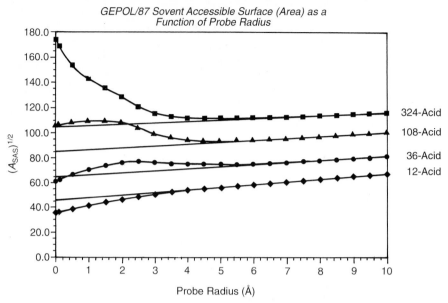

Figure 2.9. Graphical depiction of porosity for generations 1 through 4 of polyamido, carboxylic acid terminated cascades.

dendrimers. Comparison of the PAMAM series, which is structurally characterized by a $1 \rightarrow 2$ *N*-branching pattern, to the $1 \rightarrow 3$ *C*-branched polyamide acid series reveals that a $1 \rightarrow 3$ branching pattern reaches fractal dendritic status more rapidly (i.e., generation 3 for the $1 \rightarrow 3$ growth series vs. generation 5 for the $1 \rightarrow 2$ series). It should be further noted that these measurements are based on static, computer generated models; as usual, experimental data are needed to fully validate, or substantiate, the findings.

2.3.3.4   Dendritic Fractality

The self-similarity (see Section 2.3.2.1) of a dendrimer is readily apparent when each generation is viewed consecutively. Figure 2.10 shows the three-dimensional, computer-generated space-filling models of the first through the fourth tier polyamido, acid-terminated dendrimers.[87, 88] The appearance of each generation is strikingly similar to the next with respect to such features as $1 \rightarrow 3$ branching, distance between branching centers, and nuclei connectivity. It should be noted that the requirement of self-similarity does not necessitate the strict use of similar monomers at each generation. Each tier is gray coded in order to visually demonstrate the openness and accessibility.

Generational self-similarity is also evident in Figures 2.11 and 2.12. The former shows a computer generated quadrant of a third tier, hydrocarbon-based, carboxylic acid terminated dendrimer[89] with the van der Waals surface added. Repetition of the branching pattern at different generations results in large, superstructure-bounded void volumes, as indicated by the atomic ruler juxtaposed along an extended arm. Figure 2.12 depicts a cross-sectioned and space-filling view of two dendrimers specifically connected[90] via a terpyridine–ruthenium–terpyridine complex centrally located in each diagram. Scale invariance at differing levels of observation is apparent within the limiting physical boundaries of the macroassembly. The diagrams (Figures 2.10–2.12) support the implications derived from SAS calculations in the previous section.

Determination of the fractal nature of a dendritic surface was first reported by Avnir and Farin.[91] Employing data obtained from solvent accessible surface area analysis (see Sect. 2.3.3.3) for the PAMAM dendrimers, a surface fractal dimension (D) was derived. Two methods were used. The first method applied the relation,

$$A \approx \sigma^{2-D/2} \tag{2.2}$$

in which A and σ correspond to the accessible dendritic surface area and the probe radius, respectively. For a generation six dendrimer (192 termini) in the A_{SAS} region of

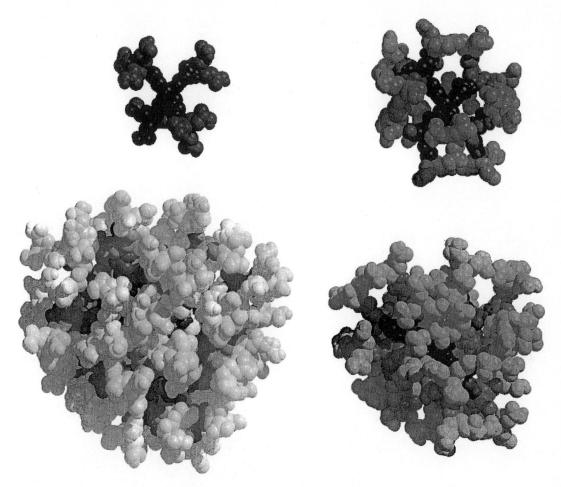

Figure 2.10. Gray coded, space filling models of dendrimers at generations 1 through 4. Internal accessibility to guests via a dynamic porosity is readily apparent.

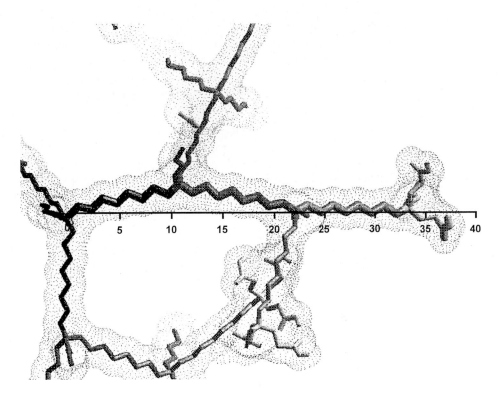

Figure 2.11. Computer generated extended quadrant of a hydrocarbon-based dendrimer illustrating segmented self-similarity.

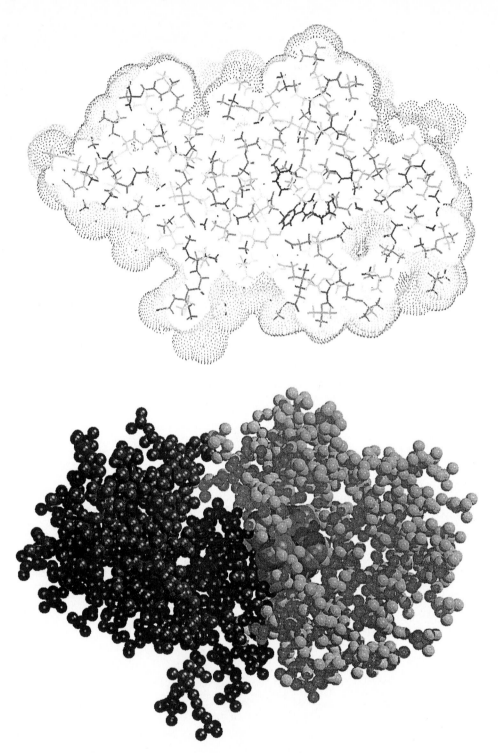

Figure 2.12. Wire-frame and space-filling representations of two metal-connected dendrimers.

probe radius 1.5–7.0 Å, $D = 2.41$ (as determined from a log–log plot of A vs. σ). This result compared favorably with that obtained via the application of another proportional relationship, where

$$A \approx d^D \tag{2.3}$$

in which d represents the size of an object with surface area A. A log–log plot of A vs. d gives $D = 2.42$.

Similar analyses of A_{SAS} data for a much more compacted pentaerythritol-based poly-ether dendrimer[85] gave $D = 1.96$. Comparison of the fractal dimensions for both series, i.e., the PAMAMs and polyethers, suggests structural differences. For $D = 1.96$, surface complexity and irregularity is minimized and it is relatively smooth. This is corroborated by the difficulty, in forcing all of the desired surface reactions to completion, when attempting to prepare higher generations; functional group surface congestion is most likely to be blamed due to short spacers, or connectors, between tetrahedrally-substituted, *C*-branching centers. For the sixth generation PAMAM dendrimer, the fractal dimension ($D = 2.42$) suggests an open and complex surface comprised of repeating peaks and valleys.

A similar analysis of the surface fractal dimension for the polyamido, acid terminated dendrimers has been conducted.[92] Analysis of the generations 1 through 4 of solvent accessible surface areas with probes of various sizes produced the GEPOL/87 curves pictured in Figure 2.9 (also see Figure 2.8). Application of Equation 2.2 resulted in the construction of a best fit straight line relation for a log–log plot of *A* vs. σ for the third tier 108-acid and the fourth tier 324-acid (Figure 2.13). A first order fit of the data afforded slopes corresponding to –2.00 and –2.17 solving for fractal dimensions (i.e., slope = $(2–D/2)$): $D = 2.40$ and $D = 2.43$, respectively.

These numbers compare well with those determined by Farin and Avnir[91] for the higher level PAMAMs. Employing the relationship $A = d^D$ (Eqn. 2.3), where *d* is represented by the log of the experimentally determined DOSY NMR[87] dendritic diameter, $D = 2.29$ (Figure 2.14).

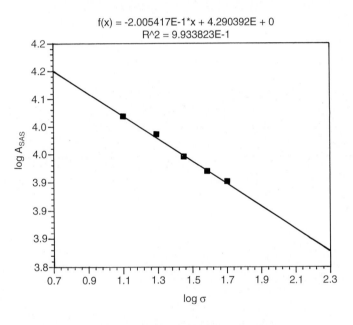

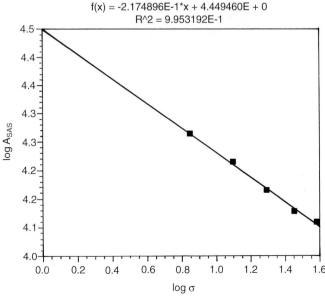

Figure 2.13. Determination of the "fractal dimension" ($D = 2.40$ and $D = 2.43$) for 3rd (top) and 4th (bottom) generation amide-based dendrimers. [See the table presented in Figure 2.8]

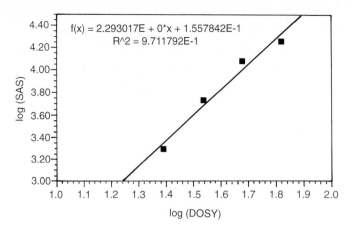

Figure 2.14. Employing the log of experimentally determined dendritic diameters the fractal dimension $D = 2.29$ was found for a series of acid terminated dendrimers. [See the table presented in Figure 2.8]

2.3.4   Fractal Summation

According to Farin and Avnir,[76] an important "contribution of fractal geometry has been a clarification of the physical meaning of *non-integer* dimensions and the creation of a continuous scale of dimensions." With respect to dendritic chemistry, fractal geometry affords a novel method of comparison of branched architecture as well as, and perhaps more importantly, a new way to envision these unique macromolecules and their properties. Dendritic surface complexity is derived from the features of the supporting superstructure which is in turn derived from the inherent nature of the building blocks used for its construction. Thus, the fractal geometry of dendritic macroassemblies implies translation, or transcription, of molecular level information to much larger superassemblies. Conveyance of information is based on topological 'scaffolding' connectivity of dendritic structures and this affords an enhanced perspective of this emerging discipline.

2.4   References

[1]   P. J. Flory, *J. Am. Chem. Soc.* **1952**, *74*, 2718.
[2]   T. Zincke, *Chem. Ber.* **1869**, *2*, 739.
[3]   C. Friedel, J. M. Crafts, *Bull. Soc. Chim.* **1885**, *43*, 53.
[4]   J. H. Gladstone, J. Tribe, *J. Chem. Soc.* **1885**, *47*, 448.
[5]   W. Ostwald, *Kolloid Z.* **1907**, *1*, 331.
[6]   W. Ostwald, *Z. Chem. Ind. Kolloide* **1908**, *3*, 28.
[7]   T. Graham, *Trans. Roy. Soc. (London)* **1861**, *151*, 183.
[8]   H. Hlasiwetz, J. Habermann, *Ann. Chem. Pharm.* **1871**, *159*, 304.
[9]   P. J. Flory, *Principles of Polymer Chemistry*, Cornell University Press, Ithaca, New York, **1953**, p. 6.
[10] W. H. Hunter, G. H. Woollett, *J. Am. Chem. Soc.* **1921**, *43*, 135.
[11] C. K. Ingold, L. C. Nickolls, *J. Chem. Soc.* **1922**, *121*, 1638.
[12] See: R. Ferdinand, *Principles of Polymer Systems*, McGraw Hill, New York, **1970**, p. 8.
[13] W. H. Carothers, *Chem. Rev.* **1931**, *8*, 353.
[14] R. A. Jacobson, *J. Am. Chem. Soc.* **1932**, *54*, 1513.
[15] W. Kuhn, *Chem. Ber.* **1930**, *63*, 1503.
[16] See p. 19 in ref. 9.
[17] A.-V. Lourenco, *Compt. Rend.* **1860**, *51*, 365.
[18] A.-V. Lourenco, *Ann. Chim. Phys.* **1863**, *67(3)*, 273.
[19] K. Kraut, *Ann.* **1869**, *150*, 1.
[20] See p. 14 in ref. 9.
[21] P. J. Flory, *J. Am. Chem. Soc.* **1941**, *63*, 3083.
[22] P. J. Flory, *J. Am. Chem. Soc.* **1941**, *63*, 3091.
[23] P. J. Flory, *J. Am. Chem. Soc.* **1941**, *63*, 3096.
[24] P. J. Flory, *J. Am. Chem. Soc.* **1942**, *64*, 132.

[25] W. H. Stockmayer, *J. Chem. Phys.* **1943**, *11*, 45.

[26] P. J. Flory, *J. Phys. Chem.* **1949**, *17*, 303.

[27] R. J. Goldberg, *J. Am. Chem. Soc.* **1952**, *74*, 5715.

[28] R. J. Goldberg, *J. Am. Chem. Soc.* **1953**, *75*, 3127.

[29] P. E. Rouse, Jr., *J. Chem. Phys.* **1953**, *21*, 1272.

[30] B. H. Zimm, *J. Chem. Phys.* **1956**, *24*, 269.

[31] E. Buhleier, W. Wehner, F. Vögtle, *Synthesis* **1978**, 155.

[32] J.-M. Lehn, "Design of Organic Complexing Agents. Strategies towards Properties" in *Structure and Bonding* (Eds. J. D. Dunitz, P. Hemmerich, J. A. Ibers, C. K. Jørgensen, J. B. Neilands, D. Reinen, R. J. P. Williams), Springer, New York, **1973**, Vol. 16, chapt. 1.

[33] D. J. Cram, R. C. Helgeson, L. R. Sousa, J. M. Timko, M. Newcomb, P. Moreau, F. de Jong, G. W. Gokel, D. H. Hoffman, L. A. Domeier, S. C. Peacock, K. Madan, L. Kaplan, *Pure Appl. Chem.* **1975**, *43*, 327.

[34] J.-M. Lehn, *Angew. Chem.* **1988**, *100*, 91; *Angew. Chem. Int. Ed. Engl.* **1988**, *27*, 89.

[35] J. M. Lehn, *Angew. Chem.* **1990**, *102*, 1347; *Angew. Chem. Int. Ed. Engl.* **1990**, *29*, 1304.

[36] H. Ringsdorf, B. Schlarb, J. Venzmer, *Angew. Chem.* **1988**, *100*, 117; *Angew. Chem. Int. Ed. Engl.* **1988**, *27*, 113.

[37] J. S. Lindsey, *New J. Chem.* **1991**, *15*, 153.

[38] R. G. Denkewalter, J. F. Kolc, W. J. Lukasavage, U. S. Pat. 4, 410, 688 (Oct. 18, **1983**).

[39] S. M. Aharoni, C. R. Crosby III, E. K. Walsh, *Macromolecules* **1982**, *15*, 1093.

[40] H. R. Kricheldorf, Q.-Z. Zang, G. Schwarz, *Polymer* **1982**, *23*, 1821.

[41] G. R. Newkome, Z.-Q. Yao, G. R. Baker, V. K. Gupta, *J. Org. Chem.* **1985**, *50*, 2003.

[42] a) D. A. Tomalia, H. Baker, J. R. Dewald, M. Hall, G. Kallos, S. Martin, J. Roeck, J. Ryder, P. Smith, *Polym. J.* **1985**, *17*, 117; b) I. Bidd, M. C. Whiting, *J. Chem. Soc., Chem Commun.* **1985**, 543.

[43] C. Hawker, J. M. J. Fréchet, *J. Chem. Soc., Chem. Commun.* **1990**, 1010.

[44] T. M. Miller, T. X. Neenan, *Chem. Mater.* **1990**, *2*, 346.

[45] J. S. Moore, Z. Xu, *Macromolecules* **1991**, *24*, 5893.

[46] P. J. Flory, *Statistical Mechanics of Chain Molecules*, Wiley, New York, **1969**.

[47] W. H. Stockmayer, in *Molecular Fluids* (Eds.: R. Balian, G. Weill), Gordon and Branch, New York, **1976**.

[48] M. Maciejewski, *J. Macromol. Sci. - Chem.* **1982**, *A17*, 689.

[49] W. Burchard, K. Kajiwara, D. Nerger, *J. Polym. Sci., Polym. Phys. Ed.* **1982**, *20*, 157.

[50] J. L. Spouge, *Proc. Roy. Soc. London* **1983**, *A387*, 351.

[51] P.-G. de Gennes, H. Hervet, *J. Phys. Lett.* **1983**, *44*, L-351.

[52] *The Fractal Approach to Heterogenous Chemistry: Surfaces, Colloids, Polymers*, (Ed.: D. Avnir), New York, Wiley, **1989**.

[53] B. B. Mandelbrot, *Les Objets Fractals: Forme, Hasard et Dimension*, Paris, Flammarion, **1975**.

[54] B. B. Mandelbrot, *Fractals: Form, Chance and Dimension*, San Francisco, Freeman, **1977**.

[55] B. B. Mandelbrot, *The Fractal Geometry of Nature*, San Franciso, Freeman, **1982**.

[56] H. Takayasu, *Fractals in the Physical Sciences*, Manchester University Press, Manchester, **1990**.

[57] A. Harrison, *Fractals in Chemistry*, New York, Oxford University Press, **1995**.

[58] Also page 34 in ref. 55.

[59] D. H. Rouvray, "Similarity in Chemistry: Past, Present, and Future", in *Topics in Current Chemistry: Molecular Similarity I*, Springer, New York, **1995**, *173*, 1.

[60] A. Blumen, H. Schnörer, *Angew. Chem.* **1990**, *102*, 158; *Angew. Chem. Int. Ed. Engl.* **1990**, *29*, 113.

[61] *Fractals, Quasicrystals, Chaos, Knots, and Algebraic Quantum Mechanics* (Eds.: A. Amann, L. Cederbaum, W. Gans), NATO ASI (C), Kluwer, Dordrecht, The Netherlands, **1988**, Vol. 235.

[62] M. Zander, "Molecular Topology and Chemical Reactivity of Polynuclear Benzenoid Hydrocarbons", in *Topics in Current Chemistry*, Springer, Berlin **1991**, *153*, 101.

[63] D. Farin, S. Peleg, D. Yavin, D. Avnir, *Langmuir* **1985**, *1*, 399.

[64] D. Farin, D. Avnir, *New J. Chem.* **1990**, *14*, 197.

[65] T. Wegner, B. Tyler, *Fractal Creations*, 2nd ed., Waite Group Press, Corte Madera, California, **1993**, p. 16.

[66] P.-G. de Gennes, *Scaling Concepts in Polymer Physics*, Cornell University Press, Ithaca, New York, **1979**.

[67] P. Pfeifer, U. Wely, H. Wippermann, *Chem. Phys. Lett.* **1985**, *113*(6), 535.

[68] M. Lewis, D. C. Rees, *Science* **1985**, *230*, 1163.

[69] M. Muthukumar, *J. Chem. Phys.* **1985**, *83*, 3161.

[70] D. J. Klein, M. J. Cravey, G. E. Hite, in *Polycyclic Aromatic Compounds*, Gordon and Breach, **1991**, Vol. 2, p. 163.

[71] S. M. Aharoni, N. S. Murthy, K. Zero, S. F. Edwards, *Macromolecules* **1990**, *23*, 2533.

[72] C. Abad-Zapatero, C. T. Lin, *Biopolymers* **1990**, *29*, 1745.

[73] G. E. P. Box, D. R. Cox, *J. Royal Stat. Soc., Series B* **1964**, *26*, 211.

[74] J. Aqvist, O. Tapia, *J. Mol. Graphics* **1987**, *5*, 30.

[75] D. Farin, D. Avnir, *J. Phys. Chem.* **1987**, *91*, 5517.

[76] D. Farin, D. Avnir, *J. Am. Chem. Soc.* **1988**, *110*, 2039.

[77] D. Farin, D. Avnir, *Colloid and Surfaces* **1989**, *37*, 155.

[78] J. Nittmann, H. E. Stanley, E. Touboul, G. Daccord, *Phys. Rev. Lett.* **1987**, *58*, 619.

[79] M. L. Mansfield, *Polymer* **1994**, *35*, 1827.

[80] B. J. West, A. L. Goldberger, *Am. Sci.* **1987**, *75*, 354.

[81] C. Rabouille, S. Cortassa, M. A. Aon, *J. Biomolec. Struct. Dynam.* **1992**, *9*, 1013.

[82] P. B. Tomlinson, *Am. Sci.* **1983**, *71*, 141.

[83] P. Prusinkiewicz, A. Lindenmayer, *The Algorithmic Beauty of Plants*, Springer, New York, **1990**.

[84] The computer-generated "trees" were constructed using the Fractint program available from the Waite Group in Fractal Creations: see ref. 65.

[85] D. A. Tomalia, A. M. Naylor, W. A. Goddard III, *Angew. Chem.* **1990**, *102*, 119; *Angew. Chem. Int. Ed. Engl.* **1990**, *29*, 138.

[86] J. L. Pascal-Ahuir, E. Silla, J. Tomasi, R. Bonaccors, *GEPOL, QCPE Program No. 554*, Quantum Chemistry Program Exchange Center, Bloomington Indiana, **1987**.

[87] G. R. Newkome, J. K. Young, G. R. Baker, R. L. Potter, L. Audoly, D. Cooper, C. D. Weis, K. F. Morris, C. S. Johnson, Jr., *Macromolecules* **1993**, *26*, 2394.

[88] J. K. Young, Ph. D. Dissertation, University of South Florida, **1993**.

[89] G. R. Newkome, C. N. Moorefield, G. R. Baker, A. L. Johnson, R. K. Behera, *Angew. Chem.* **1991**, *103*, 1205; *Angew. Chem. Int. Ed. Engl.* **1991**, *30*, 1176.

[90] G. R. Newkome, R. Güther, C. N. Moorefield, F. Cardullo, L. Echogoyen, E. Pèrez-Cordero, H. Luftmann, *Angew. Chem.* **1995**, *107*, 2159; *Angew. Chem. Int. Ed. Engl.* **1995**, *34*, 2023.

[91] D. Farin, D. Avnir, *Angew. Chem.* **1991**, *103*, 1408; *Angew. Chem. Int. Ed. Engl.* **1991**, *30*, 1379.

[92] G. R. Baker, C. N. Moorefield, J. K. Young, G. R. Newkome, unpublished results (USF), **1995**.

3   Nomenclature

3.1   Background on Trivial and Traditional Names

When Vögtle and his coworkers[1] described the first synthetic examples of discrete, branched, polyfunctional molecules prepared via an iterative, step-wise "cascade synthesis," he opened the door to a new vista of discrete meso- and macromolecules, whose structures can be easily envisioned, but are nearly impossible to name based on current nomenclature systems. Thus, researchers in the field resorted to naming their new materials with trivial names, such as: arborols,[2] cascadol,[3] cauliflower polymers,[4] crowned arborols,[5] dendrimers,[6] molecular fractals,[7, 8] polycules,[9] silvanols,[10] and "starburst" dendrimers.[6] Reliance on the IUPAC or *Chemical Abstracts* nomenclature resulted in names longer, in most cases, than the associated experimental details and were impossible for researchers to use for retrieval purposes. For example, one of the first macromolecules to be reported was [27]-arborol (Figure 3.1);[2] the *Chemical Abstracts* name is 1,19-dihydroxy-N,N',N'',N'''-tetrakis[2-hydroxy-1,1-bis(hydroxymethyl)ethyl]-10-[[4-[[2-hydroxy-1,1-bis(hydroxymethyl)ethyl]amino]-3,3-bis[[[2-hydroxy-1,1-bis(hydroxymethyl)-ethyl]amino]carbonyl]-4-oxobutoxy]methyl]-2,2,18,18-tetrakis(hydroxymethyl)-4,16-dioxo-10-pentyl-8,12-dioxa-3,17-diazanonadecane-5,5,15,15-tetra-carboxamide. From the name one cannot readily ascertain the terminal (or surface) groups (the 27 alcohol units), the branching multiplicity (three (3)) or the initiator core (a 1,1,1-trisubstituted hexane moiety). Application of the following rules[11, 12] for Figure 3.1 leads to 27-Cascade:hexane[3-1,1,1]:(4-oxapentylidyne):(3-oxo-2-azapropylidyne):methanol.

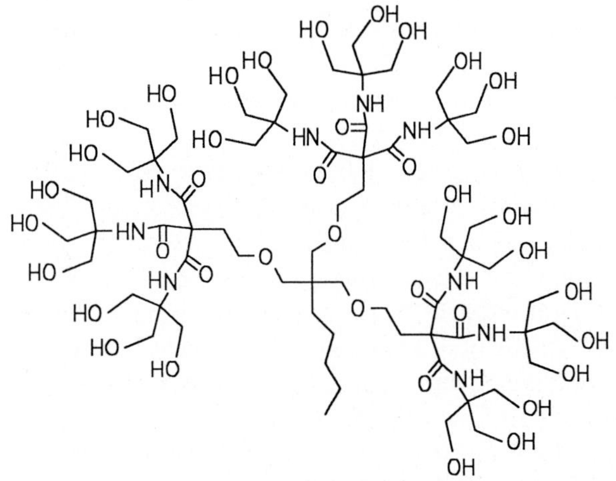

Figure 3.1.   27-Cascade:hexane[3-1,1,1]:(4-oxapentylidene):(3-oxo-2-azapropylidene):methanol.

3.2   Definition of a Cascade Polymer

A cascade polymer at the nth generation has the general formula:

$$C[R_1(R_2(...R_1(...R_n(T)N_{b_n}...)N_{b_i}...)N_{b_2})N_{b_1}]N_c \qquad (3.1)$$

where C is the formula for the core moiety; R_i is the formula for the repeat or branch unit; T is the formula for the terminal moieties; N_{b_i} is the branch multiplicity of the ith repeat unit or generation; and N_c is the multiplicity of the branching from the central core. The number of terminal moieties (Z) is calculated:

$$Z = N_c \prod_{i=1}^{n} N_{b_i} \tag{3.2}$$

If the branching multiplicity remains constant throughout the macromolecule, i. e. $N_{b_1} = N_{b_2} = \cdots = N_{b_n}$, then the relationship simplifies to give: $Z = N_c N_b^G$. For cascade polymers possessing the same branch unit throughout the structure, the line formula may be more simply represented by:

$$[\text{Core unit}][\text{Repeat Unit})_{N_b}^{G}(\text{Terminal Unit})]_{N_c} \tag{3.3}$$

The proposed nomenclature was derived from these line notations.

3.3 Proposed Cascade Nomenclature Rules[11, 12]

   1. In the cascade name, components (names of the units) are separated from each other by colons and are cited in sequence from the core unit out to the terminal units.
   2. The name begins with a numeral corresponding to the number of terminal functionalities, followed in succession by "Cascade" (to denote this class of molecules) and names of the core unit, repeat intermediate unit(s), and finally the terminal unit(s).
   3. The combination of core and terminal unit names resembles conjunctive nomenclature. The multiplicity of branching (cascading) from the core unit is indicated by a bracketed numeral immediately following the name of the core unit; if locants are necessary, they are also enclosed within the brackets, following and separated by a hyphen from the multiplicity numeral.
   4. A repeat intermediate unit consists of the molecular fragment extending from (but not including) one branch atom (or group) through the next cascade branching site.
   5. The parent chain of an intermediate or terminal unit always terminates at a cascade branching site.
   6. A superscript on a parenthetical unit denotes the number of successive repetitions of that unit.
   7. Numbering of chains of units, including the core and terminal ones, is in descending order in the direction core → terminal unit. This order preserves the well established IUPAC rules that the locant for example -CO_2H is one (1) and that for the point of attachment of the alkyl and related groups is also one (1).
   8. Repetition of combinations of repeat units is indicated by enclosure of the component repeat names (inside parentheses and separated by colons) within brackets. A superscript on the bracketed unit denotes the number of repetitions of that sequence.
   9. When the repeat unit is composed of nonequivalent branches extending from a cascade branching site, the name of that unit includes a name for each branch. Within the parenthetical name of the repeat unit, the branch names are cited by increasing parent chain length and then alphabetically, as well as are separated by colons. This situation usually results in the attachment of nonequivalent terminal groups at the last cascade branch point; in such cases, the terminal unit names are cited in sequence of increasing parent chain length and separated by colons.
   10. For cascade molecules with different arms emanating from the core, e. g., a segmental block cascade polymer, the name of each cascade segment, exclusive of the core, is given within square brackets. The segment names are separated by a hyphen, arranged in alphabetical order, and proceeded by core locants and group multipliers, when necessary.

3.4 General Patterns

3.4.1 Similar Repeat Internal Units

Z-Cascade:Core[N_c]:(Internal Units)^n:Terminal Unit

The simplest cascade structure[13] possesses symmetrical branches and identical repeat units. The second generation Micellanoic Acid^{TM} dendrimer has the structure shown in Figure 3.2 and the cascade name is given below the structure. The IUPAC name for this structure is 4,4,40,40-tetrakis(propylcarboxy)-13,13,31,31-tetrakis[12-carboxy-9,9-bis(3-propylcarboxy)dodecyl]-22,22-bis[21-carboxy-18,18-bis(propylcarboxy)-9,9-bis[12-carboxy-9,9-bis(propylcarboxy)dodecyl]heneicosyl]tritetracontanedioic acid. The cascade name is derived in several steps. First, the number of the terminal groups is determined by calculating the product of bracket subscripts (i.e., $3 \times 3 \times 4$); this number prefixes the class designation in the name (i.e., 36-Cascade). From the line formula representation, the initiator core is a tetrasubstituted methane (i.e., methane[4]). The next two molecular fragments are an identical nine (9) carbon chain (i.e., (nonylidyne)^2). Finally, the termini are propanoic acid groups.

$$C[(CH_2)_8C[(CH_2)_8C[CH_2CH_2CO_2H]_3]_3]_4$$

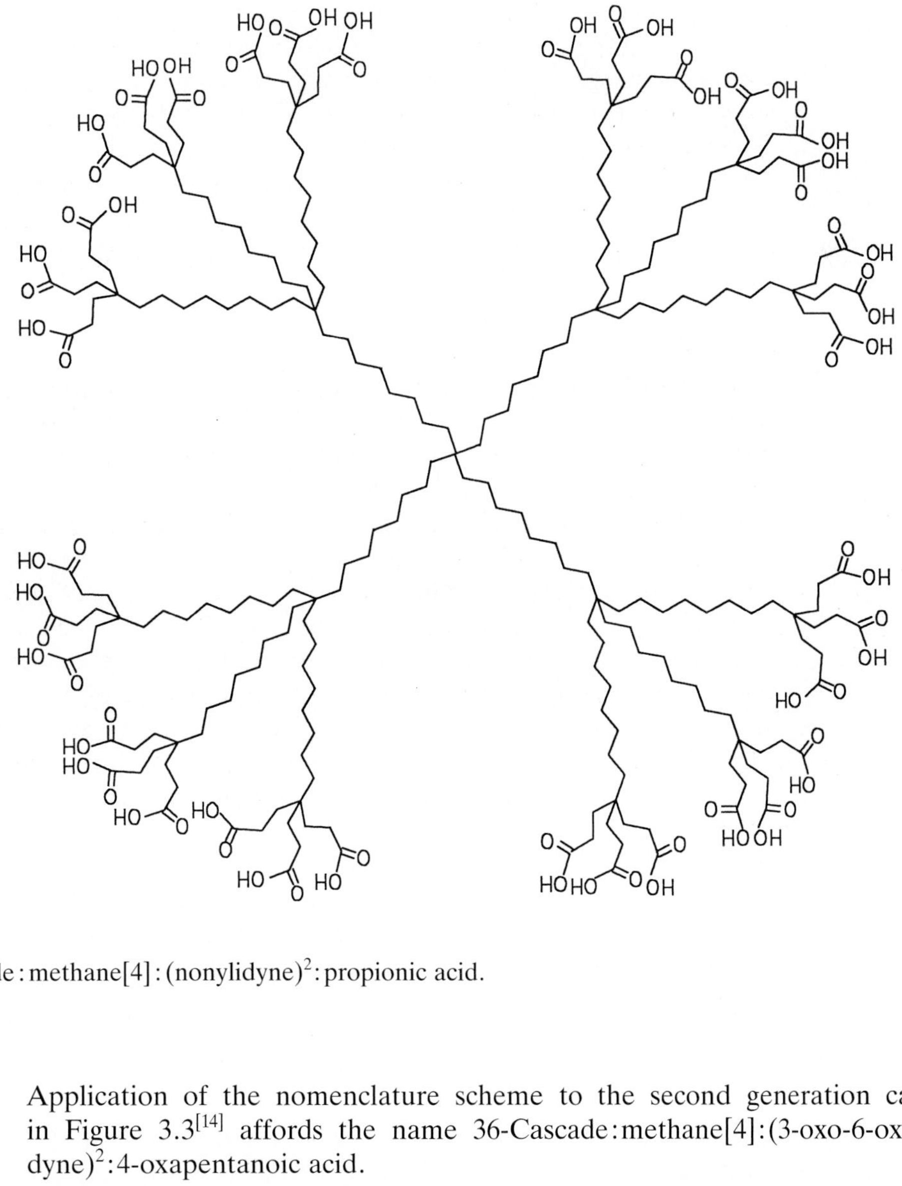

Figure 3.2. 36-Cascade:methane[4]:(nonylidyne)^2:propionic acid.

Application of the nomenclature scheme to the second generation cascade, shown in Figure 3.3[14] affords the name 36-Cascade:methane[4]:(3-oxo-6-oxa-2-azaheptylidyne)^2:4-oxapentanoic acid.

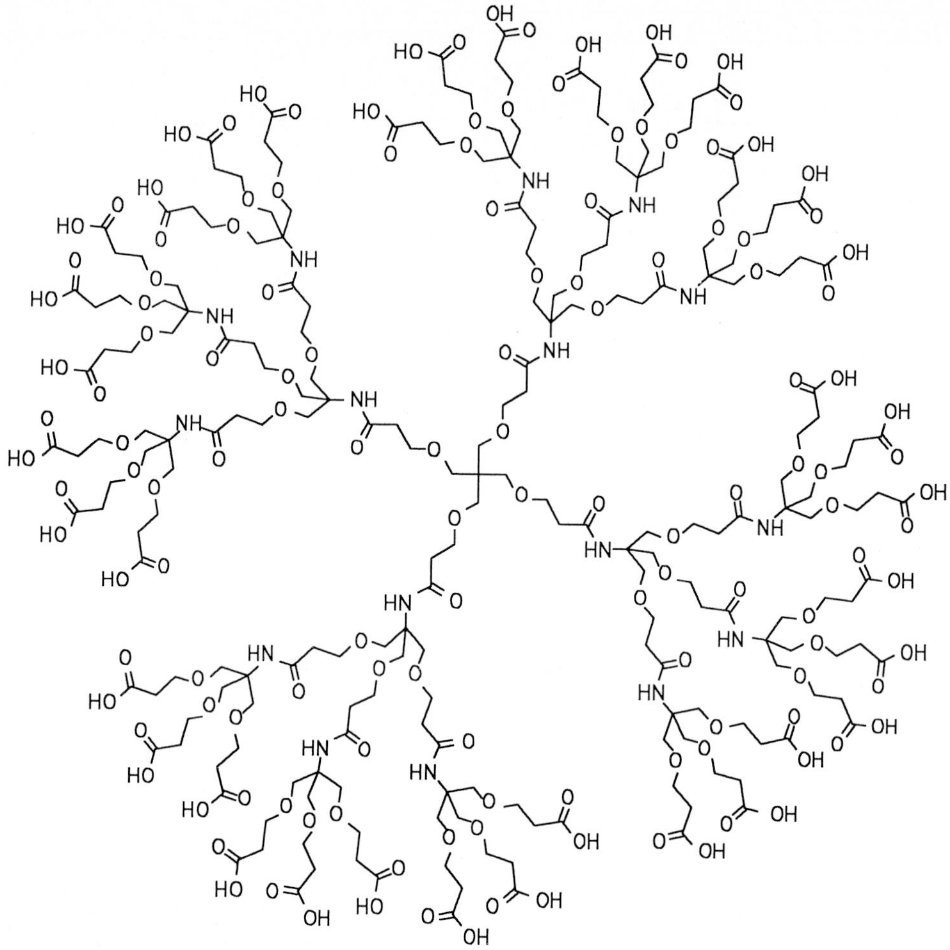

Figure 3.3. 36-Cascade : methane[4] : (3-oxo-6-oxa-2-azaheptylidyne)^2 : 4-oxapentanoic acid.

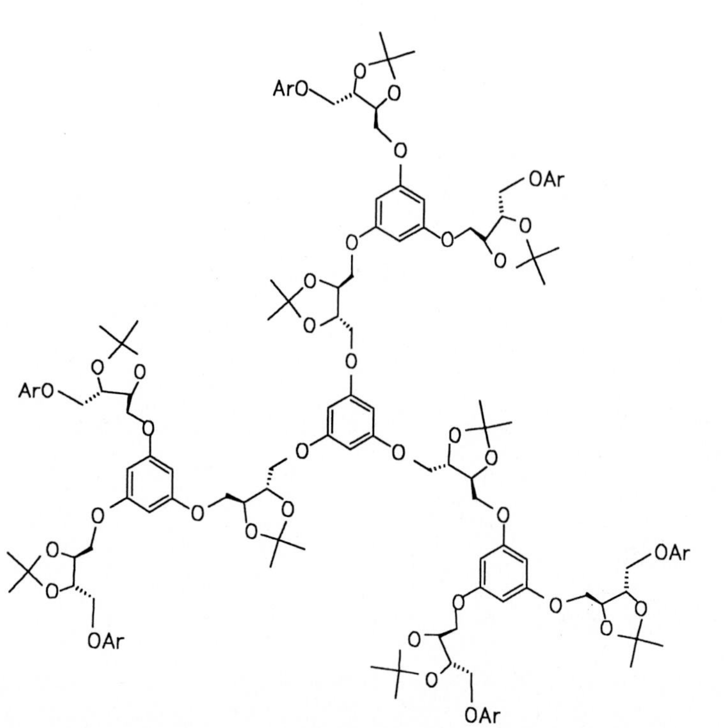

Figure 3.4. 6-Cascade : benzene[3-1,3,5] : (5-(3*S*,4*S*-1,6-dioxa-3,4-*O*-isopropylidene-3,4-dihydroxyhexyl)-1,3-phenyl-ene) : 4-(3*S*,4*S*-1,6-dioxa-3,4-*O*-isopropylidene-3,4-dihydroxyhexyl)-*tert*-butylbenzene.

The introduction of chirality into dendritic assemblies can be easily addressed as shown for Figure 3.4,[15] which would be named: 6-Cascade:benzene[3-1,3,5]:(5-(3*S*,4*S*-1,6-dioxa-3,4-*O*-isopropylidene-3,4-dihydroxyhexyl)-1,3-phenylene):4-(3*S*,4*S*-1,6-dioxa-3,4-*O*-isopropylidene-3,4-dihydroxyhexyl)-*tert*-butylbenzene.

3.4.2   Similar Arms with Dissimilar Internal Units

Z-Cascade:Core[*N$_c$*]:(Internal Units A)^m:(Internal Units B)^n:Terminal Units

The cascade name for Figure 3.5[16] is derived by the process described above, except that the attachments to the adamantane core (i.e., 1,3,5,7) must be noted after the core multiplicity. The repeat units are readily named via *replacement nomenclature*, thus;

$$\leftarrow \text{Core (Internal Unit A) Terminus} \rightarrow$$

$$\text{adamantane--}\overset{3}{C}(=O)\text{--}\overset{2}{N}H\text{--}\overset{1}{C}R_3$$

$$\leftarrow \text{(Internal Unit A) (Internal Unit B) Terminus} \rightarrow$$

$$C\text{--}[\overset{5}{C}H_2\overset{4}{C}H_2\overset{3}{C}(=O)\overset{2}{N}H\overset{1}{C}R_3']_3$$

$$\leftarrow \text{(Internal Units A \& B) Terminus} \rightarrow$$

$$C\text{--}[CH_2CH_2CO_2H]_3$$

Replacement nomenclature emphasizes the length of the repeat unit but masks the functionality (i.e., amide) in these units. Alternatively, a repeat unit name (such as, (propanamido)methylidyne), which indicates the functionality but obscures the chain length, may be preferred. The style chosen for the internal (or core or terminal) unit name(s) does not affect the general form of the proposed cascade nomenclature. Thus, the cascade in Figure 3.5 is 36-Cascade:tricyclo[3.3.1.1^{3,7}]decane[4-1,3,5,7]:(3-oxo-2-azapropylidyne):(3-oxo-2-azapentylidyne):propanoic acid.

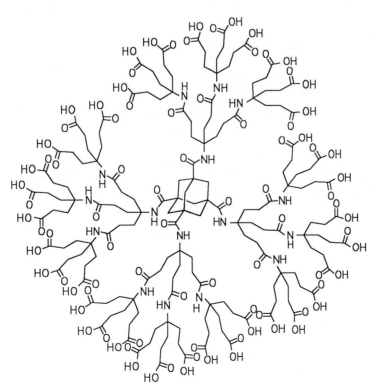

Figure 3.5. 36-Cascade:tricyclo[3.3.1.1^{3,7}]decane[4-1,3,5,7:(3-oxo-2-azapropylidyne):(3-oxo-2-azapentylidyne):propionic acid.

3.4.3 Dissimilar Arms with Similar Internal Branches

Z-Cascade:Core[N_c]:(Internal   Units)^m:Terminal   Unit]:[(Internal   Units)^n:Terminal Unit]:[(Internal Units)^p:Terminal Unit]

The chiral dendrimer (Figure 3.6[17]) in its racemic form represents an example of the class, its name is: *rac*-14-Cascade:benzyloxyethane[3-2,2,2,2]:[(5-(2-oxapropyl)-1,3-phenylene):(2-oxaethyl)benzene]:[(5-(2-oxapropyl)-1,3-phenylene):(5-(2-oxaethyl)-1,3-phenylene):(2-oxaethyl)benzene]:[(5-(2-oxapropyl)-1,3-phenylene):(5-(2-oxaethyl)-1,3-phenylene)^2:(2-oxaethyl)benzene].

An alternative name, which is based on methane as the core, would be: *rac*-15-Cascade:methane[4]:[(2-oxapropyl)benzene]:[(5-(2-oxapropyl)-1,3-phenylene):(2-oxaethyl)benzene]:[(5-(2-oxapropyl)-1,3-phenylene):(5-(2-oxaethyl)-1,3-phenylene):(2-oxaethyl)benzene]:[5-(2-oxapropyl)-1,3-phenylene):(5-(2-oxaethyl)-1,3-phenylene)^2:(2-oxaethyl)benzene].

3.4.4 Dissimilar Arms with Dissimilar Internal Branches or Terminal Groups

Z-Cascade:Core[N_c]:[(Internal Units A)^a:(Internal Units B)^b:Terminal Units X)]^m-[(Internal Units C)^c:(Internal Units D)^d:Terminal Units Y)]^n

The cascade depicted in Figure 3.7[18] possesses dissimilar arms with the same internal segments; its name is (where, X = H): 48-Cascade:ethane[3-1,1,1]:bis[(5-(2-*p*-phenyl-2-oxaethyl)-1,3-phenylene):(5-(2-oxaethyl)-1,3-phenylene)^3:(2-oxaethyl)benzene]-[(5-(2-*p*-phenyl-2-oxaethyl)-1,3-phenylene):(5-(2-oxaethyl)-1,3-phenylene)^3:4-(2-oxaethyl)-1-bromobenzene]; or where (X = Br): 48-Cascade:ethane[3-1,1,1]:[(5-(2-*p*-phenyl-2-oxaethyl)-1,3-phenylene):(5-(2-oxaethyl)-1,3-phenylene)^3:(2-oxaethyl)benzene]-bis[(5-(2-*p*-phenyl-2-oxaethyl)-1,3-phenylene):(5-(2-oxaethyl)-1,3-phenylene)^3:4-(2-oxaethyl)-1-bromobenzene].

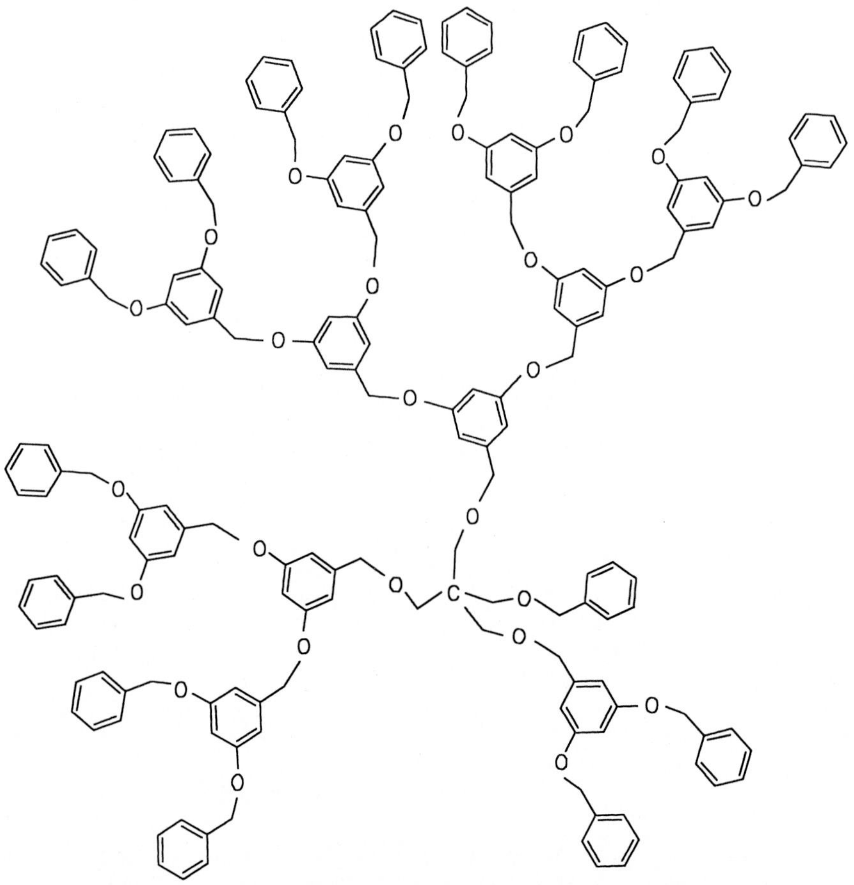

Figure 3.6. *rac*-14-Cascade:benzyloxyethane[3-2,2,2,2]:[(5-(2-oxapropyl)-1,3-phenylene):(2-oxaethyl)benzene]:[(5-(2-oxapropyl)-1,3-phenylene):(5-(2-oxaethyl)-1,3-phenylene):(2-oxaethyl)benzene]:[(5-(2-oxapropyl)-1,3-phenylene):(5-(2-oxaethyl)-1,3-phenylene)^2:(2-oxaethyl)benzene].

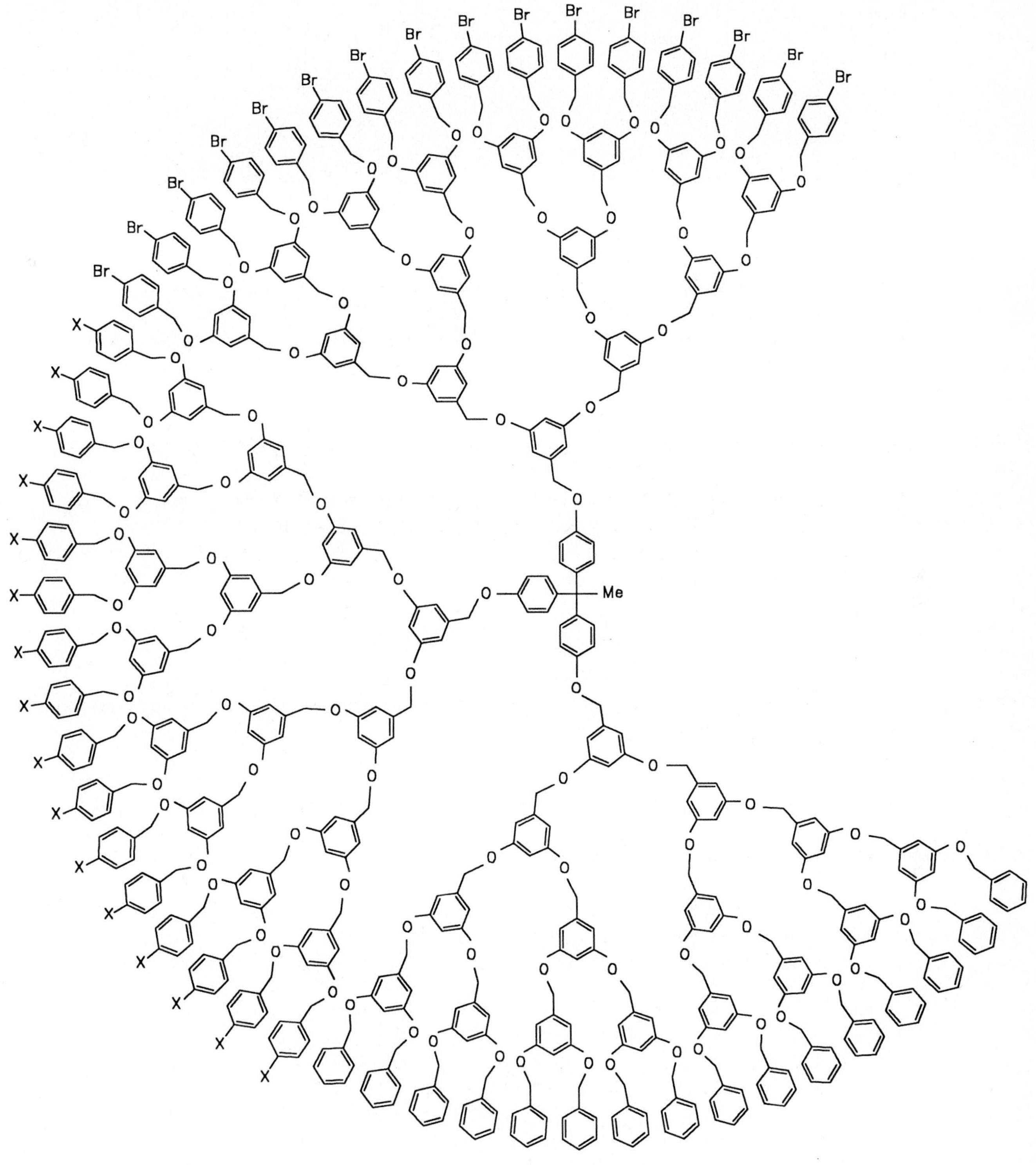

Figure 3.7. 48-Cascade:ethane[3-1,1,1]:*bis*[(5-(2-*p*-phenyl-2-oxaethyl)-1,3-phenylene):(5-(2-oxaethyl)-1,3-phenyl-ene)^3:(2-oxaethyl)benzene]-[(5-(2-*p*-phenyl-2-oxaethyl)-1,3-phenylene):(5-(2-oxaethyl)-1,3-phenylene)^3:4-(2-oxaethyl)-1-bromobenzene] (where, X = H); or 48-Cascade:ethane[3-1,1,1]:[(5-(2-*p*-phenyl-2-oxaethyl)-1,3-phenylene):(5-(2-oxa-ethyl)-1,3-phenylene)3:(2-oxaethyl)benzene]-*bis*[(5-(2-*p*-phenyl-2-oxaethyl)-1,3-phenylene):(5-(2-oxaethyl)-1,3-phenyl-ene)^3:4-(2-oxaethyl)-1-bromobenzene] where (X = Br).

The cascade shown in Figure 3.8 is a hypothetical dendrimer composed of known building blocks (including the core), which possesses different arms comprised of various internal units. The termini also differ. Its name is 36-Cascade:methane[4]:bis[(3-oxo-6-oxa-2-azaheptylidyne)^2:propylamine]-bis[(3-oxo-6-oxa-2-azaheptylidyne):(3-oxo-2-aza-pentylidyne):propanoic acid].

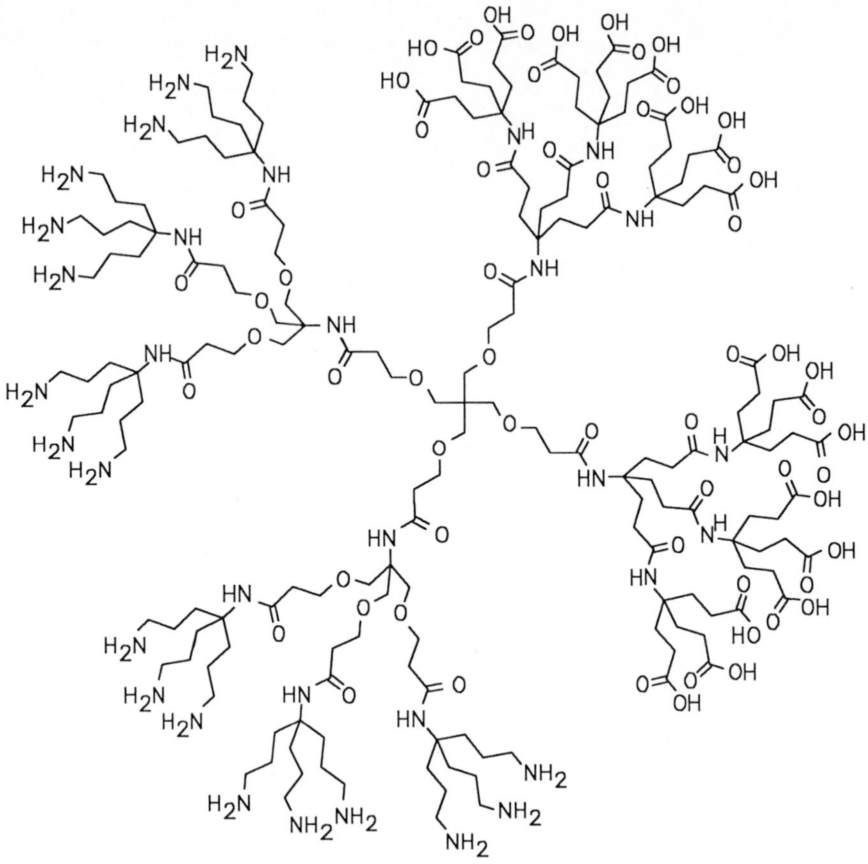

Figure 3.8. 36-Cascade : methane[4] : *bis*[(3-oxo-6-oxa-2-azaheptylidyne)² : propylamine]-*bis*[(3-oxo-6-oxa-2-azaheptylidyne) : (3-oxo-2-azapentylidyne) : propanoic acid].

3.4.5 Unsymmetrically Branched Cascades

The cascade in Figure 3.9,[19] based on lysine, possesses unsymmetrical branches, its name is: 8-Cascade : *N*-(diphenylmethyl)acetamide[2-2,2] : (2-oxo-3-azapropylidene : 2-oxo-3-azaheptylidene)² : *N*-(*tert*-butoxycarbonyl)amine : *N*-(*tert*-butoxycarbonyl)butylamine.

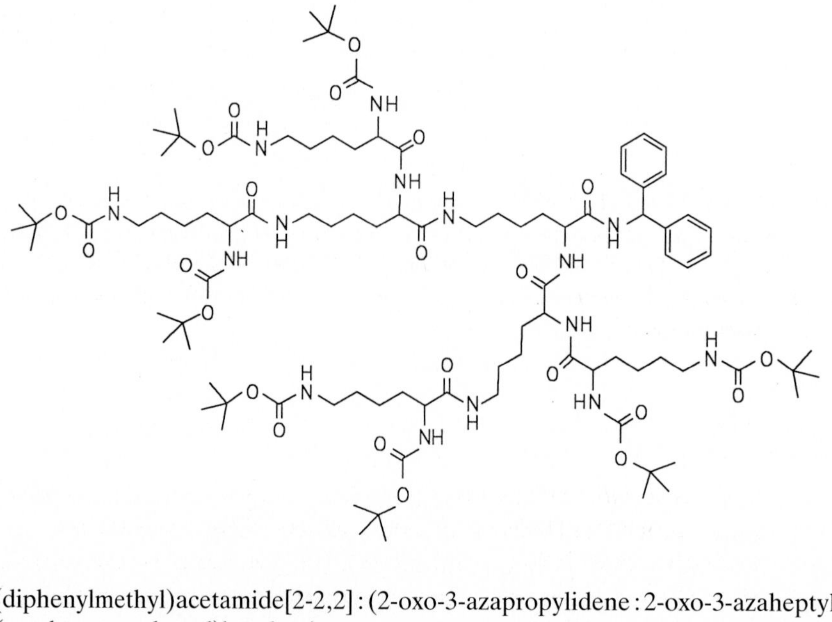

Figure 3.9. 8-Cascade : *N*-(diphenylmethyl)acetamide[2-2,2] : (2-oxo-3-azapropylidene : 2-oxo-3-azaheptylidene)² : *N*-(*tert*-butoxycarbonyl)amine : *N*-(*tert*-butoxycarbonyl)butylamine.

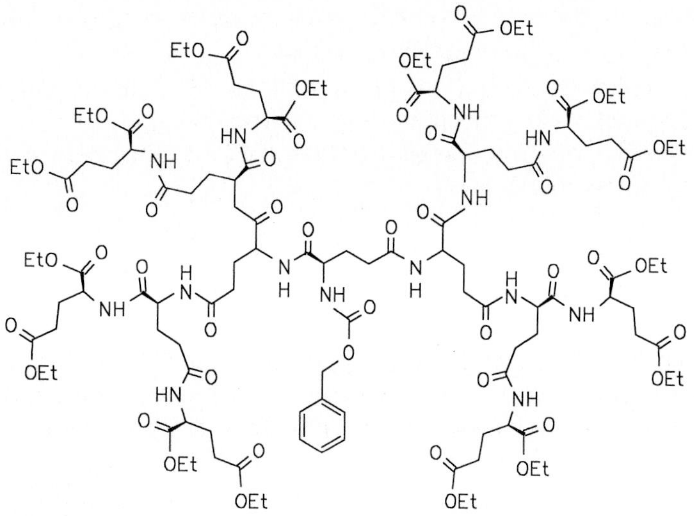

Figure 3.10. 16-Cascade: (*N*-benzyloxycarbonyl)methylamine[2-1,1]: (1*S*-3-oxo-2-azapropylidene: 1*S*-3-oxo-2-azapentylidene)[3]: ethyl formate: ethyl propanoate.

A hypothetical example of an unsymmetrically branched chiral dendrimer[20] is shown in Figure 3.10, which has the name: 16-Cascade:*N*-benzyloxycarbonyl)methylamine[2-1,1]: (1*S*-3-oxo-2-azapropylidene: 1*S*-3-oxo-2-azapentylidene)[3]: ethyl formate: ethyl propanoate.

3.5 Fractal Notation

A fractal notation has been proposed by Mendenhall[7, 8] to address the complexity of cascade macromolecules, since they possess a high degree of molecular symmetry. This highly condensed description of a symmetrical molecule possessing a fractal architecture can be generalized in the following form:

<u>periphery units</u>   <u>subscription</u>   <u>connectors</u>   <u>core</u>
(terminal group)f$_{(\text{outermost hub ... innermost hub})}$(n,n',n'', ... n^z)core
        fractal notation

Although this notation can be applied to diverse structures, the following rules and examples have been confined to very simple dendritic systems.

3.5.1 General Rules for the Proposed Fractal Notation

1. The *terminal* or *peripheral group(s)* is (are) left blank if hydrogen, otherwise it denotes the specific terminal functionality. With diverse functionality, $<X>_n <Y>_m$, where n and m are statistical average values of groups $<X>$ and $<Y>$, respectively. The number of terminal groups (R) can be denoted by a subscript, e.g., (R)$_\#$. If the molecule has a cyclic structure, the terminal group is absent or incorporated into the core, this is signified by "Ø" in the terminal position of the notation.

2. *f* denotes the fractal notation.

3. *Subscripts* refer to the junctures or branching points in the molecule. If the branch point is a tetrahedral carbon, the subscript is omitted. When the branch point is identical throughout the structure, the atom (or group) is only cited once as a subscript. If, however, there are multiple branch points, they are sequentially listed for the outermost → innermost and separated by a period.

4. *Connector groups* are listed again from the outermost → innermost and enclosed in parentheses as well as separated by periods.

5. *Core* is denoted without parentheses at the far right and can be omitted if it is identical to the juncture or branching atom or group.

Thus, a summary of the basic fractal notation is: (peripheral groups)$_{number}$f$_{junctures\ (coordination\ number)}$(connectors)core.

3.5.2   Examples Using the Fractal Rules

The original [27]-Arborol[2] (see Figure 3.1) would be named by this procedure as: the *terminal groups* ≡ (HO)$_{27}$; "f"; *junctures* ≡ tetrahedral carbon, therefore omitted; *connectors* (from the outside in) ≡ (CH$_2$.NHCO.CH$_2$CH$_2$OCH$_2$); and the *core* ≡ C–C$_5$H$_{11}$–*n*.

Thus, the complete fractal notation for the structure would be:

$$HOf(CH_2.NHCO.CH_2CH_2OCH_2)C–C_5H_{11}–n$$

It was noted[8] that further simplification could be made by a simple number replacement for consecutive methylene groups; therefore, this complete fractal notation can be simplified (or condensed) to:

$$HOf(1.NHCO.2O1)C5H$$

and for practical utilization the number of terminal groups is appended to afford:

$$(HO)_{27}f(1.NHCO.2O1)C5H$$

For Tomalia's dendritic PAMAM amine, depicted in the line notation as:

$$N[CH_2CH_2CONHCH_2CH_2N(CH_2CH_2CONHCH_2CH_2N[CH_2CH_2$$
$$CONHCH_2CH_2N(CH_2CH_2CONHCH_2CH_2NH_2)_2]_2]_3$$

it would be greatly simplified to

$$H_2Nf_N(2NHCO2.)_4$$

where the 4 signifies the fourth generation.

For applications to more complex macromolecules and the use of the additional rules, not cited herein, associated with the fractal notation, one should consult the Mendenhall article.[8]

3.6   References

[1]   E. Buhleier, W. Wehner, F. Vögtle, *Synthesis* **1978**, 155.

[2]   G. R. Newkome, V. K. Gupta, G. R. Baker, Z.-Q. Yao, *J. Org. Chem.* **1985**, *50*, 2003.

[3]   A. F. Bochkov, B. E. Kalganov, V. N. Chernetskii, *Izv. Akad. Nauk. SSSR Ser. Khim.* **1989**, 2394 (*Chem. Abstr.* **1990**, *112*, 216174p).

[4]   P.-G. de Gennes, H. Hervet, *J. Phys. Lett.* **1983**, *44*, 351.

[5]   T. Nagasaki, M. Ukon, S. Arimori, S. Shinkai, *J. Chem. Soc., Chem. Commun.* **1992**, 608.

[6]   D. A. Tomalia, H. Baker, J. Dewald, M. Hall, G. Kallos, S. Martin, J. Roeck, J. Ryder, P. Smith, *Polym. J.* **1985**, *17*, 117.

[7]   G. D. Mendenhall, S. X. Liang, E. H.-T. Chen, *J. Org. Chem.* **1990**, *55*, 3697.

[8]   For a detailed presentation see: G. D. Mendenhall, in *Mesomolecules From Molecules to Materials* (Eds.: G. D. Mendenhall, A. Greenberg, J. F. Lieberman), Chapman & Hall, New York, **1995**, pp. 181–194.

[9]   O. L. Chapman, J. Magner, R. Ortiz, *Polym. Preprints* **1995**, *36*(1), 739.

[10] G. R. Newkome, Y. Hu, M. J. Saunders, F. R. Fronczek, *Tetrahedron Lett.* **1991**, *32*, 1133.

[11] G. R. Newkome, G. R. Baker, J. K. Young, J. G. Traynham, *J. Polym. Sci.: A: Polym. Chem.* **1993**, *31*, 641.

[12] G. R. Newkome, G. R. Baker, *Polym. Preprints* **1994**, *35*, 6.

[13] G. R. Newkome, C. N. Moorefield, G. R. Baker, A. L. Johnson, R. K. Behera, *Angew. Chem.* **1991**, *103*, 1205; *Angew. Chem. Int. Ed. Engl.* **1991**, *30*, 1176.

[14] G. R. Newkome, X. Lin, *Macromolecules* **1991**, *24*, 1443.

[15] H.-F. Chow, L. F. Fok, C. C. Mak, *Tetrahedron Lett.* **1994**, *35*, 3547.

[16] G. R. Newkome, A. Nayak, R. J. Behera, C. N. Moorefield, G. R. Baker, *J. Org. Chem.* **1992**, *57*, 358.

[17] J. A. Kremers, E. W. Meijer, *J. Org. Chem.* **1994**, *59*, 4262.

[18] K. L. Wooley, C. J. Hawker, J. M. J. Fréchet, *J. Chem. Soc., Perkin Trans. 1* **1991**, 1059.

[19] R. G. Denkewalter, J. Kolc, W. J. Lukasavage, U. S. Pat. 4, 289, 872 (Sept. 15, **1981**).

[20] L. J. Twyman, A. E. Breezer, J. C. Mitchell, *Tetrahedron Lett.* **1994**, *35*, 4423.

4 Synthetic Methodologies: Divergent Procedures

4.1 General Concepts

Divergent dendritic construction results from sequential monomer addition beginning from a core and proceeding outward toward the macromolecular surface. This methodology is illustrated in Scheme 4.1. To a respective core representing the zeroth generation and possessing one or more reactive site(s), a generation or layer of monomeric building blocks is covalently connected. The number of building blocks that can be added is dependent on the available reactive sites on the particular core assuming parameters, such as monomer functional group steric hindrance and core reactive site accessibility, are not a concern. Repetitive addition of similar, or for that matter dissimilar, building blocks (usually effected via a protection–deprotection scheme) affords successive generations. A key feature of the divergent method is the exponentially increasing number of reactions that are required for the attachment of each subsequent tier (layer or generation).

Faultless growth is realized by the complete reaction of all the available reactive groups and thus results in the attachment of the maximum number of monomers possible. Defective growth, or incomplete reaction, results in branch errors, which, if it occurs in the early stages of growth, is generally more problematic than that occuring at higher generations.

Branching is dependent on building block valency (this includes the cores since they are a special class of building block). Thus, a core possessing one reactive moiety, such as a primary amine, is divalent and will accommodate two monomers assuming a neutral trisubstituted amine product. Branching therefore proceeds in a $1 \rightarrow 2$ manner, or with three monomers, the resultant product is an ammonium salt, in which branching proceeds by a $1 \rightarrow 3$ route. For neutral amine products, conversion of the new terminal groups (e.g., cyano moieties) to primary amines and repetition of monomer addition procedure (e.g., amine alkylation) results in a general $1 \rightarrow 2 \rightarrow 4 \rightarrow 8 \rightarrow 16 \rightarrow 32 \rightarrow \cdots$ branching pattern. A tetravalent, four-directional core that reacts with four equivalents of a $1 \rightarrow 2$ branching monomer will result in a progression $4 \rightarrow 8 \rightarrow 16 \rightarrow 32 \rightarrow 64 \rightarrow 128 \rightarrow \cdots$ (Scheme 4.1 (a)); whereas employing the same core with a $1 \rightarrow 3$ branching monomer will give a dendritic series with an increasing peripheral multiplicity of $4 \rightarrow 12 \rightarrow 36 \rightarrow 108 \rightarrow 324 \rightarrow 972 \rightarrow \cdots$ (Scheme 4.1 (b)).

4.2 Early Procedures

In 1978, Vögtle and coworkers in Bonn reported[1] the first preparation, separation, and mass spectrometric characterization of simple dendritic structures via a divergent iterative methodology (Scheme 4.2). They described this *cascade synthesis* as "reaction sequences which can be carried out repeatedly." The key features of polyamine **4** include the trigonal *N*-centers of branching and the critical distance imposed by the -$(CH_2)_3$- linkage between branching centers. Treatment of the primary amine **1** with acrylonitrile using a Michael-type addition afforded the desired dinitrile **2**, which was reduced to the terminal diamine **3**. After purification, **3** was subjected to the same reaction sequence to generate heptaamine **4**, the structure of which was fully characterized. The original reduction conditions have proven unreliable in certain laboratories; recent repetition[2] of this sequence, but using di*iso*butylaluminum hydride, supports the original[1] structural conclusions. Technically, heptaamine **4** is not a macromolecule, but rather an example of a dendritic meso-[3] or medio-molecule. The most important aspect of this synthetic

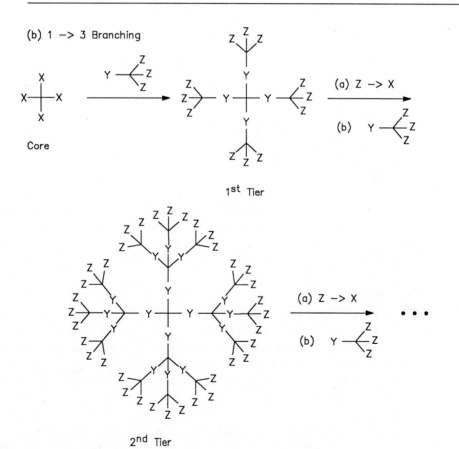

Scheme 4.1. Examples of divergent synthetic procedures for the macromolecular construction.

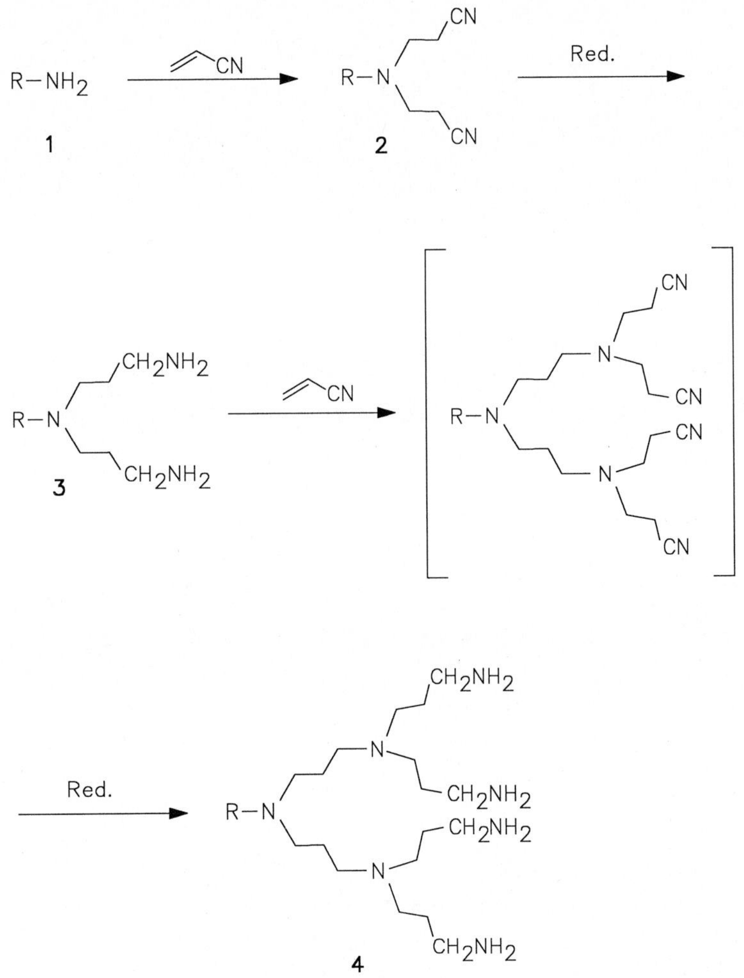

Scheme 4.2. Vögtle et al.'s[1] original repetitive synthesis.

sequence was that Vögtle presented the first application of an iterative reaction sequence and thus the birth of cascade construction.

In 1981, Denkewalter et al. reported[4] in a patent the first divergent preparation of dendritic polypeptides utilizing the protected amino acid, *N,N'*-bis(*tert*-butoxycarbonyl)-L-lysine, as the monomeric building block. The two-directional, asymmetric core (**5**) was constructed from L-lysine and benzhydrylamine. The coupling procedure was accomplished by the use of an activated *p*-nitrophenyl ester (**6**), followed by removal of the *tert*-butoxycarbonyl (*t*-BOC) protecting groups; see Scheme 4.3. The now free polyamine moieties (**7**) were then available for the construction of the next generation. Repetition of the simple coupling and deprotection sequences, utilizing ester **6**, led to lysine polymers containing 1023 terminal (*t*-BOC) protected lysine groups, and the creation of a series of generations possessing increasing sizes and molecular weights. These biopolymers were inherently asymmetrical, used peptide coupling technology, and contained a 1 → 2 *C*-branching pattern, which resulted in an approximate generational doubling of the molecular weight through the 10th generation (molecular weight at the 10th generation: 233,600 amu). The original[4] and related patents[5, 6] however, afforded little insight into the purification, characterization, or physical/chiral properties of these macromolecules.

In 1985, Newkome et al.[7] and Tomalia et al.[8] published different divergent routes to 1 → 3 *C*-branching arborols (*Arbor*: Latin: tree) and 1 → 2 *N*-branching dendrimers (*Dendro*-: Greek: tree-like), respectively (Scheme 4.4). These authors described the construction of polyfunctional molecules that possessed multiple branching centers *and* offered spectral characterization supporting the structural assignments. These and all subsequent divergent routes will be described in this Chapter under the appropriate branching pattern and mode of monomer connectivity.

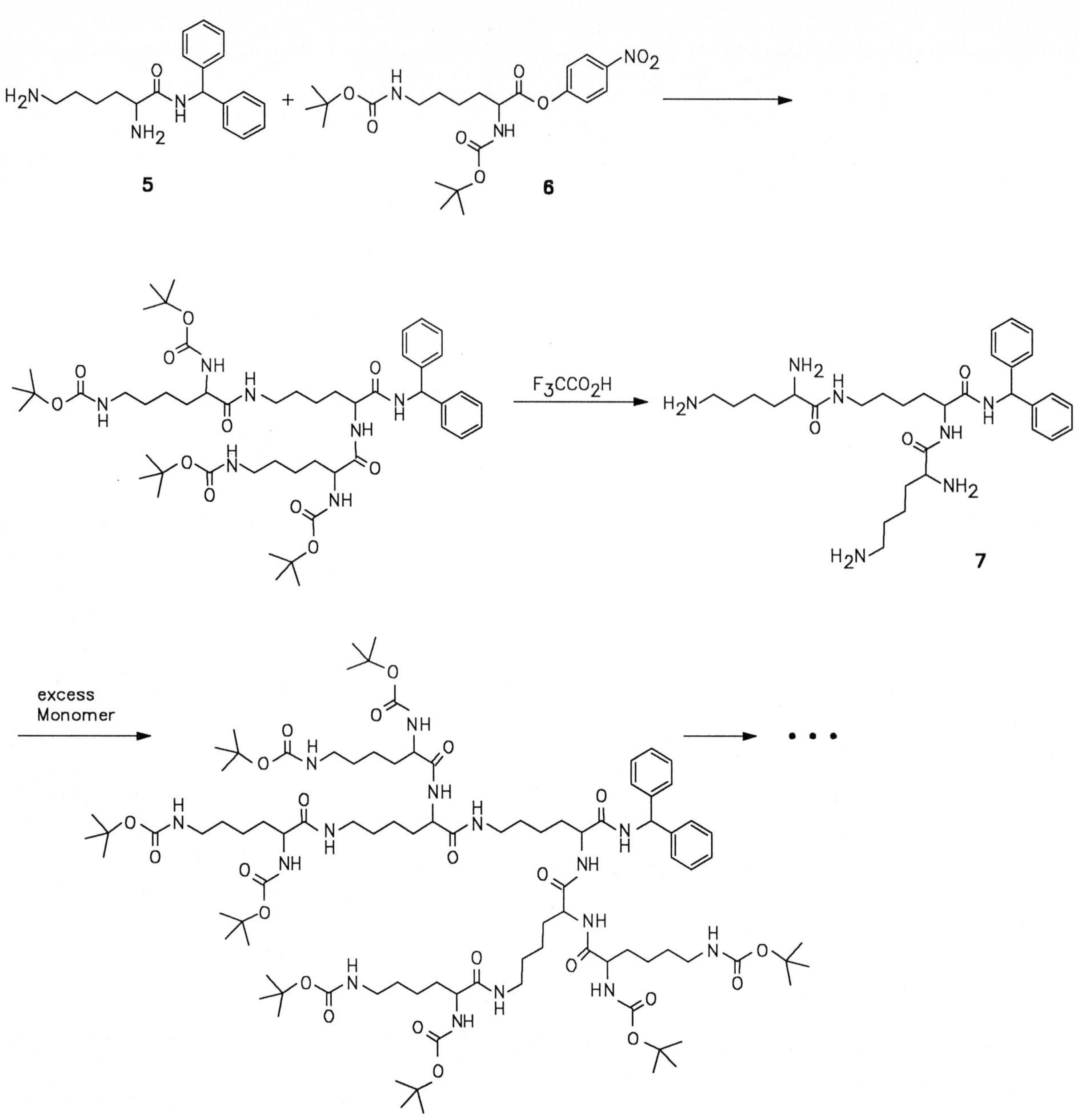

Scheme 4.3. Denkewalter et al.'s[4] preparation of dendritic polypeptides.

4.3   1 → 2 *N*-Branched

4.3.1   1 → 2 *N*-Branched and Connectivity

In their quest for large molecular cavities capable of binding molecules or ionic guests, Vögtle and coworkers[1] synthesized non-cyclic polyaza compounds via a repetitive stepwise process (one-directional case: Scheme 4.2). The use of two-directional cores (Scheme 4.5), e.g., ethylenediamine, 2,6-di(aminomethyl)pyridine (**8**) or 1,3-di(aminomethyl)benzene, with acrylonitrile afforded the tetranitrile **9**, which was reduced with Co(II)-catalyzed borohydride to produce hexaamine **10**. This terminal tetraamine was subsequently treated with excess acrylonitrile to generate the octanitrile **11**; although the process was terminated at this stage, a foundation of dendrimer construction was established.

Wörner and Mülhaupt[9] improved Vögtle's procedure by modifications that do not utilize excess reagents and complicated purifications. They first improved the nitrile reduction step by the use of Raney nickel[10] at ambient temperature (8 bar H$_2$) with a trace of sodium hydroxide in an ethanolic solvent. They further found that the cyanoethylation,

General Construction

1→2 N—Branching (Tomalia)

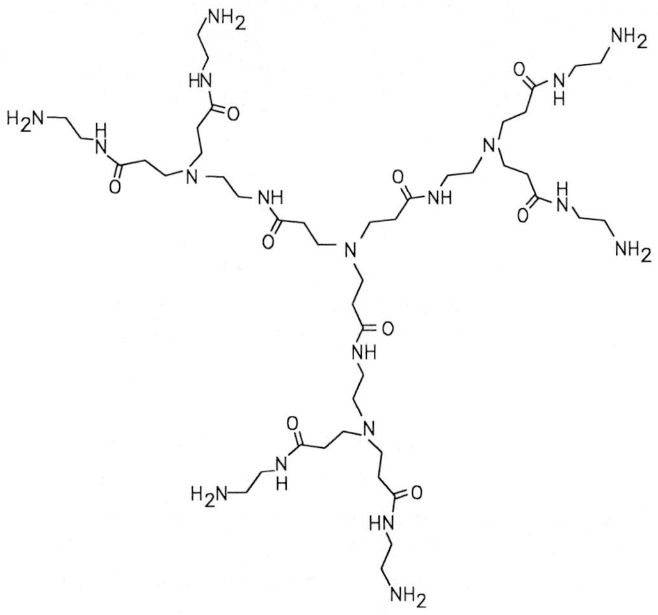

1→3 C—Branching (Newkome)

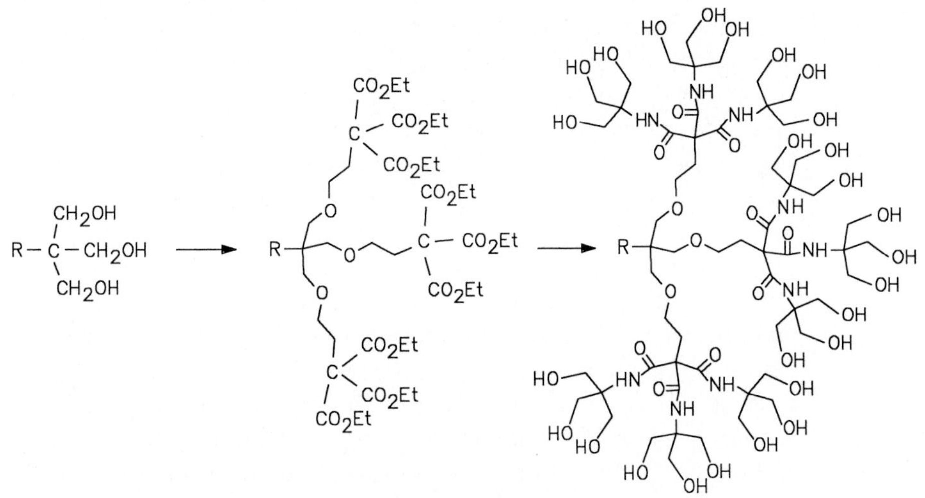

Scheme 4.4. Tomalia et al.'s[8] (1 → 2 branching) and Newkome et al.'s[7] (1 → 3 branching) original motifs.

when conducted in methanol, occurred without monosubstituted side products. By application of these synthetic modifications, the original route of Vögtle[1] could be extended to the fifth generation nitrile in excellent overall yields at most stages.

De Brabander-van den Berg and Meijer[11] also reported (in back-to-back manuscripts with Wörner and Mülhaupt[9]) procedures for the large scale preparation of poly(propylene imine) dendrimers using a sequence of reactions analogous to those employed by Vögtle et al.[1] in his original cascade synthesis. Thus, repetitive addition of a primary amine to two equivalents of acrylonitrile, followed by hydrogenation of the resulting nitrile moieties with Raney cobalt in a hydrogen atmosphere, afforded the desired polyamine terminated macromolecules (Scheme 4.6). ^{15}N NMR spectra of these dendrimers confirm their highly branched and well-defined structures.[12] Starting with 1,4-diaminobutane (**12**), the fifth tier polynitrile (**13**) was produced in kilogram quantities.[13] These nitrile-terminated dendrimers were subsequently converted to the corresponding acid-terminated series by treatment with hydrochloric acid.

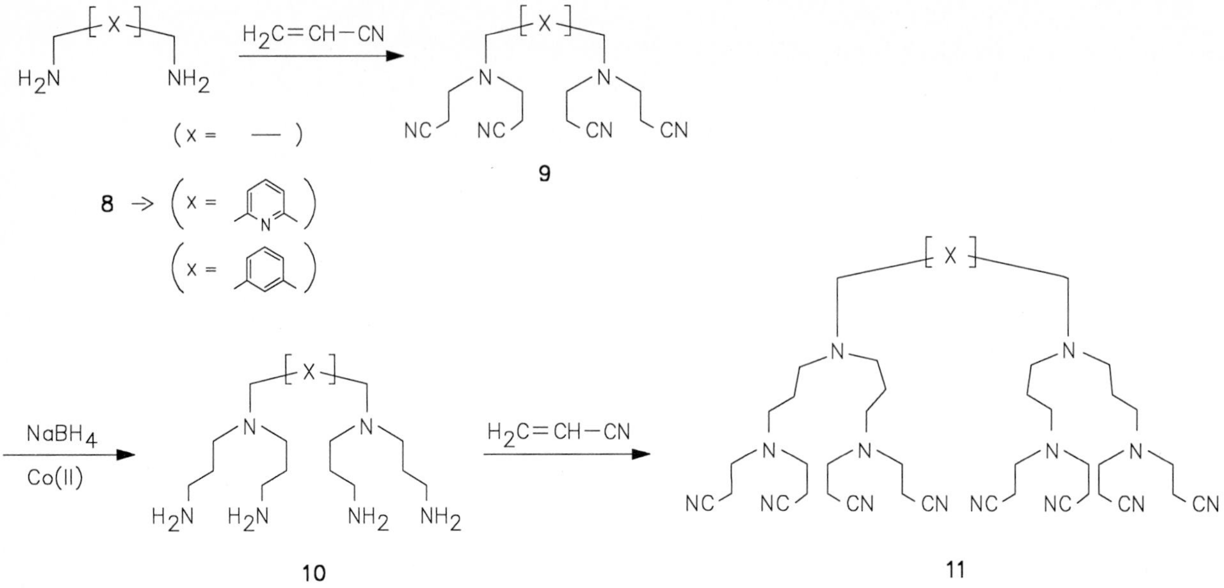

Scheme 4.5. Repetitive *N*-alkylation and reduction for the construction of polyamines and polynitriles.

In an alternative scheme for the preparation of $1 \rightarrow 2$ amine branched macromolecules possessing $-(CH_2)_2-$ spacer moieties, Tomalia et al. reported[14] a series of cascade polymers that were called "starburst" poly(ethylenimine)s (SPEI), which were constructed by nucleophilic addition of an amine to a mesylate activated aziridine (i.e., **14**). Scheme 4.7 shows the addition of excess **14** to a simple branched tetraamine (**15**) to generate the protected polyamine **16**. Deprotection of **16** afforded terminal primary amines, which can be further treated with excess *N*-methylaziridine (**14**) to create successive generations (e.g., **17**). This dendrimer series would be called *Z*-Cascade:ammonia[3]:(1-aza-propylidene)^G:ethylamine. Due to the diminished distance between branching centers in these SPEI dendrimers as compared with their PAMAM series (see Section 4.3.2), the resultant cascades reached dense packing limits at lower generations. Tomalia asserted that in this SPEI series, "divergence from branching ideality becomes significant as one approaches generation 3 or 4 and especially at generation 5."[14]

Meijer et al.[15] utilized the fifth generation polyamine (abbreviated as **18**) generated from the reduction of the corresponding nitrile for the construction of dendritic 'boxes' (Scheme 4.8) that allowed the sterically-induced entrapment of guest molecules. The concept of "trapping topologically by shell molecules" was considered theoretically by Maciejewski[16] and was previously reported by Denkewalter et al.[4–6] with the synthesis of lysine-based dendrimers therein described as "non-draining" spheres. Meijer's dendritic boxes (**20**), with molecular weights up to 24,000 amu, were prepared[15] by surface modification of the poly(propylenimine) cascade (PPI) with activated chiral esters, for example, the *N*-hydroxysuccinimide ester of a *tert*-butoxycarbonyl (*t*-BOC) protected amino acid (phenylalanine shown; **19**); further detail concerning chirality is presented in Chapter 7. The concept of a dendritic box has been expanded[17a] by the introduction of other surface amino acid groups, e.g. L-alanine, L-*t*-Bu-serine, L-tyr-cysteine, and L-*t*-Bu-aspartic ester.

Polypropylenimine dendrimers have been constructed on one end of a polystyrene (M_n = 3.2×10^3) core thereby forming macromolecular surfactant amphiphiles. Dendritic growth was continued through five generations. For polystyrene-poly(propyleneimine) hybrids possessing multiple CO_2H termini, a pH-dependent[17b] aggregation was observed (e.g., PS-dendri-$(CO_2H)_8$ at high pH formed "worm-like" micelles; PS = polystyrene, dendri = branched head group). Amine coated head groups (i.e., PS-dendri-$(NH_2)_n$) imparted generation-dependent aggregation;[17c] aqueous solutions were observed to possess micellar spheres, rods, and vesicles for $-(NH_2)_{8, 16, and 32}$, respectively.

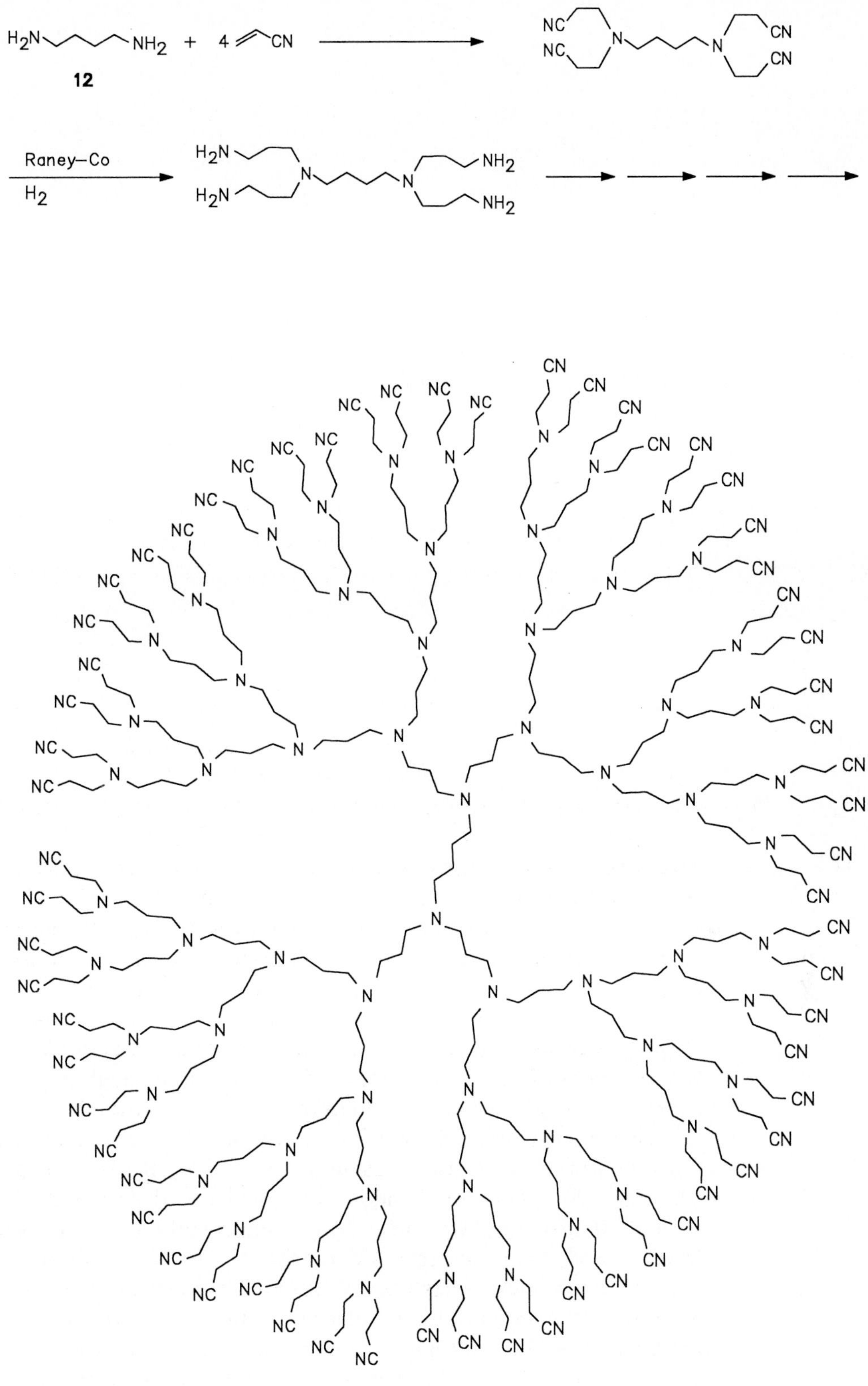

Scheme 4.6. Procedure for the large scale synthesis of poly(propylene imine) dendrimers.

A simple series of 1 → 2 *N*-branching cascade molecules based on 4,4',4''-tris(*N*,*N*-diphenylamino)triphenylamine has been prepared and characterized.[18] The ESR spectrum of the cationic triradical of the related 1,3,5-tris(diphenylamino)benzene has been studied[19] in detail.

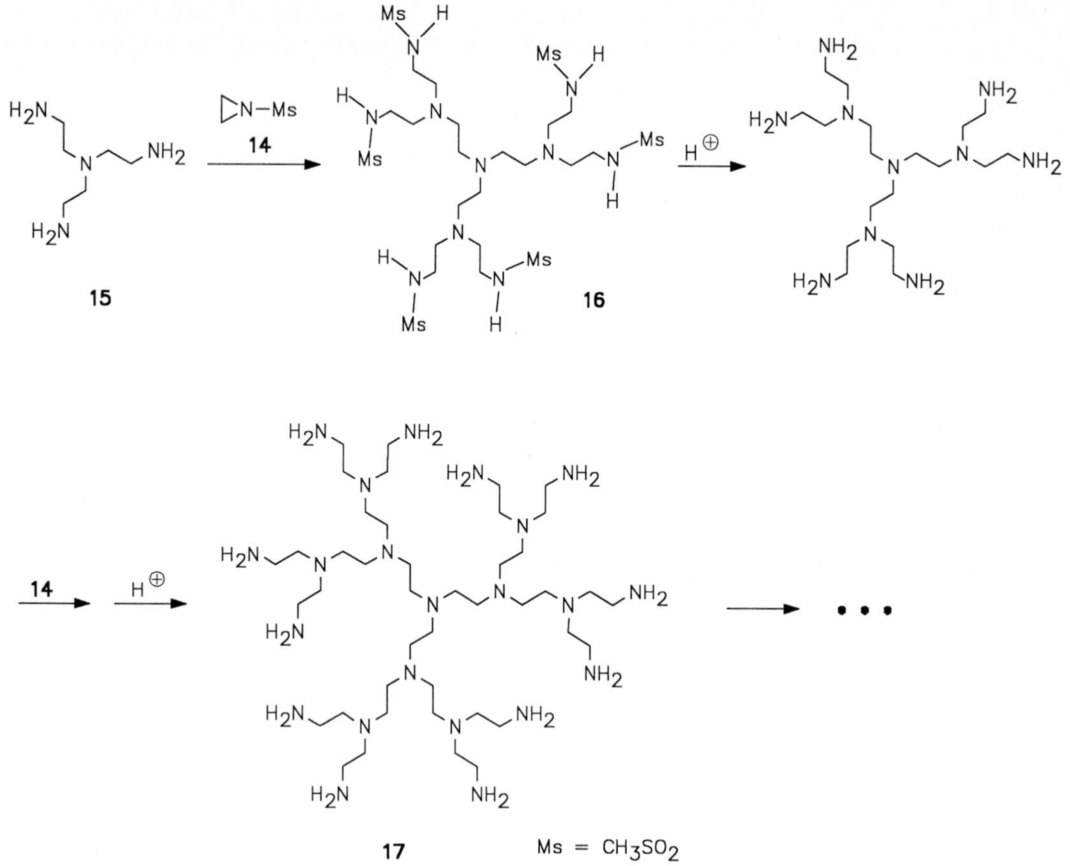

Scheme 4.7. Method for the construction of "starburst" poly(ethylene imine) cascades using *N*-mesylaziridine.

4.3.2 1 → 2 *N*-Branched, *Amide* Connectivity

In 1985 and 1986, Tomalia et al. described[8, 20] the preparation of polyamidoamine, termed "starburst polymers", or "dendrimers", which were generated from a three-directional core (ammonia) and possessed three-directional *N*-branching centers as well as amide connectivity (Scheme 4.9). Each generation was synthesized by exhaustive Michael-type addition of amines (e.g., for an ammonia core, **21**) with methyl acrylate to generate a *β*-amino acid ester (e.g. **22**), followed by amidation with excess ethylenediamine producing a new branched polyamine **23**. The general procedure was repeated to create the higher generations (e.g., **24**). Similar dendrimers were prepared by employing related cores, such as ethylenediamine as well as aminoalcohols and other functionalizable groups, such as amino and thiol moieties.[21] This procedure is applicable to most primary amines, resulting in the 1 → 2 branching pattern. It was noted that aryl esters could be utilized as the initiator core.[22, 23]

In order to achieve a high degree of monodispersity in these dendrimers, Tomalia and his coworkers minimized potential synthetic problems associated with amine additions to esters (e.g., *intra*molecular cyclization (lactam formation), *retro*-Michael reactions, incomplete addition, and *inter*molecular coupling) by the use of excess diamine, maintaining a moderate (< 80 °C) reaction temperature, and avoiding aqueous solvents.[8, 24] With optimized conditions, defects produced by these undesired reactions can be, for the most part, suppressed. This work was especially noteworthy because, for the first time, an iterative synthesis was reported, allowing the preparation of high molecular weight cascade polymers, herein denoted using the series terminology *Z*-cascade:ammonia[3]:(5-oxo-1,4-diazaheptylidene)^G:(4-oxo-3-azahexylamine). Structural, computer simulated,[25] comparative and electron microscopy,[26] and physical characterization[27, 28] of these macromolecules included standard spectroscopic methods, e.g., ^{1}H, ^{2}H,[29] and [13]C[30] NMR,[31] IR, as well as mass spectrometry (electrospray),[32] HPLC, GPC, DSC, TGH, and intrinsic viscosity.[33] The dependence of hindered diffusion for PAMAMs and linear poly(styrene)s in porous glasses have been evaluated,[34] dendrimers are in agreement with the hydrodynamic theory for a hard sphere in a cylindrical pore.

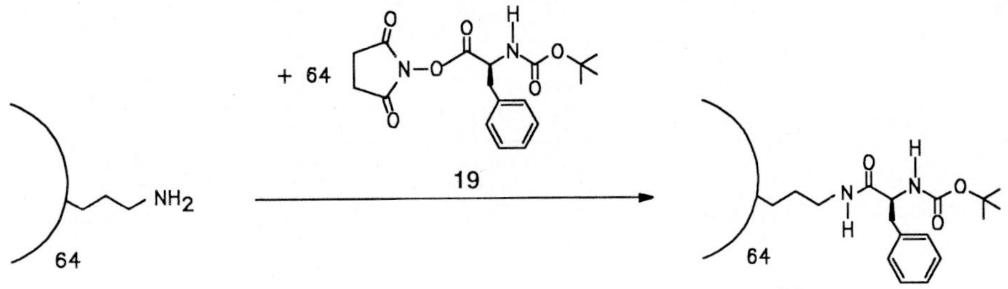

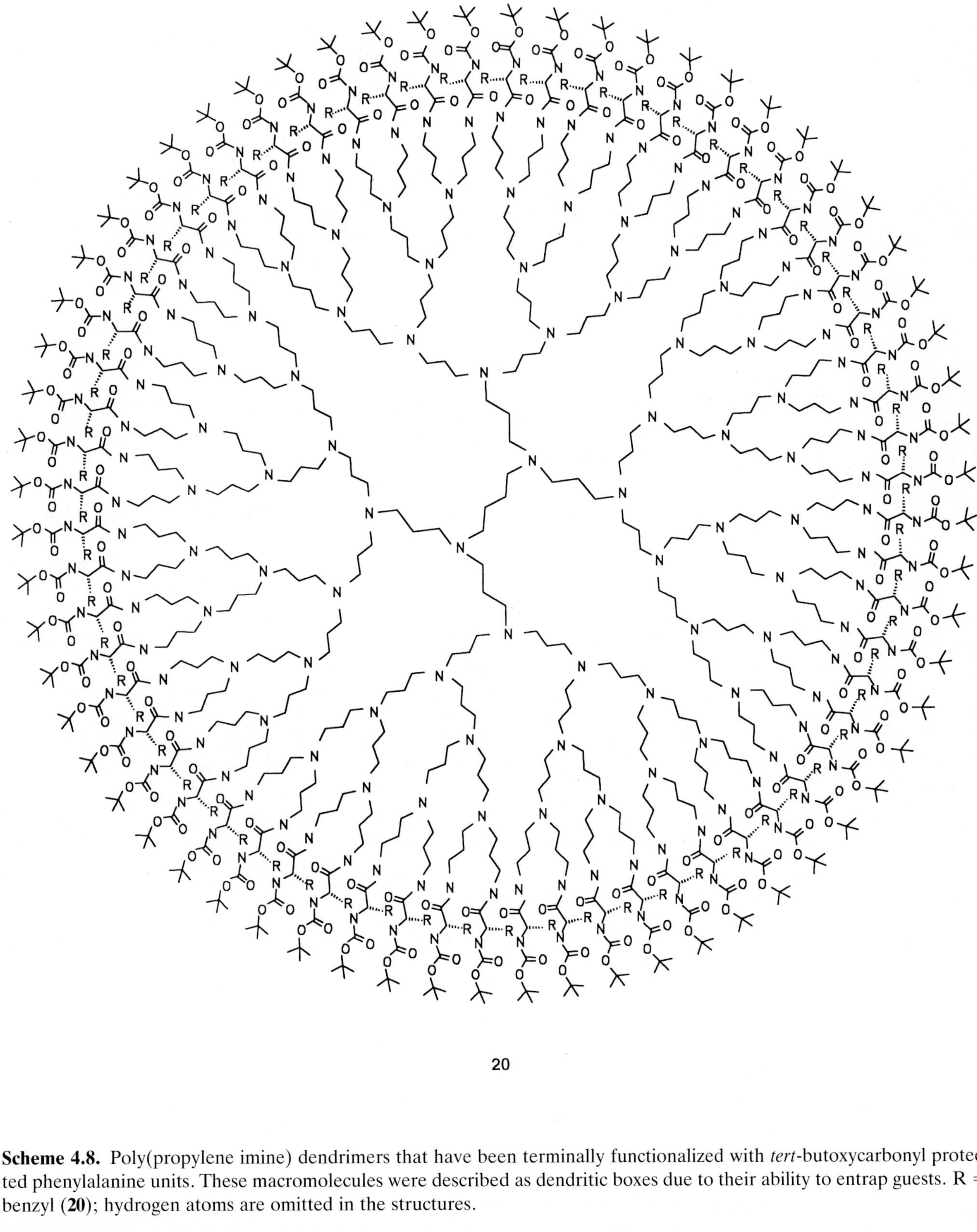

20

Scheme 4.8. Poly(propylene imine) dendrimers that have been terminally functionalized with *tert*-butoxycarbonyl protected phenylalanine units. These macromolecules were described as dendritic boxes due to their ability to entrap guests. R = benzyl (**20**); hydrogen atoms are omitted in the structures.

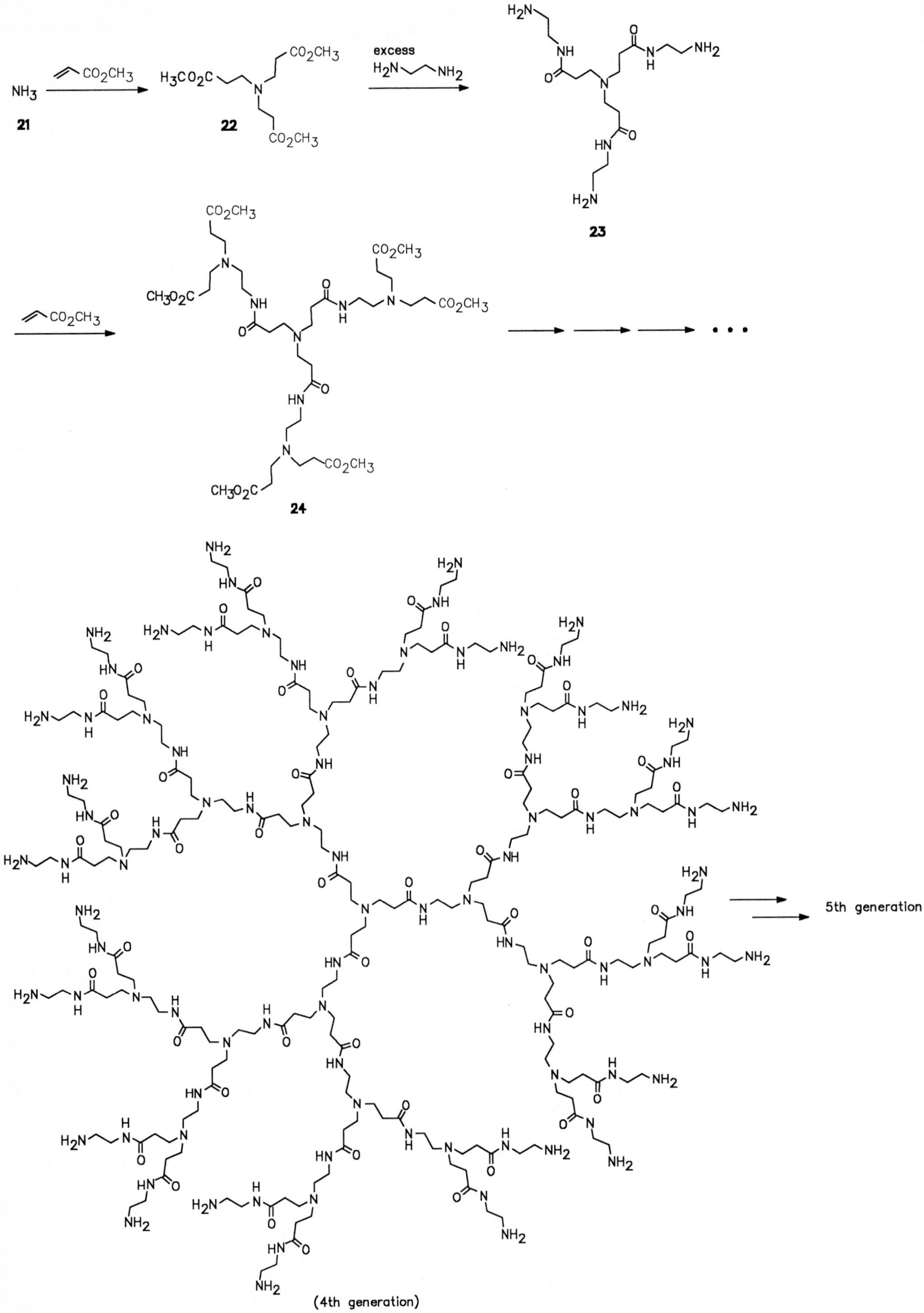

Scheme 4.9. Procedure for the preparation of PAMAM (*polyamidoam*ine) starburst polymers.

Related dendrimers possessing primary amines were synthesized[35] from hexaacrylonitrile, a by-product in acrylonitrile polymerization, which was converted to the corresponding hexaester. Aminolysis with ethylenediamine and Michael addition with ethyl acrylate follow Tomalia's process; the sequence was repeated to generate the higher generations.

Turro et al.[36, 37] have characterized the PAMAM structures utilizing fluorescence spectroscopy in which pyrene was used as the photoluminescence probe to evaluate the internal hydrophobic sites. Their results were consistent with the theoretically predicted morphology of this family of dendrimers, at generations 0.5–3.5 (carboxylate surface) the structures are open, but at generations 4.5–9.5 they are closed, and the surface becomes congested with increasing generations. Photoinduced electron transfer between species associated with the surface carboxylate groups also support the structural change at the 3.5th generation.[38] The dynamics of electron-transfer quenching of photoexcited $Ru(phen)_3^{2+}$ using methyl viologen in solution with the anionic PAMAM dendrimers were evaluated.[39] The results support the structural change at generation 3.5 and despite the structural differences with micellar aggregates striking similarities were demonstrated. Reviews[40, 41a] of their studies have appeared and should be consulted for details.[41b]

Ottavianna et al.[42] have conducted extensive ESR studies on half generation PAMAMs, those possessing carboxylate surface functionality. Positively charged nitroxide radicals, attached to carbon chains with variable lengths, were used to evaluate the hydrophobic and hydrophilic binding loci. Mobility (τ_c) and polarity (A_n) parameters as a function of pH demonstrated electrostatic interactions in binding at the dendritic surface–water interface. The radical chain was demonstrated to intercalate in the inner regions of the dendrimer and interact at hydrophobic sites. Activation energies for the rotational motion of the probe were also determined.

Miller et al.[43] described the coating of the surface of the PAMAM series with aryl diimides capable of forming anion radicals.[44, 45] The vis-IR spectra of the generated anion radicals in this series suggest the formation of diimide "π-stacks," and cyclic voltammograms further indicated a diimide aggregation phenomena. Subsequent investigations[45b] indicated that the dendrimer-based π-stacked diimide network established a mode of electrical conductivity, as opposed to ionic. Films cast at 60 °C possessed conductivity values ($\sigma = 2 \times 10^{-3}$ S cm^{-1}) ten times greater than those cast at 120 °C suggesting that stacking is improved at lower temperatures. It was postulated that the "3-D" features of dendrimers make the "isotropic nature of these films of particular interest."

De Gennes and Hervet[46] statistically found that these "cauliflower polymers"[8] (Tomalia's PAMAMs) exhibit restricted idealized growth, also known as 'dense packing', when the number of generations $m = m_1$, where $m_1 \cong 2.88$ (ln P + 1.5). This relates in space to the limiting radius R_1, which increases linearly with P monomers. Below this limit, the radius $R(M)$ of the dendrimer, when plotted as a function of molecular weight (M), should increase ($m^{0.2}$); whereas above this limit ($R > R_1$), compact structures ($R \approx M^{0.33}$) should result.

These starburst dendrimers have been subjected[47] to two different fractal analyses: [48, 49] (a) $A \approx \sigma^{(2-D)/2}$, where A is the surface area accessible to probe spheres possessing a cross-sectional area, σ, and the surface fractal dimension, D, which quantifies the degree of surface irregularity; and (b) $A \approx d^D$, where d is the object size. Both methods give similar results with $D = 2.41 \pm 0.04$ (correlation coefficient = 0.988) and 2.42 $\pm$ 0.07 (0.998), respectively. Essentially, the dendrimers at the larger generations are porous structures with a rough surface. For additional information on dendritic fractality, see Section 2.3.

Watanabe and Regen reported[50] the use of these PAMAM dendrimers in the preparation of Iler-like arrays.[51] These arrays were constructed on a (3-aminopropyl)triethoxysilane activated silicon wafer, via sequential exposure to K_2PtCl_4, rinsing, a solution of the dendrimer, K_2PtCl_4, and rinsing. Multilayers were constructed by the repetition of the last three steps. Examination of a multilayer coating after 5-cycles by atomic force microscopy demonstrated that the surface is smooth at the molecular level, with an average roughness of 7.1 Å.

Yu and Russo[52a] reported the fluorescence photobleaching recovery and dynamic light

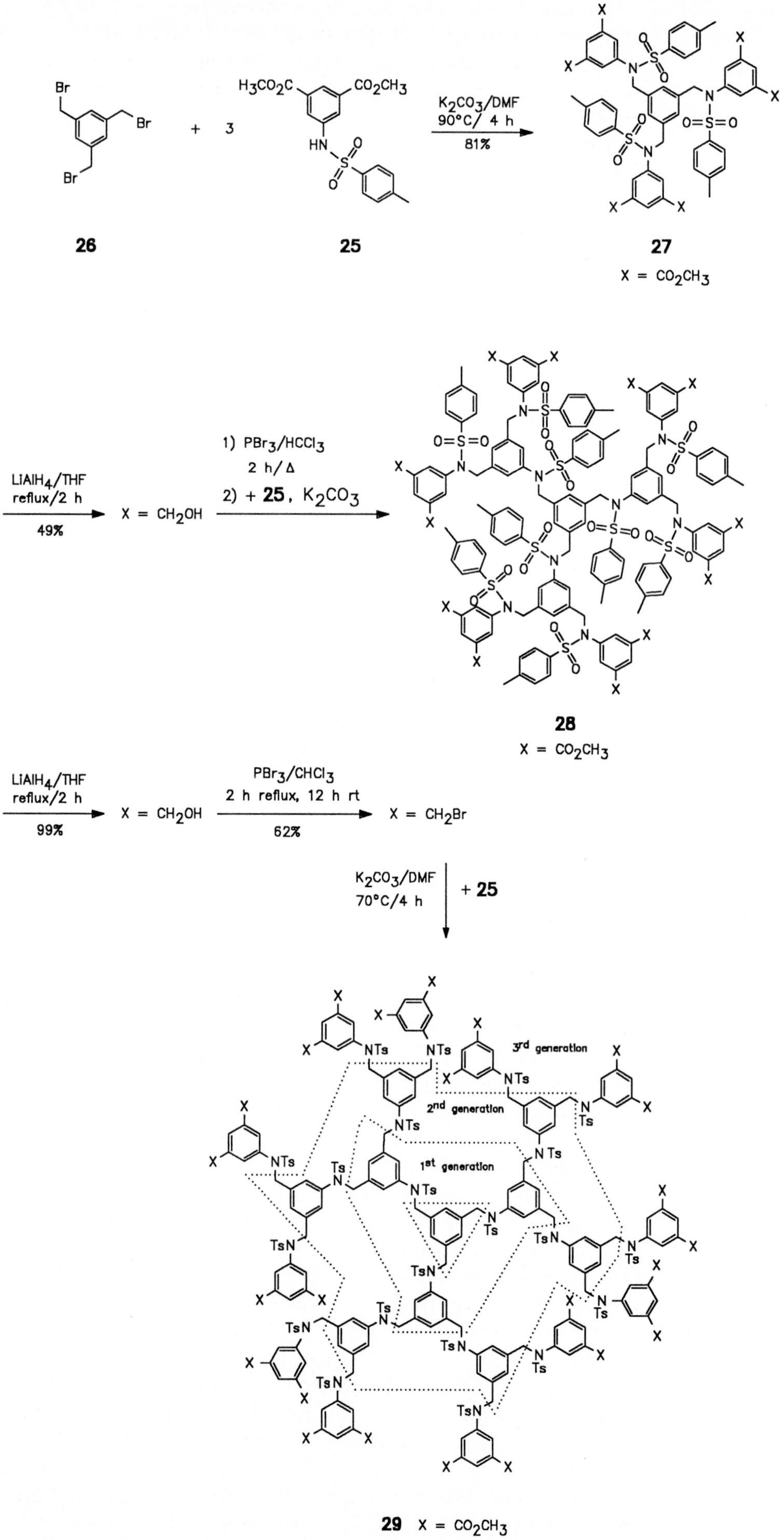

Scheme 4.10. Synthesis of "bulky" dendrimers via the sequential addition of a *N*-tosylated aminoisophthalate diester building block or monomer.

scattering characterization of these polyamidoamine cascade polymers (PAMAMs).[8] Agreement of the two techniques suggests that the attachment of a fluorescent dye does not significantly change the diffusion coefficient of the fifth generation dendrimer. At high salt concentrations, the measured hydrodynamic diameters obtained via the Stokes–Einstein equation, are close to the reported diameters as determined by size exclusion chromatography (SEC). In an unrelated report, Shinkai et al.[52b] have coated PAMAM dendrimers with anthracene-phenylboronic acid units; these dendrimers act as "saccharide sponges." Crooks and Wells[52c] have attached PAMAM dendrimers to *self-assembled monolayers* (SAMs) and subsequently demonstrated their usefulness in constructing *surface acoustic wave* (SAW) devices. Generation 4 was determined to be the most useful as a mass balance detector due to globular architecture and readily accessible interior endoreceptors.

4.4   1 → 2 *Aryl*-Branched, *Amide* Connectivity

Vögtle et al.[53–56] devised a simple route to very bulky cascade molecules by employing the *N*-tosylate of dimethyl 5-aminoisophthalate (**25**; Scheme 4.10). Utilizing 1,3,5-tris (bromomethyl)benzene[57] (**26**) as the core and tosylate **25** as the building block, they were able to generate hexaester **27** (X = CO_2CH_3), which was reduced and transformed to the hexakis(bromomethyl) derivative, and then treated with 6 equivalents of **25**, generating the corresponding dodecaester **28**. This three-step divergent reaction sequence was repeated, ultimately achieving three generations, e. g., **29**. An X-ray structure (Figure 4.1) of the intermediate hexaester **27** was obtained, from which an appreciation of the internal rigidity can be acquired. The structural homogeneity associated with further tier construction beyond the third tier (**29**) could be problematic due to the steric demands of the bulky monomers (**25**) and the diminished spatial region available for reaction of each peripheral bromomethyl moiety.

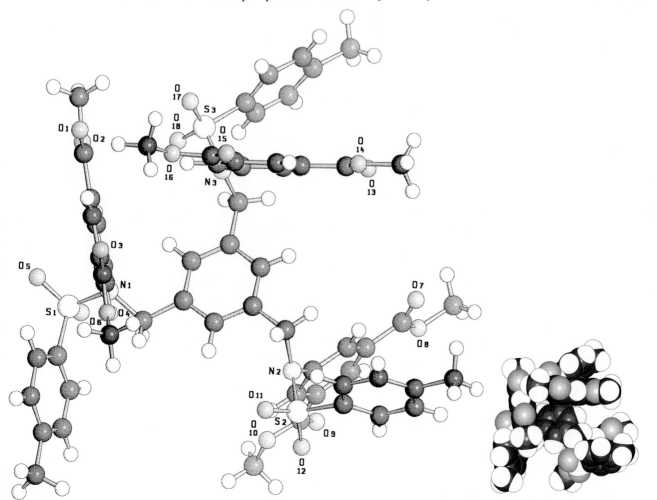

Figure 4.1. X-ray structure of the 1st generation tri-*N*-tosylated dendrimer (**27**; Scheme 4.10.).[55]

4.5   1 → 2 *Aryl*-Branched, *Ester* Connectivity

Haddleton et al. [58] reported the determination of molecular masses of two series of dendritic aryl esters by matrix-assisted laser desorption ionization (MALDI) mass spectrometry. Their divergent mode of construction utilized either phloroglucinol (1,3,5-trihydroxybenzene) or hydroquinone (**30**) as the core and the benzyl protected 3,5-dihydroxybenzoic acid (**31**), as the building block (Scheme 4.11). Treatment of diol **30** with two equivalents of acid **31** in the presence of dicyclohexylcarbodiimide (DCC) gave the bisester **32**, which was deprotected by catalytic hydrogenolysis to liberate tetrahydroxy diester **33**. Repetition of the esterification process gave rise to hexaester **34**; three tiers were constructed and characterized by MALDI mass spectral studies which showed no evidence for dimer or trimer formation either during the synthesis or within the mass spectrometer.

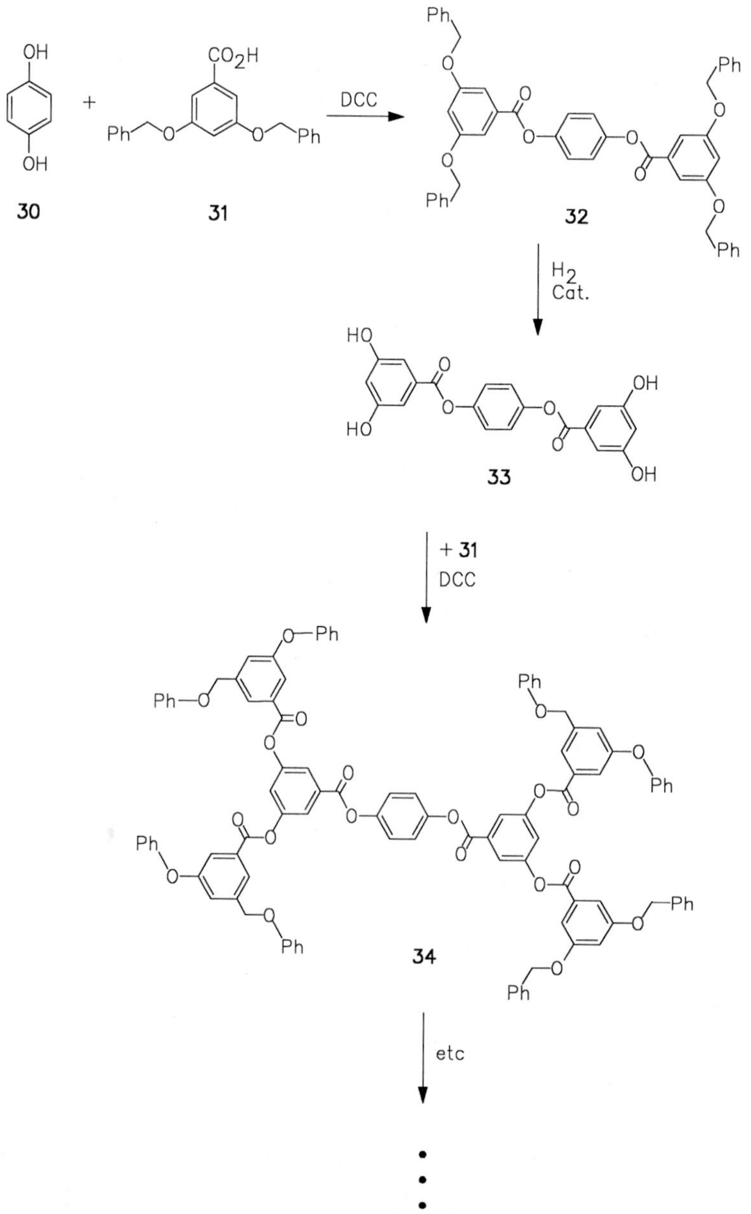

Scheme 4.11. Construction of 1 → 2 aryl-branched poly esters via phenol acylation.

4.6   1 → 2 *C*-Branched

4.6.1   1 → 2 *C*-Branched, *Amide* Connectivity

Denkewalter et al.[4] synthesized a series of *tert*-butoxycarbonyl-protected poly(α, ε-L-lysine) macromolecules (see above, Scheme 4.3), whose molecular models suggested them to be globular, dense spheres and whose molecular weight distribution was determined to be very narrow ($M_w/M_n \cong 1.0$). Since each generation in this series was synthesized in a stepwise manner, each member was predicted to have a monodisperse molecular weight.

Aharoni and coworkers characterized[59] Denkewalter's cascade macromolecules[4] by employing classical polymer techniques: viscosity determinations, photo correlation spectroscopy (PCS), and size exclusion chromatography (SEC). It was concluded that at each tier (2 through 10) these globular polymers were, in fact, monodisperse and behaved as nondraining spheres. The purity of these molecules was not ascertained and the dense packing limits were either not realized or simply not noted.

Tam and coworkers[60–65] coupled amino acids by means of a Merrifield-type peptide synthesizer using lysine-based amine acylation technology, similar to that reported by Denkewalter,[4] to produce octopus-immunogens (**35**) and scaffolding, or core matrices, for a *multiple antigen peptide* (MAP core, **36**) (Scheme 4.12). Recently, the synthesis of peptide dendrimers utilized a novel tetravalent glyoxylyl–lysinyl core peptide, {[(OHCCO)$_2$Lys]$_2$Lys-AlaOH} in the formation of oxime, hydrazone, or thiazolidine linkages.[66] In general, a branched lysine core matrix with an aldehyde ligated four copies of the unprotected peptides possessing bases such as aminoxy, hydrazide or cysteine 1,2-aminothiol groups at their *N*-termini forming the synthetic proteins. The use of the mutually reactive weak base and aldehyde pair will provide a convenient route to peptide dendrimers and artificial proteins.[67]

Pessi et al.[68] prepared a different multiple antigen peptide employing the continuous-flow polypeptide procedure[69] for the solid-phase peptide synthesis. The resultant one-directional, octaantigen, polypeptide cascade **37** is depicted in Figure 4.2 and was characterized by gel permeation chromatography (GPC), FAB MS, and amino acid ratio analysis.

Roy and coworkers[70, 71] described the solid state preparation of the first four generations (e.g., **38**) of the dendritic sialoside inhibitors of influenza A virus haemagglutinin

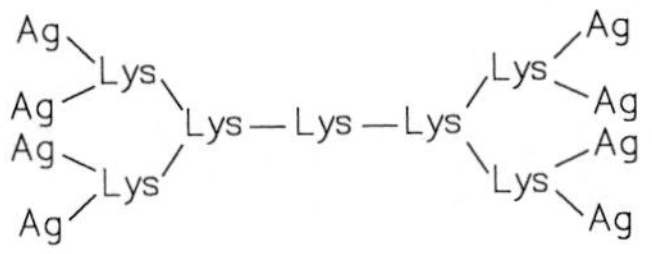

35

An Octopus–Immunogen

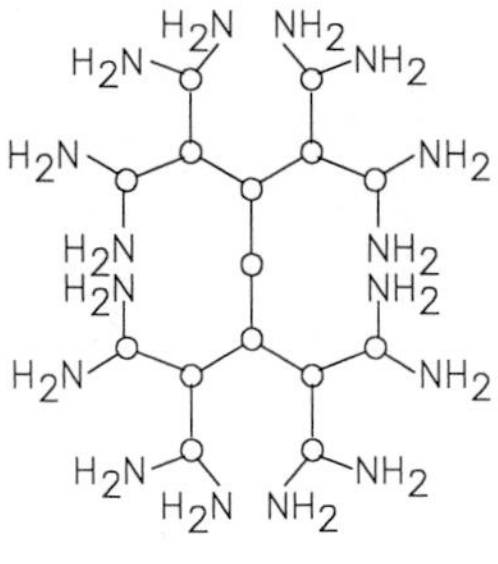

36   (O ≡ Lys)

A Lysine–base core matrix used
for the preparation of a
Multiple–Antigen–Peptide

Scheme 4.12. Lysine-based dendritic "scaffolding" used for the preparation of polyimmunogens.

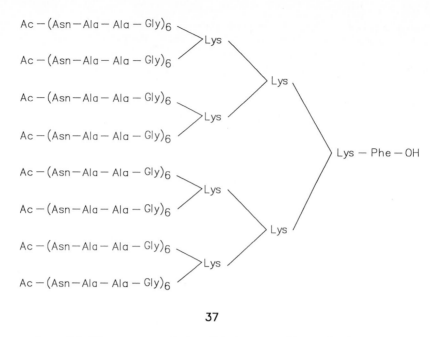

37

Figure 4.2. Octaantigen polypeptide **37** was prepared via solid phase peptide synthesis.

(Scheme 4.13 (a)). They used a similar backbone to that found in the Denkewalter lysine-based series,[4] but employed $N^{\alpha},N^{\varepsilon}$-di(fluorene-9-ylmethoxycarbonyl)-L-lysine benzotri-azole (**39**) as the activated ester building block for tier construction (Scheme 4.13 (b)). After acylation and subsequent deprotection, the two-step process was repeated. Each tier was terminally functionalized with the preformed benzotriazole ester of chloroace-tylglycylglycine, followed by reaction with peracetylated 2-thiosialic acid. Removal from the solid state support, followed by hydrolysis, afforded the desired inhibitor (e.g., **38**) in nearly quantitative yield.

Chapman et al.[72a] reported the synthesis of poly(ethylene oxide) supported dendritic *t*-BOC-poly(α,ε-L-lysines). Methoxy terminated poly(ethylene oxide) was esterified with BOC-protected glycine (**40**), then deprotected (TFA) to liberate the free amine, follow-ed by the divergent construction (Scheme 4.14) of the poly(α,ε-L-lysine) (**41**). Employing a procedure related to that of Denkewalter,[4] pentafluorophenyl N^{α}, N^{ε}-di(*t*-BOC)-L-lysinate (**42**) was repetitively coupled, then deprotected. These 'chimerical' molecules with multiple terminal hydrophobic *tert*-butyl groups were coined as hydroamphiphiles, since in water they were observed to form foams possessing good temporal stability.

Scrimin et al.[72b] have reported the preparation of a small, three-directional, polypep-tide that is useful for the modulation of membrane permeability. Decapeptide fragments were attached to a TREN (tris(2-aminoethyl)amine) core.

4.6.2   1 → 2 *C*-Branched and Connectivity

Hart and his coworkers[72-76] reported the construction of numerous polyaromatic, all-hydrocarbon cascades termed "iptycenes", which are extended triptycenes using bicy-clo[2.2.2]octane moieties as the branching groups. Connection of the bicyclic moieties with benzene units results in a three-directional pattern. Although their construction is not strictly iterative, multiple Diels–Alder transformations with bis(9,10-anthradiyl) sub-stituted butadiene monomer (**43**) were employed (Scheme 4.15). Treatment of the highly substituted chlorobutadiene **44** with quinone **45** afforded the bis-triptycene intermediate **46**, which generated dione **47** when treated with diene **43**. Transformation of the core dione moiety of **47** to the anthracene nucleus **48** was conducted in three-steps (NBS; LiAlH₄; DDQ); treatment of the tetrakis-triptycene **48** with 1,2-dichloroethene, follow-ed by Li, and subsequent Diels–Alder reaction with diene **43** afforded an intermediate (not shown) that was finally aromatized to generate "superiptycene" **49**. These highly rigid superstructures possess very unique molecular cavities as evidenced by the fact that

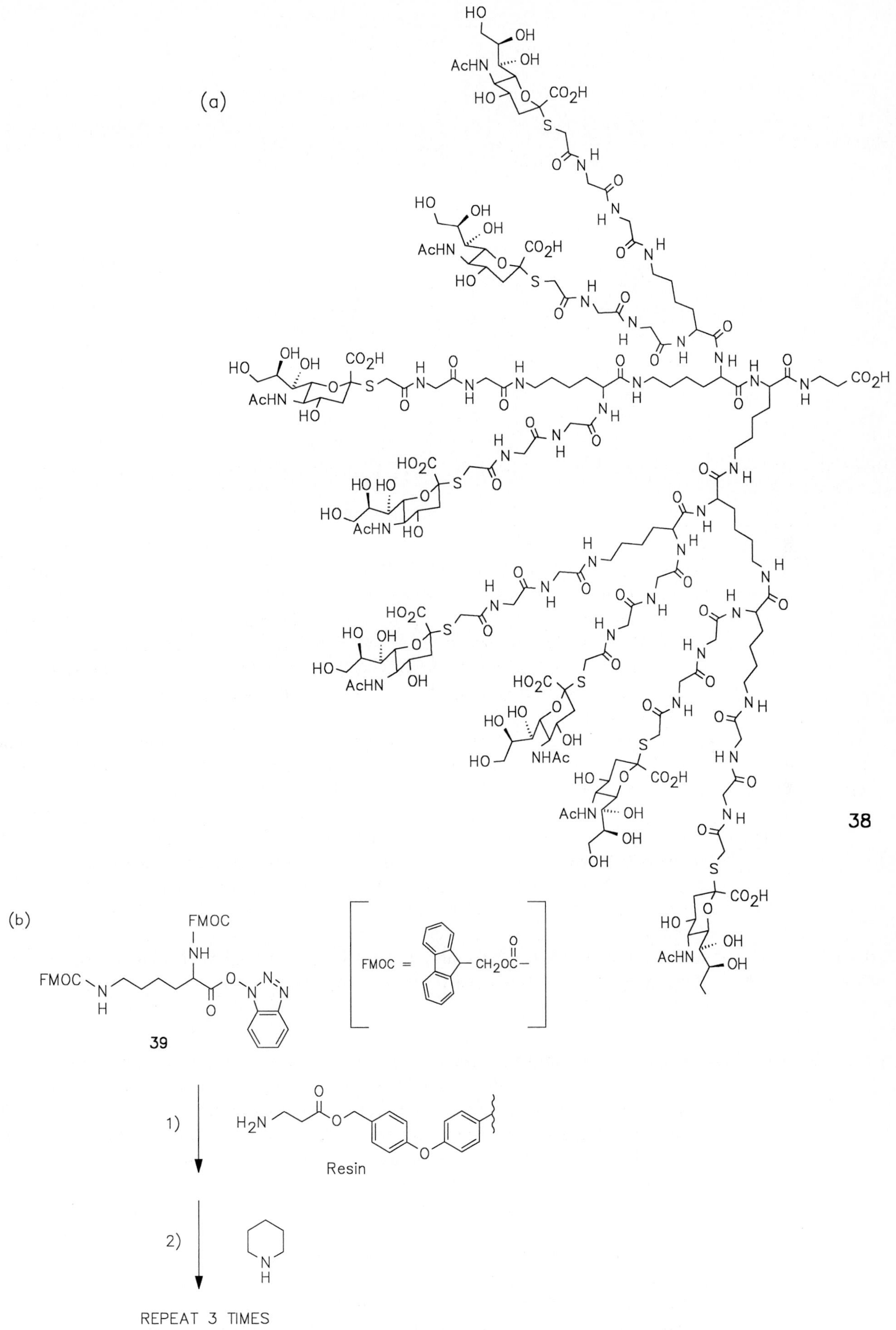

Scheme 4.13. Dendritic sialoside inhibitors of influenza A virus haemagglutinin constructed via the solid phase synthesis of a lysine-based superstructure.

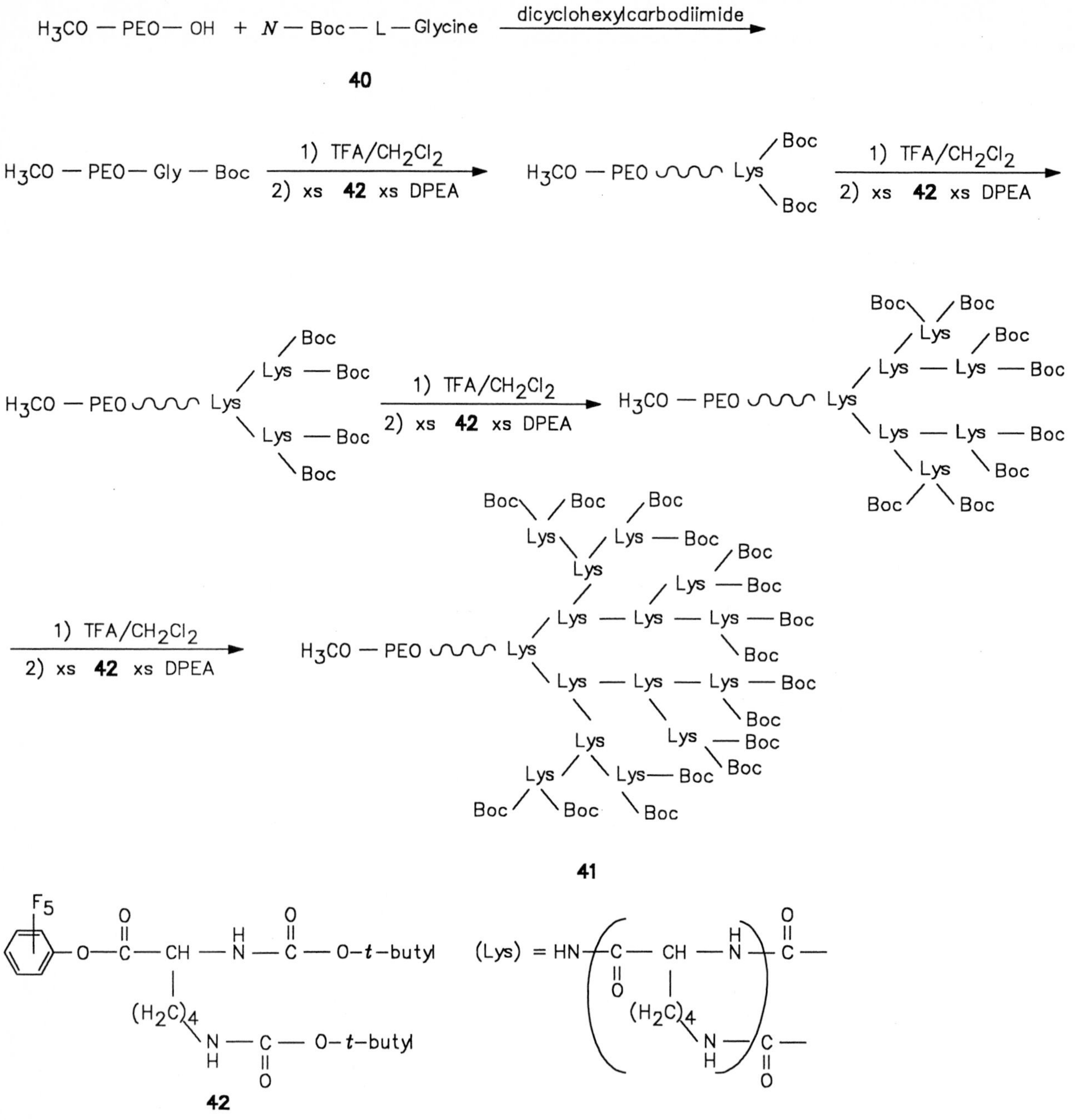

$H_3CO — PEO — OH + N — Boc — L — Glycine \xrightarrow{\text{dicyclohexylcarbodiimide}}$

40

$H_3CO — PEO — Gly — Boc \xrightarrow[\text{2) xs }\textbf{42}\text{ xs DPEA}]{\text{1) TFA/CH}_2\text{Cl}_2}$ H₃CO — PEO ∿ Lys $\genfrac{}{}{0pt}{}{\diagup \text{Boc}}{\diagdown \text{Boc}}$ $\xrightarrow[\text{2) xs }\textbf{42}\text{ xs DPEA}]{\text{1) TFA/CH}_2\text{Cl}_2}$

H₃CO — PEO ∿ Lys with branches:
Lys — Boc (Boc)
Lys — Boc
Boc

$\xrightarrow[\text{2) xs }\textbf{42}\text{ xs DPEA}]{\text{1) TFA/CH}_2\text{Cl}_2}$

H₃CO — PEO ∿ Lys (expanded dendrimer with Lys — Lys — Boc branches terminated in Boc groups)

$\xrightarrow[\text{2) xs }\textbf{42}\text{ xs DPEA}]{\text{1) TFA/CH}_2\text{Cl}_2}$

H₃CO — PEO ∿ Lys (fully expanded dendrimer **41** with multiple Lys — Lys — Lys — Boc branches terminated in Boc groups)

41

Structure **42**: pentafluorophenyl ester — F_5 (C₆ ring) — O — C(=O) — CH — N(H) — C(=O) — O—*t*—butyl, with side chain (H₂C)₄ — N(H) — C(=O) — O—*t*—butyl

42

$(\text{Lys}) = \text{HN} \left(\begin{array}{c} \text{C} — \text{CH} — \text{N}(\text{H}) — \text{C} — \\ \| \quad | \quad \quad \| \\ \text{O} \quad (\text{H}_2\text{C})_4 \quad \text{O} \\ \quad \quad \text{N} — \text{C} — \\ \quad \quad \text{H} \quad \| \\ \quad \quad \quad \text{O} \end{array} \right)$

Scheme 4.14. Lysine-based, dendritic terminated poly(ethylene oxide) was described as "hydroamphiphiles" due to the formation of stable foams in water.

crystals of iptycene **49** "include" solvent molecules within these cavities; solvent inclusion was suggested as a factor affecting X-ray structure determination.

Hart et al.[77, 78] also reported the preparation of symmetric iptycenes possessing benzene cores, for example, "nonadecaiptycene" (**50**) via bicyclic vinyl halide trimerization of triptycene dimer **51**, as well as the synthesis of the related asymmetric iptycenes[78] (Scheme 4.16).

Although the initial crystallographic study of the simple heptiptycene[79] did not afford a structural model,[80] recently the structure of the crystalline 1:1 heptiptycene–chlorobenzene clathrate, in which the solvent molecule was packed in the channels between ribbons of the heptiptycene, was ascertained.[81] The calculated molecular geometry via Hartree–Fock (6-31G(D)) and local density methods compared well with the X-ray data.

Webster[82] described the preparation of a water-soluble tritriptycene and examined the ¹H NMR chemical shift changes of various substrates due to interactions with the aromatic ring currents. For example, a D₂O solution of the tritriptycene and *p*-toluidine exhibited an up-field shift ($\Delta v = 55$ Hz) for the substrate methyl group absorption.

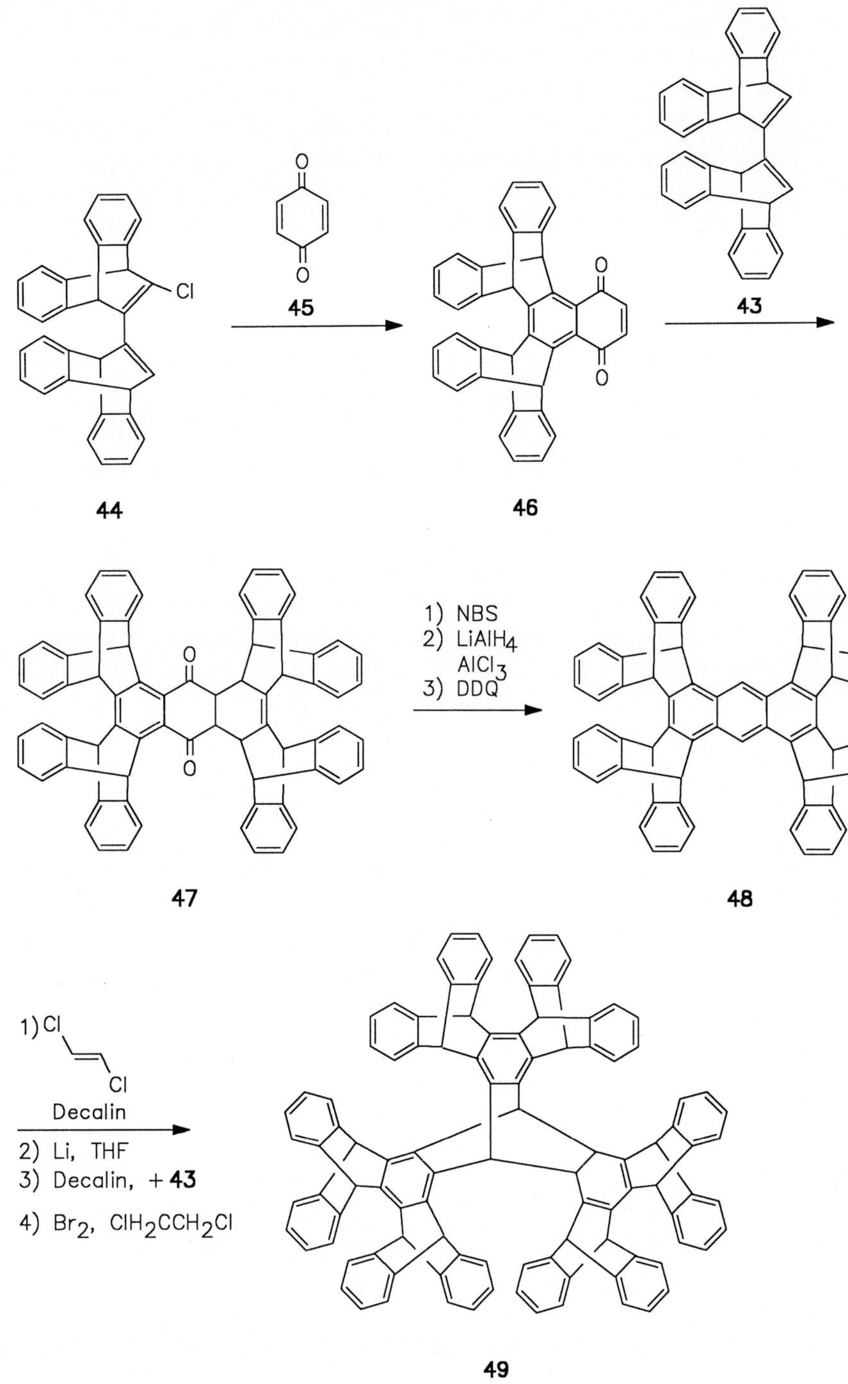

Scheme 4.15. Hart et al.'s[76] "iptycene" construction.

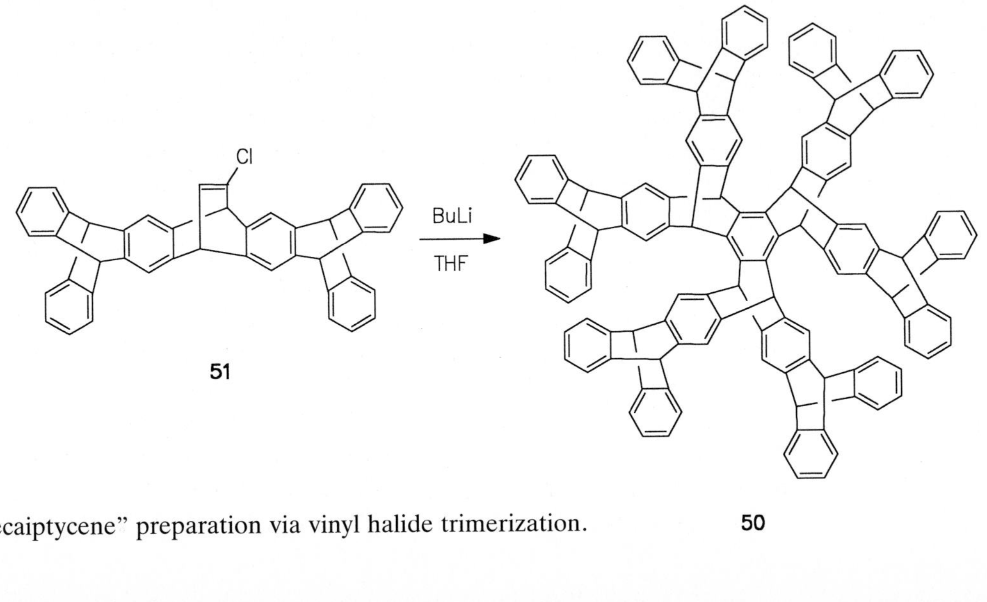

Scheme 4.16. "Nonadecaiptycene" preparation via vinyl halide trimerization.

Zefirov et al.[83] described a general synthetic strategy for the preparation of *branched triangulanes* or spiro-condensed polycyclopropanes (Scheme 4.17). Key features of this strategy include chloromethylcarbene addition to methylenecyclopropanes,[84] followed by dehydrohalogenation. Thus, treatment of bicyclopropylidene **52** with the chloromethylcarbene generated from dichloride **53** gave the tricyclopropane **54**. Dehydrochlorination followed by cyclopropanation (CH_3CHCl_2, *n*-BuLi) and alkene formation again (*tert*-BuOK, DMSO) gave the unsaturated ether **55**. After $Pd(OAc)_2$ mediated methyl carbene addition (CH_2N_2), acidic alcohol deprotection (HCl), bromination (Ph_3PBr_2, pyr) and β-elimination (*tert*-BuOK, DMSO), alkene **56** was converted to the hexakis(spirocyclopropane) (**57**).

The preparation of other branched triangulanes with varying symmetries has also been reported. A notable feature of this series of small hydrocarbon cascades is that the framework is composed entirely of quaternary, tetraalkyl-substituted carbons. This unique architecture closely resembles or is at least reminiscent of Maciejewski's[16] proposed cascade molecule comprised of an all $1 \rightarrow 3$ *C*-branched interior framework (i.e., without spacers between branching centers).

de Meijere et al.[85] reported the preparation of symmetrical branched triangulane constructed of 10 fused cyclopropane moieties (Scheme 4.18). Although the synthesis was not repeating, or iterative, the resulting structurally rigid spirocyclopropane **59** was finally obtained (14 %) by the reaction of nitrosourea **58** with bicyclopropylidene and NaOMe. Unequivocal structure determination of **59** was provided by X-ray crystallography, which demonstrated its D_{3h} molecular symmetry. [10]Triangulane **59** showed high thermal stability even though its thermal strain energy ($\approx$ 1130 kJ mol^{-1}) indicates that it is more strained than cubane.[86, 87a] The authors speculated on the potential for a carbon network based on spiro-linked, three-membered rings. Strain energies in [*n*]triangulanes and spirocyclopropanated cyclobutanes have been determined.[87b]

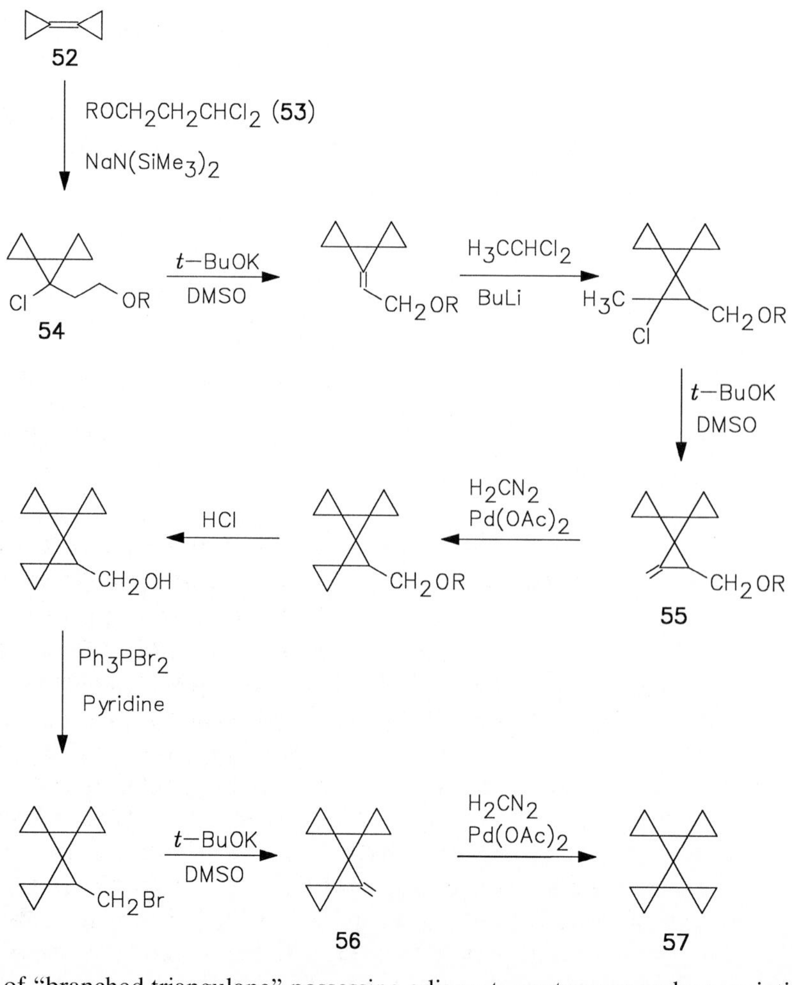

Scheme 4.17. Synthesis of "branched triangulane" possessing adjacent, quaternary carbon moieties. R = Me or THP.

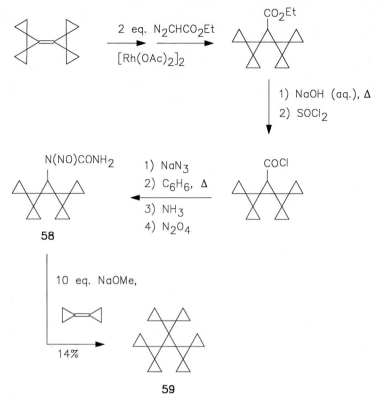

Scheme 4.18. Non-iterative preparation of a [10]triangulane possessing D_{3h} symmetry.

4.7 1 → 2 *C & Aryl* Branched and Connectivity

Veciana et al.[88–92] have reported the synthesis of perchlorinated polyradicals through two generations (Scheme 4.19). The first tier triradical **64** was obtained by subjecting 1,3,5-trichlorobenzene **60** to dihalomethylation (AlCl₃, CHCl₃) to give the Reimer–Tiemann intermediate like tris(α,α-dichloromethylbenzene) **(61)**, which on treatment with pentachlorobenzene in the presence of AlCl₃ afforded the polychlorinated heptaaryl radical precursor **62**. Deprotonation at the triphenylmethane sites (n-Bu₄N⁺OH⁻, 35 days) gave trianion **63**, which was converted to the corresponding triradical **64** by treatment with excess p-chloranil.

Triradical **64** was isolated in two isomeric forms possessing D_3 and C_2 symmetry. Due to steric shielding provided by the chloro groups, the polyradicals showed exceptional stability in the solid state at temperatures up to 250 °C. The second generation (**65**; Figure 4.3) of these perchlorinated polyradicals was prepared, although the authors reported that "several structural defects disrupted some of the desired ferromagnetic couplings."

Iwamura et al.[93] have reported the preparation of a "branched-chain" nonacarbene possessing a nonadecet ground state with the goal of constructing superparamagnetic polycarbenes. The preparation of three-directional, diradicals was reported by Bock et al.[94] and a tetrahedral, tetraradical has been synthesized by Kirste et al.[95] The synthesis, characterization, and physical properties of the perchloro-2,6-bis(diphenylmethyl)-pyridine-α,α'-ylene biradical have also been reported.[96]

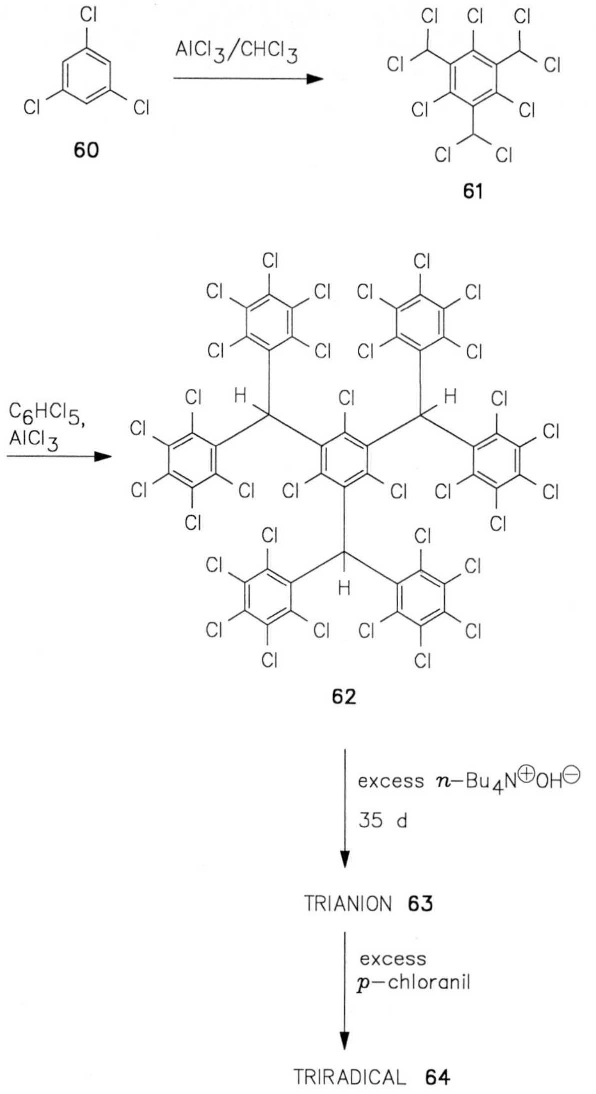

Scheme 4.19. Preparation of a perchlorinated triradical via AlCl₃ activated arylation. The triradical was isolated in two isomeric forms, D₃ and C₂.

4.8 1 → 2 *Aryl* Branched, *N*-Connectivity

Hall and Polis[97] prepared a series of polyarylamines using an aromatic nucleophilic substitution–reduction sequence (Scheme 4.20). Hence, 2,4-dinitrofluorobenzene **66** was treated with *p*-diaminobenzene to afford tetranitro **67**, which was reduced to give the corresponding first generation, diamine **68**. Repetition of this sequence afforded the second generation tetradecaamine **69**. These 'starburst' polyarylamines were complexed with iodine to form semiconducting materials and were the first dendrimers to be examined by cyclic voltammetry.

4.9 1 → 2 *Ethano*-Branched, *Ether* Connectivity

Wong et al.[98] have reported the solid-phase preparation of branched glycopeptides. A key step in the synthesis included the connection of β-1,4-galactosyl monomeric units via treatment with β-1,4-galactosyltransferase. Enzymatic fucosylation after α-chymotrypsin-mediated cleavage of the linear glycopeptide from the silica-based solid phase, afforded the sialyl Lewis X[99] glycopeptide (**70**; Figure 4.4).

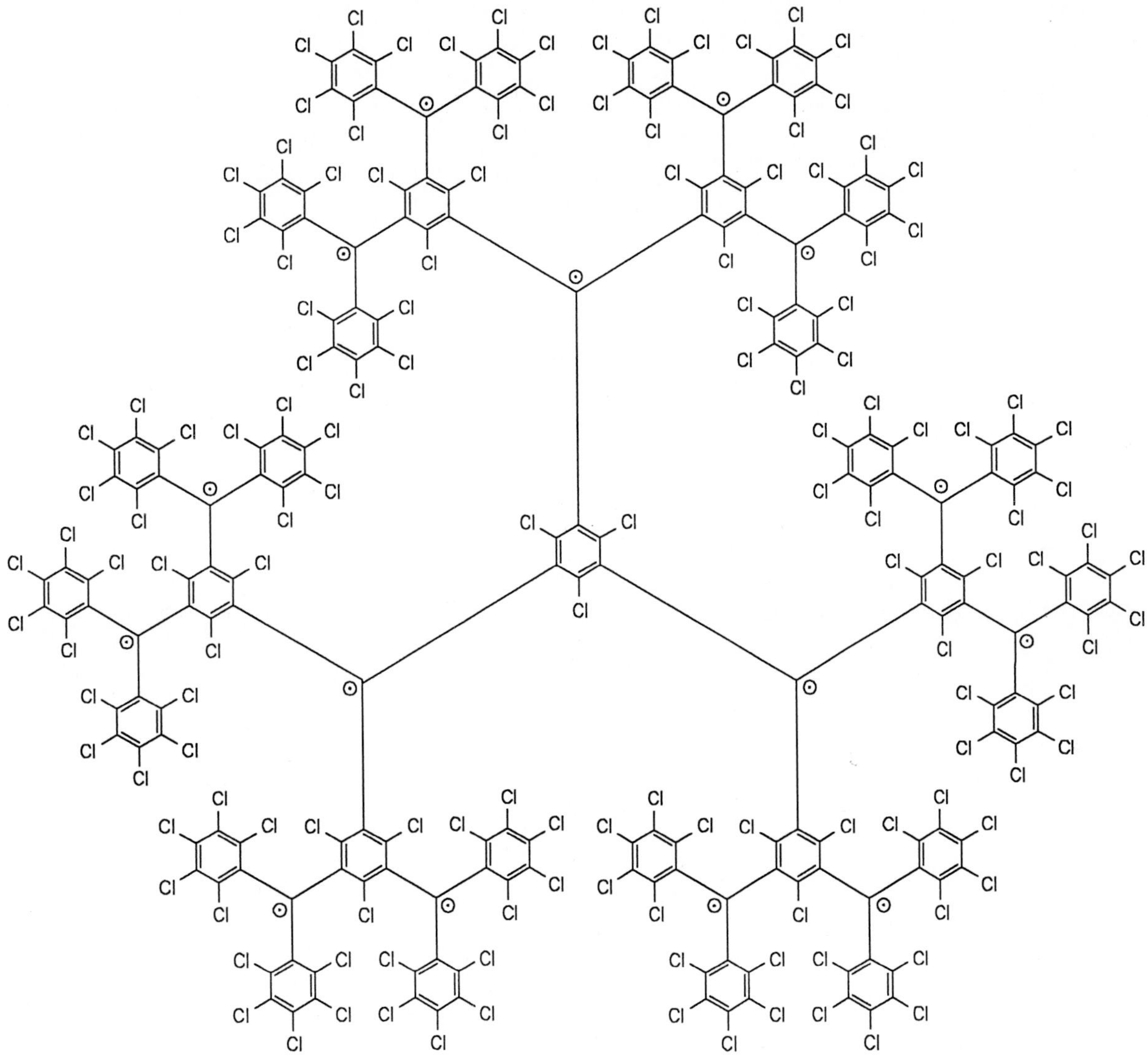

65

Figure 4.3. A 2nd generation, perchlorinated polyradical constructed to investigate solid state radical stability.

4.10 1 → 2 *Si*-Branched and Connectivity

Masamune et al.[100] reported the preparation of the first series of high molecular weight, silicon-branching macromolecules by means of the procedure shown in Scheme 4.21. Their iterative procedure utilized two differently branched synthetic equivalents: a trifunctional, hydrido-terminated core **71** and a trigonal monomer **72**. Syntheses of the polysiloxane core **71** and building block **72** were each accomplished by the treatment of trichloromethylsilane with three or two equivalents of the siloxane oligomers, HO[Si-(Me)$_2$O]$_5$Si(Me)$_2$H and HO[Si(Me)$_2$O]$_3$Si(Me)$_2$H, respectively.

Repetitive silicon-based transformations were then employed for dendritic construction. Palladium-catalyzed silane hydroxylation of the core **71** afforded triol **73**, which was then treated with three equivalents of monochloropolysiloxane **72** to generate the hexahydrido, first generation dendrimer **74**. Further application of the Pd-mediated hydroxylation, followed by attachment of monochloro-monomer **72**, led to the second (**75**) and third (**76**) generation polysiloxane cascades.

Morikawa, Kakimoto, and Imai[101] employed divergent methodology for the construction of a series of siloxane-based dendrimers possessing dimethylamino, phenyl, benzyl, or hydroxy peripheral moieties. Sequential tier addition included transforming phenyl

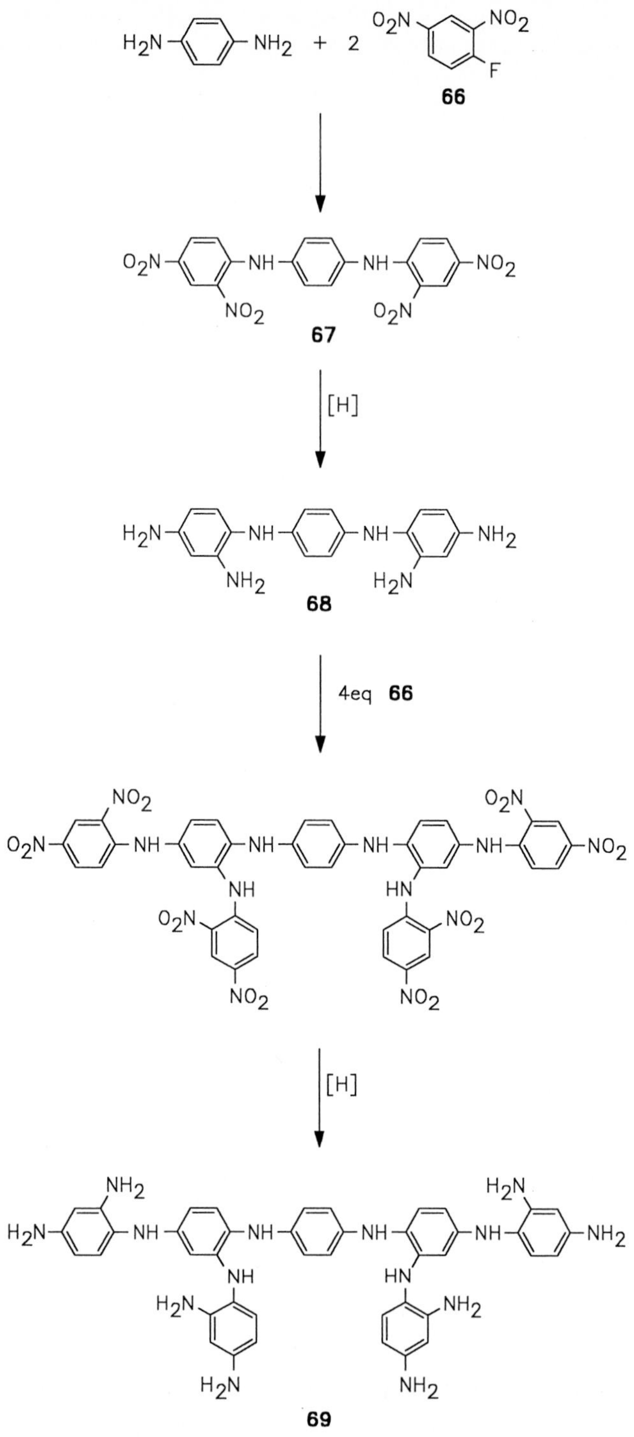

Scheme 4.20. Synthesis of poly(aryl amines) via aromatic nucleophilic displacement of fluoride ion.

silane termini to the corresponding silyl bromide (Br$_2$), treatment with HNEt$_2$ to generate the silylamine moiety, and hydroxysilyl monomer displacement of the amino group. Characterization via gel permeation chromatography and ^{1}H NMR is discussed. Polydispersity indices were determined to range between 1.30 for the second generation, phenyl terminated polysiloxane to 1.71 for the third generation, hydroxyl terminated dendrimer. The water-soluble, poly(hydrochloride) salt of the dimethylsilylamine terminated dendrimer (third tier) was compared to a unimolecular micelle due to structural similarities.

A series of carbosilane dendrimers was synthesized by Roovers et al.[102] by employing the Pt-catalyzed addition of methyldichlorosilane (**77**) to an alkene, followed by nucleophilic substitution with vinylmagnesium bromide at the terminal dichlorosilane moieties (Scheme 4.22), as the iterative method. Thus, using tetravinylsilane[103–105] (**78**) as the ini-

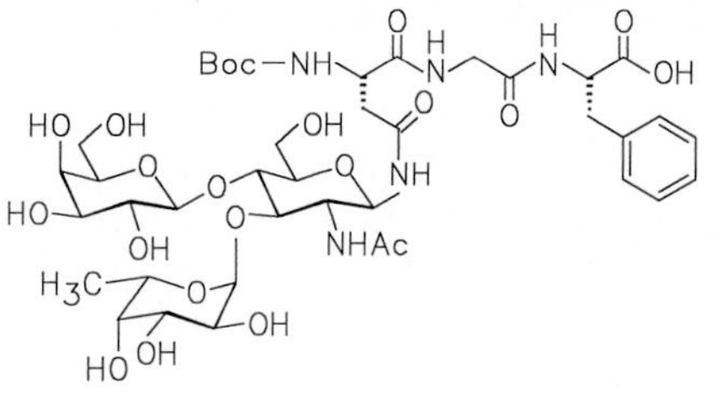

70

Figure 4.4. Solid phase synthesis of branched glycopeptides afforded this "sialyl Lewis X" glycopeptide **70**.

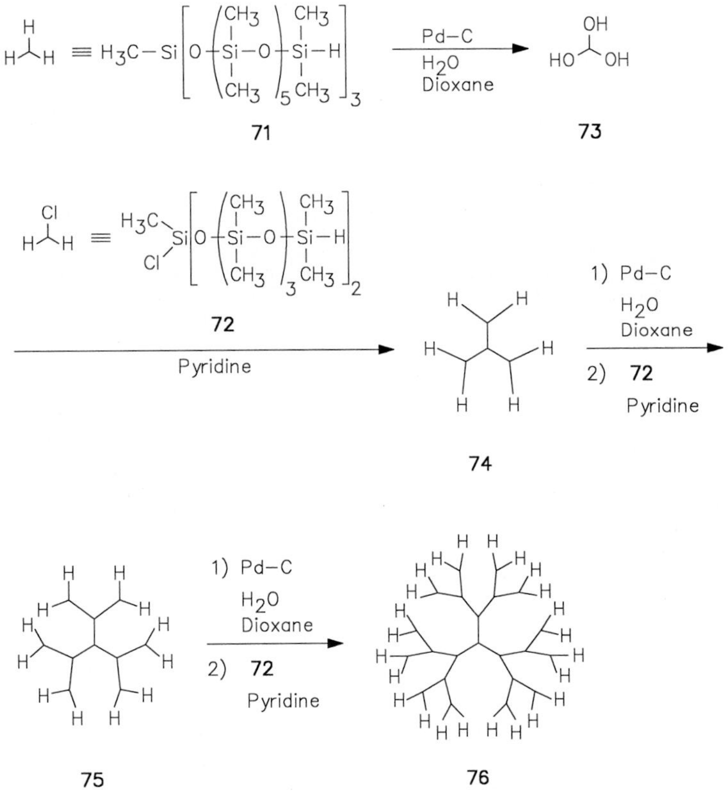

Scheme 4.21. Construction of polysiloxane dendrimers prepared by an iterative silane hydroxylation and chloride displacement at silicon.

tial tetrafunctional core, the first generation tetrakis(methyldichlorosilane) **79** was generated after addition of four equivalents of monomer **77**. Reaction of eight equivalents of vinylmagnesium bromide with pentasilane **79** generated octaolefin **80**. Continued iteration gave rise to the polyalkene **81**, possessing a molecular weight of 6,016 amu at the fourth tier and 64 terminal vinyl groups. These dendritic carbosilanes[106, 107] with 64 and 128 surface Si–Cl bonds were used as coupling reagents for monodisperse poly(butadienyl)lithium. Two series of regular star polymers with molecular weights between 6,400 and 72,000 amu were prepared; these are good models for polymeric micelles.

Morán et al.[108] reported the utilization of a similar[102] procedure (Scheme 4.23) except that allyl spacers were incorporated. Thus, tetra(allyl)silane[109] (**82**), as the core, and a simple allyl Grignard reagent were used. When silane **82** was hydrosilylated with methyldichlorosilane (**77**) under Pt-catalyzed conditions, the pentasilane **83** was generated. Subsequent branching was accomplished by reaction of the octachlorosilane (**83**) with allylmagnesium bromide to afford octaene **84**, which was hydrosilylated with dimethylchlorosilane to give the capped chlorosilane **85**. Treatment of this polychlorosilane with either lithio- or aminoethyl-ferrocene gave the corresponding *Si*-dendrimers coated with ferrocenyl moieties (**86** and **87**, respectively), which were described as noninteracting redox centers.

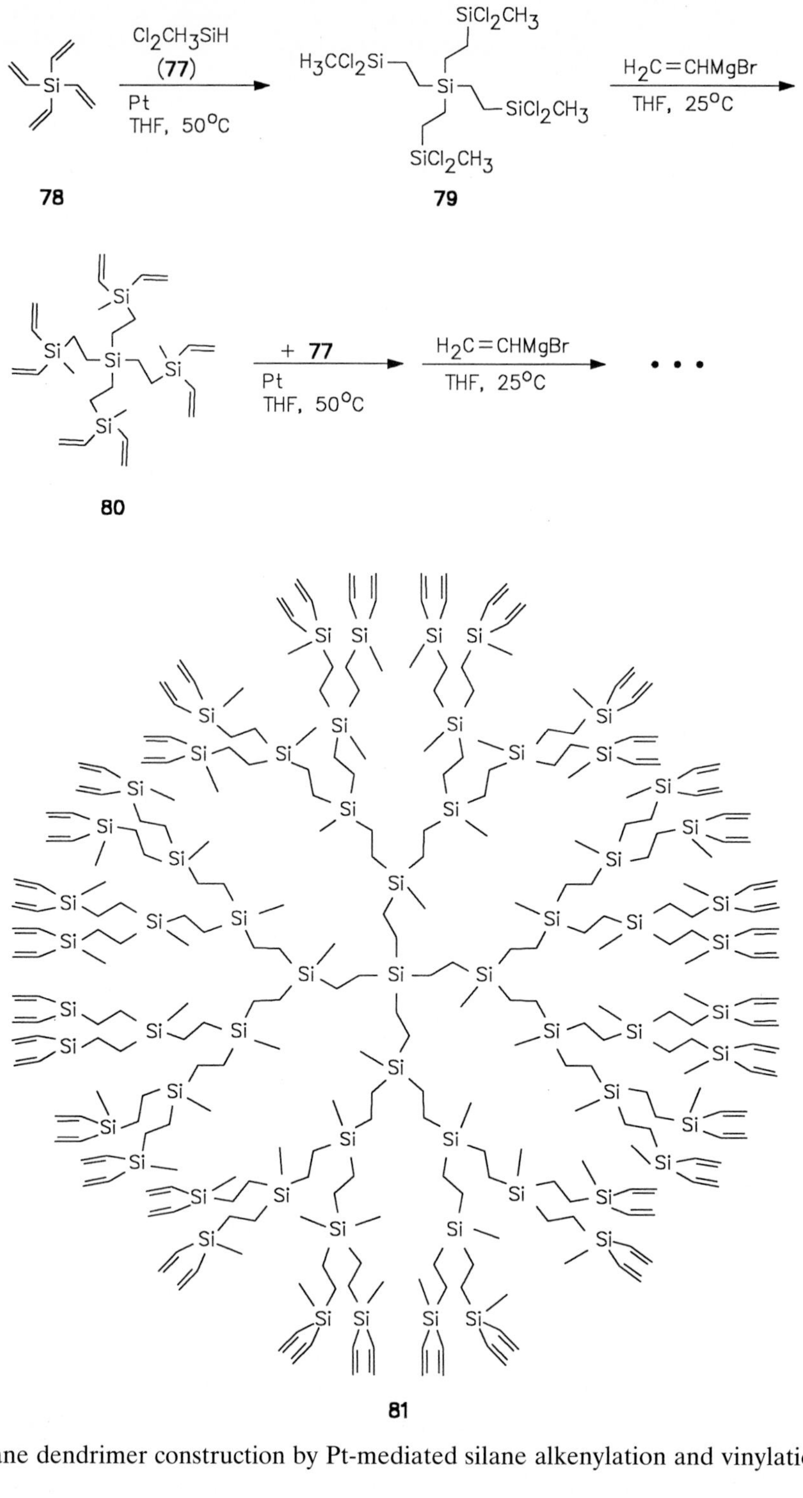

Scheme 4.22. Alkyl silane dendrimer construction by Pt-mediated silane alkenylation and vinylation.

Construction of a new series of polysilane dendrimers in which the structure of the second generation product was unambiguously confirmed by X-ray diffraction has been reported.[110] The divergent procedure for the synthesis of polysilane dendrimer **88** is shown in Scheme 4.24. The treatment of PhMe$_2$SiCl with highly inflammable silyllithium **89**, prepared (80%) by the reaction of bis(1,3-diphenylpentamethyltrisilanyl)mercury and lithium, afforded tetrasilane **90** as colorless crystals, which upon treatment with trifluoromethanesulfonic acid, followed by reaction with monomer **89** generated the next higher level dendrimer **91**. The permethylated polysilane **88** was then prepared (29%) from tridecasilane **91** by a similar two-step sequence, utilizing the capping reagent, 2-lithioheptamethyltrisilane [(Me$_3$Si)$_2$Si(Me)Li] (**92**).

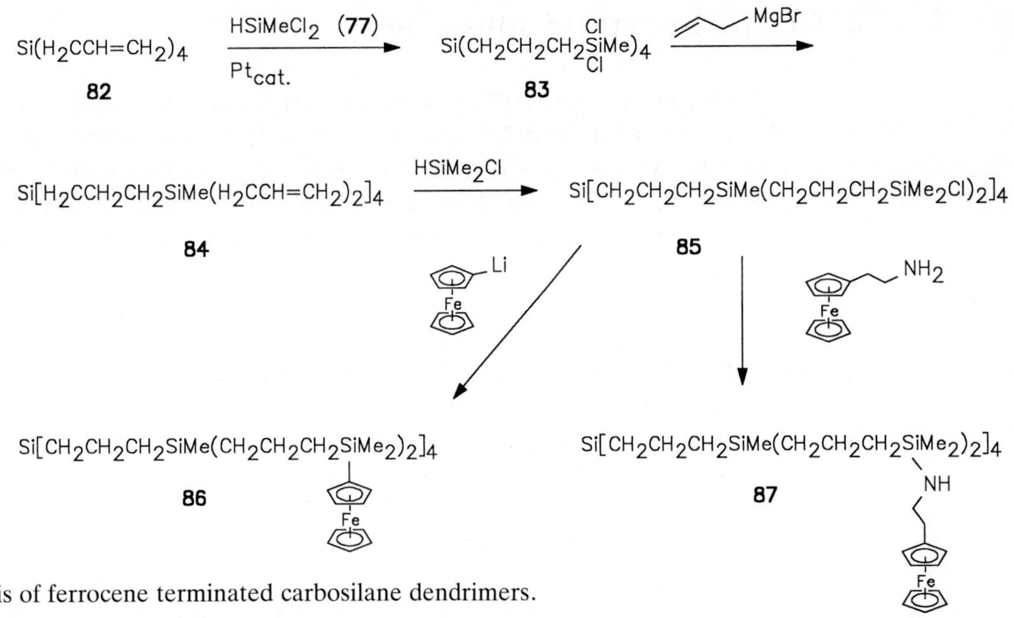

Scheme 4.23. Synthesis of ferrocene terminated carbosilane dendrimers.

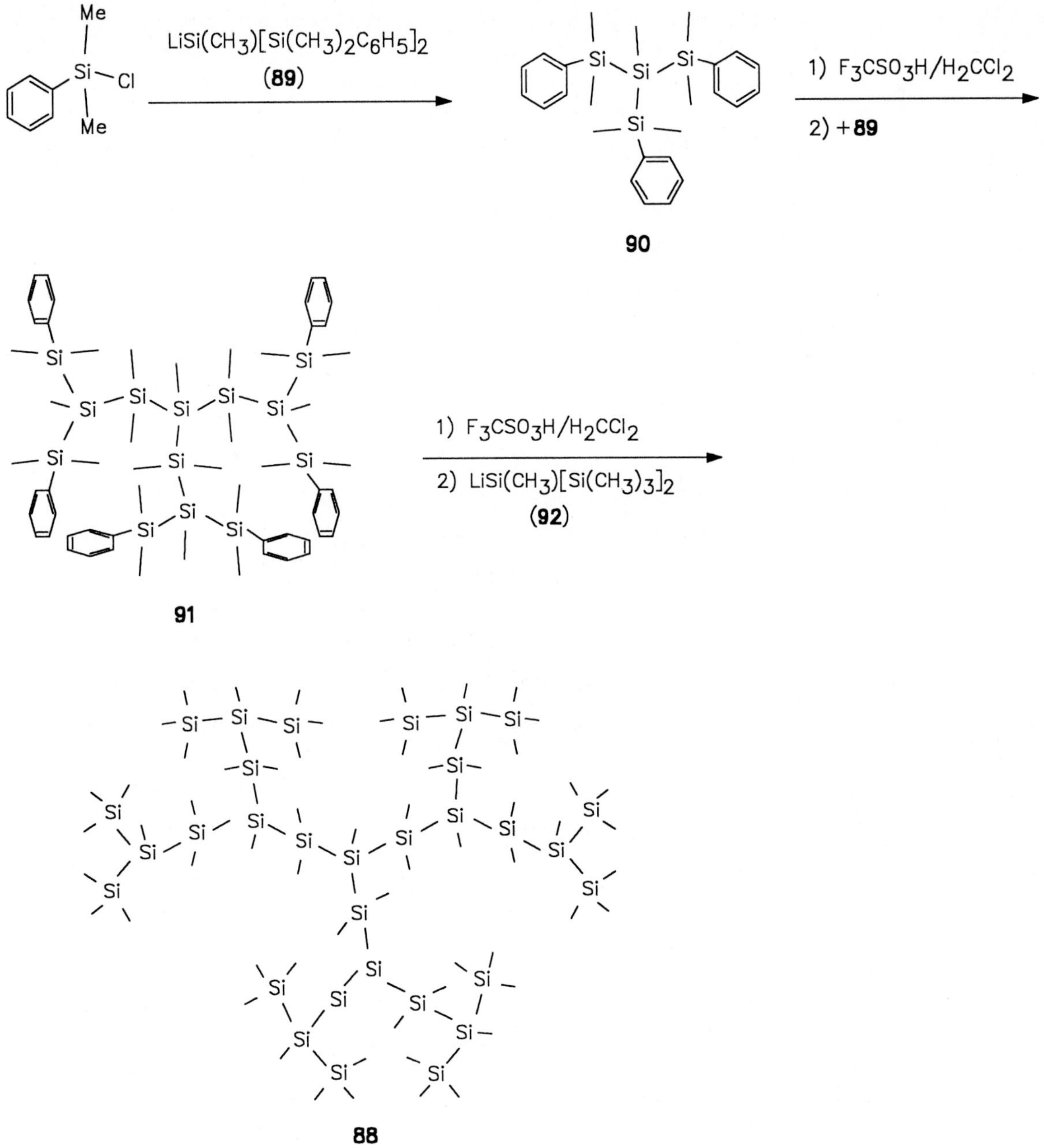

Scheme 4.24. Preparation of Si-based dendrimers with contiguous Si – Si connectivity.

4.11   1 → 2 *P*-Branched and Connectivity

DuBois et al. reported[111] the simple construction of phosphine dendrimers by the free radical addition of primary phosphines to diethyl vinylphosphonate, followed by LiAlH$_4$ reduction of the phosphonate moieties to give the corresponding polyphosphines (Scheme 4.25). Thus, triphosphine (**93**), prepared by known procedures,[112, 113] was subjected to this method to generate heptaphosphine **94** or was treated with vinyldiphenylphosphine to afford tetrakis(diphenylphosphine) **95**. This termination procedure was also used on phosphine **94** to give the phenyl-capped **96**; whereas treatment of the corresponding ethyl terminated dendrimer **97** with five equivalents of [Pd(MeCN)$_4$](BF$_4$)$_2$ resulted in pentametallation. This site specific metal complex is described in more detail in Chapter 8.

The application of this approach to phosphorus dendrimers is readily applied[111] to the creation of a tetrahedral series via the use of a four-directional silane core (Scheme 4.26). The treatment of tetravinylsilane (**78**) with phosphine **98** quantitatively gave the desired "small" dendrimer **99**, which can be transformed to the tetrakis(square planar palladium) complex; see Chapter 8.

Majoral et al.[114] reported the facile divergent synthesis of a novel *P*-dendrimer series (Scheme 4.27). Treatment of the sodium salt of 4-hydroxybenzaldehyde (**100**) with trichlorothiophosphorus(V) gave the trialdehyde **101**, which, when treated with three equivalents of the hydrazine derivative **102**, quantitatively afforded the first generation dendrimer **103** possessing six P–Cl bonds juxtaposed for repetition of the sequence. Construction of the second, third, and fourth (**104**) generations followed this iterative sequence. Key features of this sequence included no protection–deprotection procedures *and* the

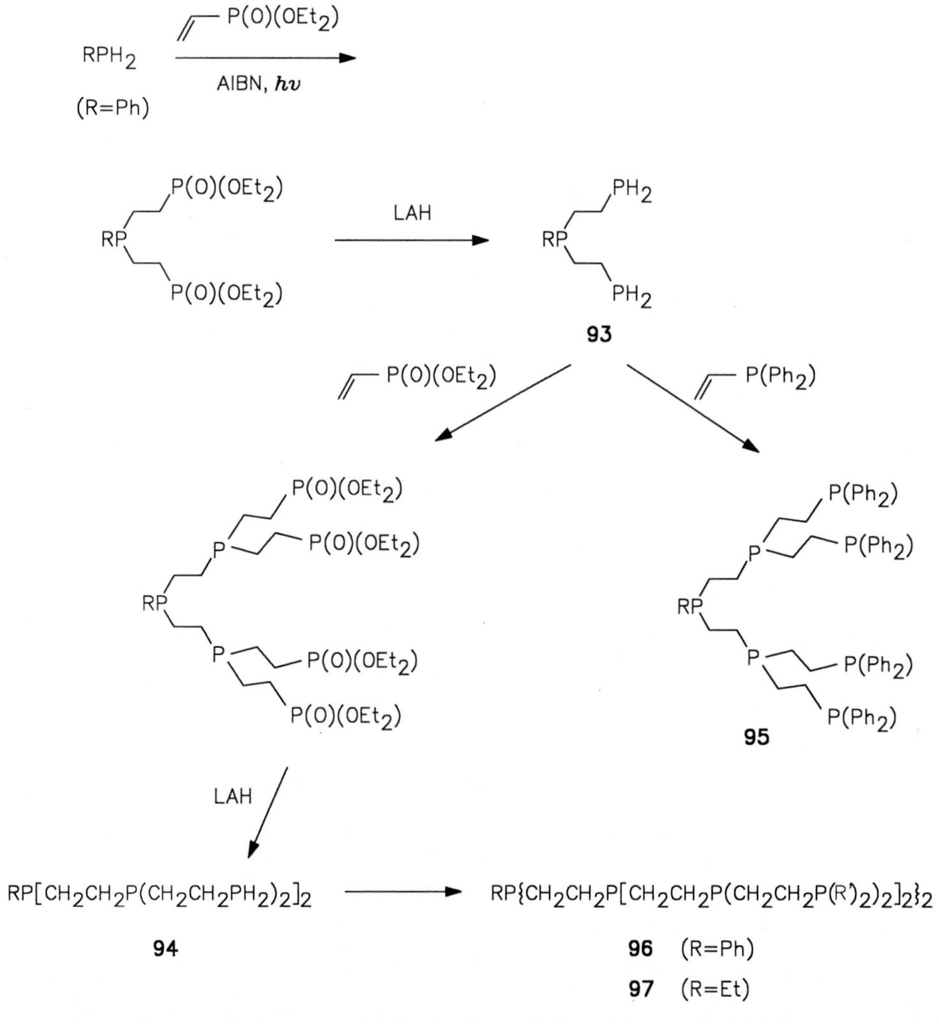

Scheme 4.25. DuBois et al.[111] have prepared *P*-based cascades for the examination of metal complexation potentials.

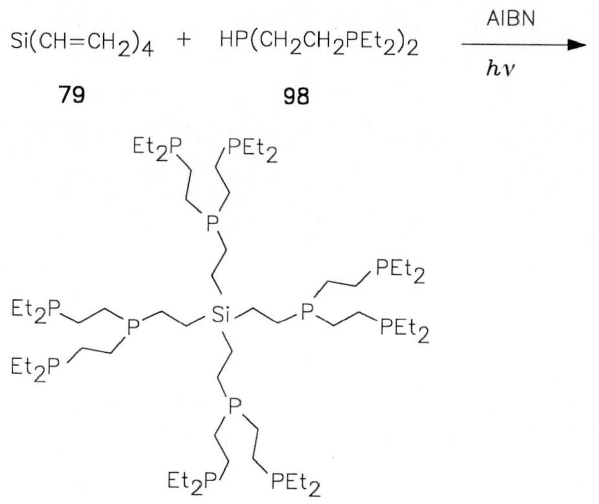

$$Si(CH=CH_2)_4 \quad + \quad HP(CH_2CH_2PEt_2)_2 \quad \xrightarrow[h\nu]{AIBN}$$

79   **98**

99

Scheme 4.26. The free-radical mediated construction of mixed *Si/P*-based dendrimers.

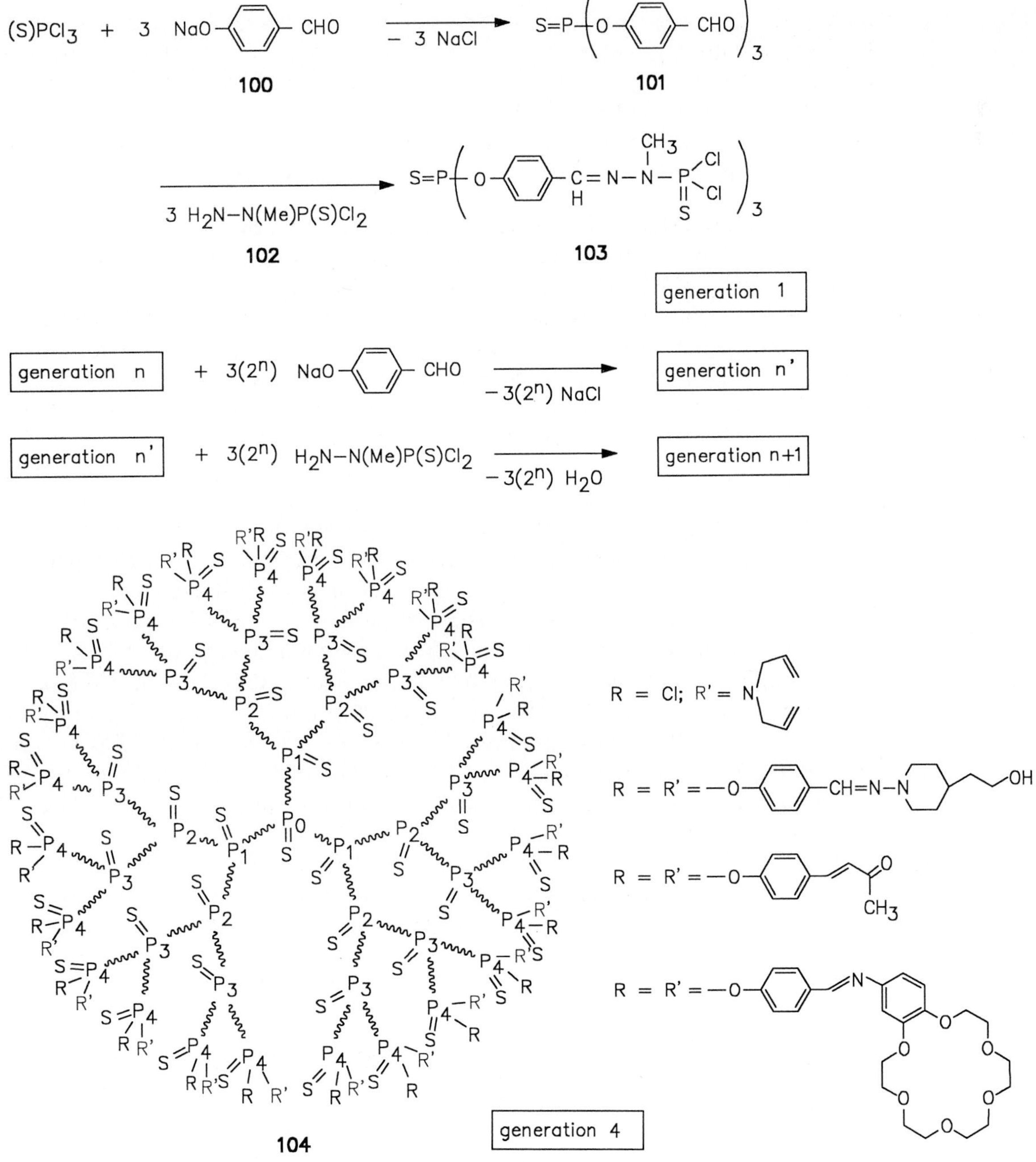

$$(S)PCl_3 \quad + \quad 3 \; NaO\!-\!\!\langle\rangle\!-\!CHO \quad \xrightarrow{-\;3\;NaCl} \quad S=P\!\!\left(O\!-\!\!\langle\rangle\!-\!CHO\right)_3$$

100   **101**

$$\xrightarrow{3\;H_2N-N(Me)P(S)Cl_2}$$

102   **103**

generation 1

generation n $+ \; 3(2^n) \; NaO\!-\!\!\langle\rangle\!-\!CHO \quad \xrightarrow{-\;3(2^n)\;NaCl}$ generation n'

generation n' $+ \; 3(2^n) \; H_2N-N(Me)P(S)Cl_2 \quad \xrightarrow{-\;3(2^n)\;H_2O}$ generation n+1

104   generation 4

$$R = Cl; \; R' = N\!\!\diagdown$$

$$R = R' = -O\!-\!\!\langle\rangle\!-\!CH=N-N\!\diagdown\!\!\langle\rangle\!-\!OH$$

$$R = R' = -O\!-\!\!\langle\rangle\!-\!CH_3$$

$$R = R' = -O\!-\!\!\langle\rangle\!-N$$

Scheme 4.27. A series of neutral pentavalent *P*-based dendrimers.

only by-products were sodium chloride and water, assuming quantitative transformations. In a subsequent paper, the expansion of these *P*-dendrimers to the fifth, sixth, and 7th generations possessing up to 384 functional groups was reported.[115] Facile functional group manipulation at the periphery allowed attachment of α,β-unsaturated ketones, crown ethers, and alcohols. Treatment of the PCl_2 moieties with bis(allyl)amine afforded the monosubstituted termini (i.e., $P(Cl)N(CH_2CH=CH_2)_2$). The surface incorporated crown ethers interestingly acted as "shields" with respect to attempted imine hydrolysis (THF, H_2O (4:1), 25 °C, 48 h). Spectral evidence for the structure included ^{31}P NMR; no overlapping resonances (^{31}P NMR) were observed up to the fifth tier.

The synthetic and spectral details to the *P*-dendrimers up through the third generation from a cyclotriphosphazene core ($N_3P_3(OC_6H_4CHO)_6$)[116, 117] possessing six formyl moieties have also recently been reported.[118] The simple procedures lead to a spherical surface with electrophilic or nucleophilic reactive moieties, such as aldehydes, hydrazones, and aminophosphines.

4.12   1 → 3 *C*-Branched

4.12.1   1 → 3 *C*-Branched, *Amide* Connectivity

4.12.1.1   1 → 3 *C*-Branched, *Amide* ('Tris') Connectivity

In 1985, Newkome et al. reported[7] the first example of divergently constructed cascade spherical macromolecules utilizing sp^3-carbon atoms as 1 → 3 branching centers. Although these initial syntheses were not strictly iterative, notable dendritic preparative features were explored and exploited. The incorporated building blocks possessed tetrahedral, tetrasubstituted *C*-branching centers, as well as maximized branching for a *C*-based system, and differential monomer layering (analogous to block copolymer construction). The molecular architecture was modeled from the Leeuwenberg model[119, 120] for trees as described by Tomlinson[121] and since this original series was terminated by hydroxyl moieties, they coined the simple descriptive term "arbor*ols*". The initial core[7] consisted of a one-directional 1,1,1-tris(hydroxymethyl)alkane (**105**) and used two readily available building blocks: trialkyl methanetricarboxylates,[122–124] or their sodium salts, (the "triester"; **106**) and tris(hydroxymethyl)aminomethane ("Tris", **107**; Scheme 4.28). The use of an appropriate spacer was found to be necessary due to steric hindrance associated with the quaternary carbon center, as subsequent studies have shown.[125] To circumvent retardation of S_N2-type chemical transformation, a three-atom distance is needed between the branch point and the reactive chemical center. Thus, triol **105** was treated with the chloroacetic acid, esterified (MeOH, H^+), reduced ($LiAlH_4$), and transformed to the tritosylate **108**, which upon treatment with the sodio anion of **106** generated the nonaester **109**. Subsequent amidation of **109** with tris (**107**) afforded the desired 27-arborol (**110**), which was fully characterized and shown to be water soluble, thus affording entrance to "unimolecular" micelles.[7]

Application of this two-step procedure (Scheme 4.29) of nucleophilic substitution of a substrate possessing an appropriate leaving group with the anion of a trialkyl methanetricarboxylate to generate a polyester, followed by amidation with tris, was extended to the preparation of two-directional (dumbbell-shaped) arborols (**111**).[126] Treatment of $1,\omega$-dibromo- or dimesyloxyalkanes (**112**) with the sodium salt of monomer **106**, followed by treatment with tris (**107**) afforded the bisnonaols (**111**), which, when $n = 8$–12, possess unique structural features permitting them to stack in an orthogonal array (Figure 4.5 (a)) resulting in the formation of spaghetti-like aggregates. These aggregates form aqueous, thermally reversible gels,[127] based on the maximization of lipophilic–lipophilic and hydrophilic–hydrophilic interactions. Fluorescence and electron microscopy, as well as light scattering experiments, provided evidence for supramolecular stacking and a rod-like micellar topology of these aggregates at low concentrations.

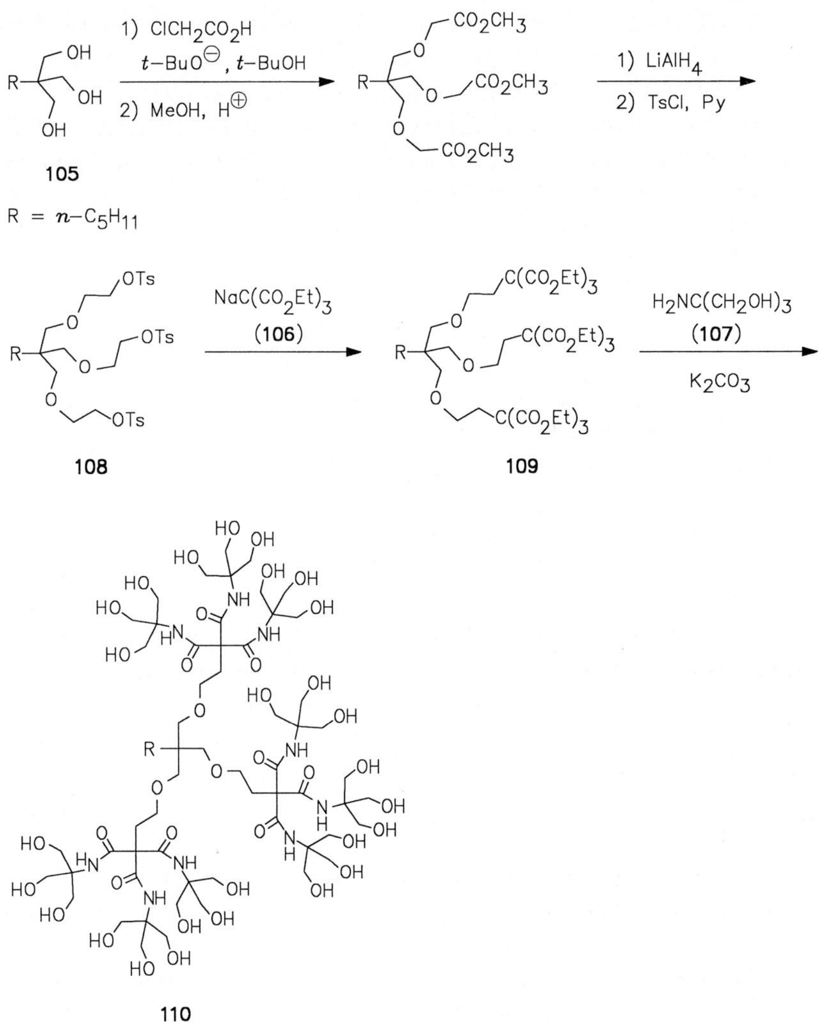

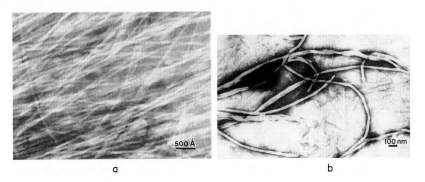

Scheme 4.28. Utilization of a methane triester and an aminotriol, as building blocks for a 27-arborol.

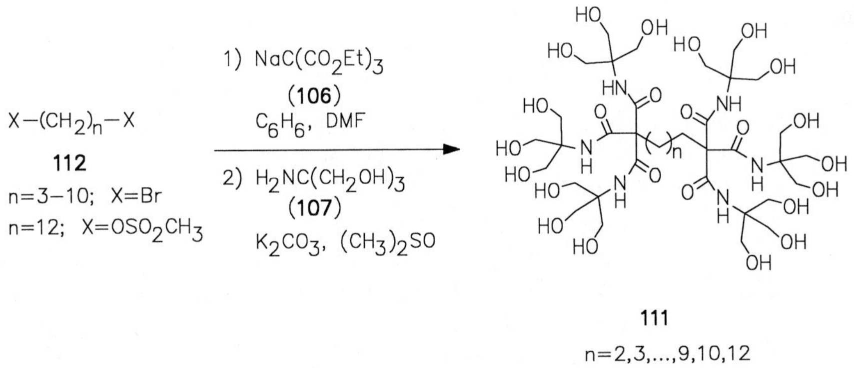

Scheme 4.29. Cascade construction of dumbbell shaped molecules that form stacked aggregates in aqueous environments.

Figure 4.5. Electron micrographs of two-directional cascades. Linear aggregates are formed with flexible alkyl bridges (4.5a); whereas, curved rope-like structures result from the incorporation of bridge structural rigidity such as an alkyne moiety (4.5b).

Since these arborols (**111**) are comprised of two hydrophilic groups connected by a hydrophobic linkage, they fit the simple definition of a bolaamphiphile; a term derived from bolaform amphiphile, originally introduced in 1951 by Fuoss and Edleson.[128] In 1984, when Fuhrhop and Mathieu[129] reported the synthesis and self-assembly of several bolaamphiphiles, these two-directional surfactant-like macromolecules represented a simple entrance to the bolaamphiphile field; this subject has recently been highlighted[130] and reviewed.[131]

Since these two-directional arborols stack in an organized manner so that the lipophilic alkyl chain moieties are orthogonally juxtaposed, a functionality incorporated on this linkage would by necessity be preorganized for subsequent interactions. The introduction of a central alkyne bond (e.g., **113**) was accomplished by transformations shown in Scheme 4.30. Application of the simple two-step procedure gave rise to either **114** or **115** depending on the ester reagents used (i.e., malonic ester or triester). Upon dissolution in water, the resultant alkyne **114** formed a gel in a manner analogous to that of alkane bridged bolaamphiphile **111**. Figure 4.5(b) shows the electron micrograph of **114** supporting the stacking phenomena, but the presence of the rigid, linear, central alkyne moiety induces a less than orthogonal chain alignment giving rise to the formation of a helical morphology.[132] The large diameters of the twisted aggregates (Figure 4.5(b)) probably result from the packing of individual rods into the groves of adjacent helical rods, or aggregates, thus producing a "super-coil" or "molecular rope". Predetermined self-assembly has been denoted as "automorphogenesis" by Lehn.[133]

Newkome et al.[134] probed the inner lipophilic region of the two-directional arborols by the incorporation of spirane and biphenyl cores, which introduced varing degrees of intermolecular interactions causing disruption to the aggregation process. Based on computationally generated pictures, a better understanding of the molecular interactions during the initial stages of preorganization prior to gelation has been proposed. Encapsulation or guest inclusion during the gelation process has been examined and also modeled.

A related three-directional member of this series[135] was the benzene [9]^3-arborol {**116**; 27-Cascade:benzene[3-1,3,5]:(ethylidene):(3-oxo-2-azapropylidene):methanol}, which was prepared by the two-step (alkylation–amidation or triester–tris) reaction sequence applied to 1,3,5-tris(bromomethyl)benzene (**26**; Scheme 4.31). Electron microscopy and

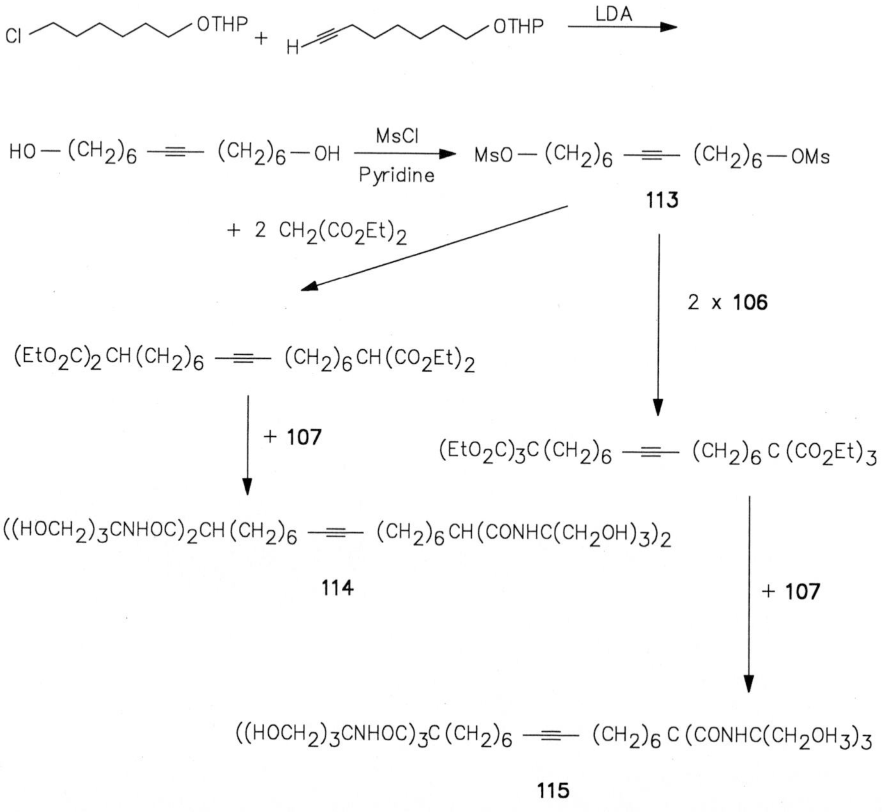

Scheme 4.30. Synthetic methodology for the incorporation of an alkyne unit within a two-directional cascade.

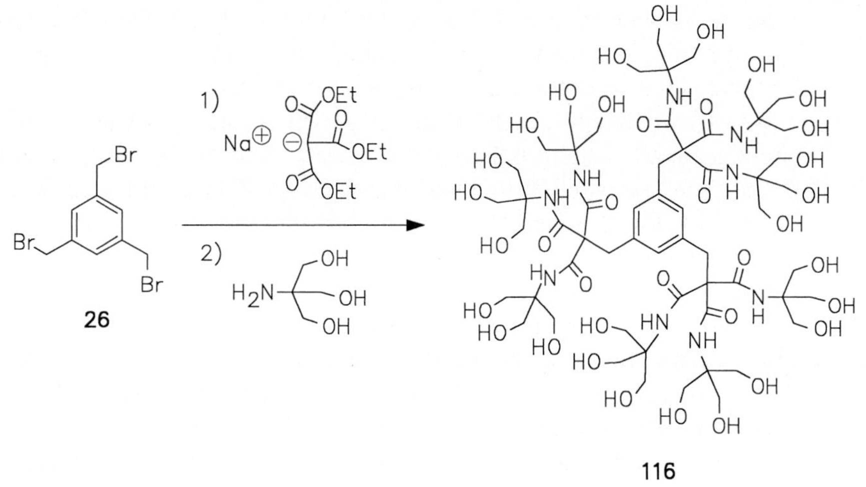

Scheme 4.31. Construction of a cascade triade.

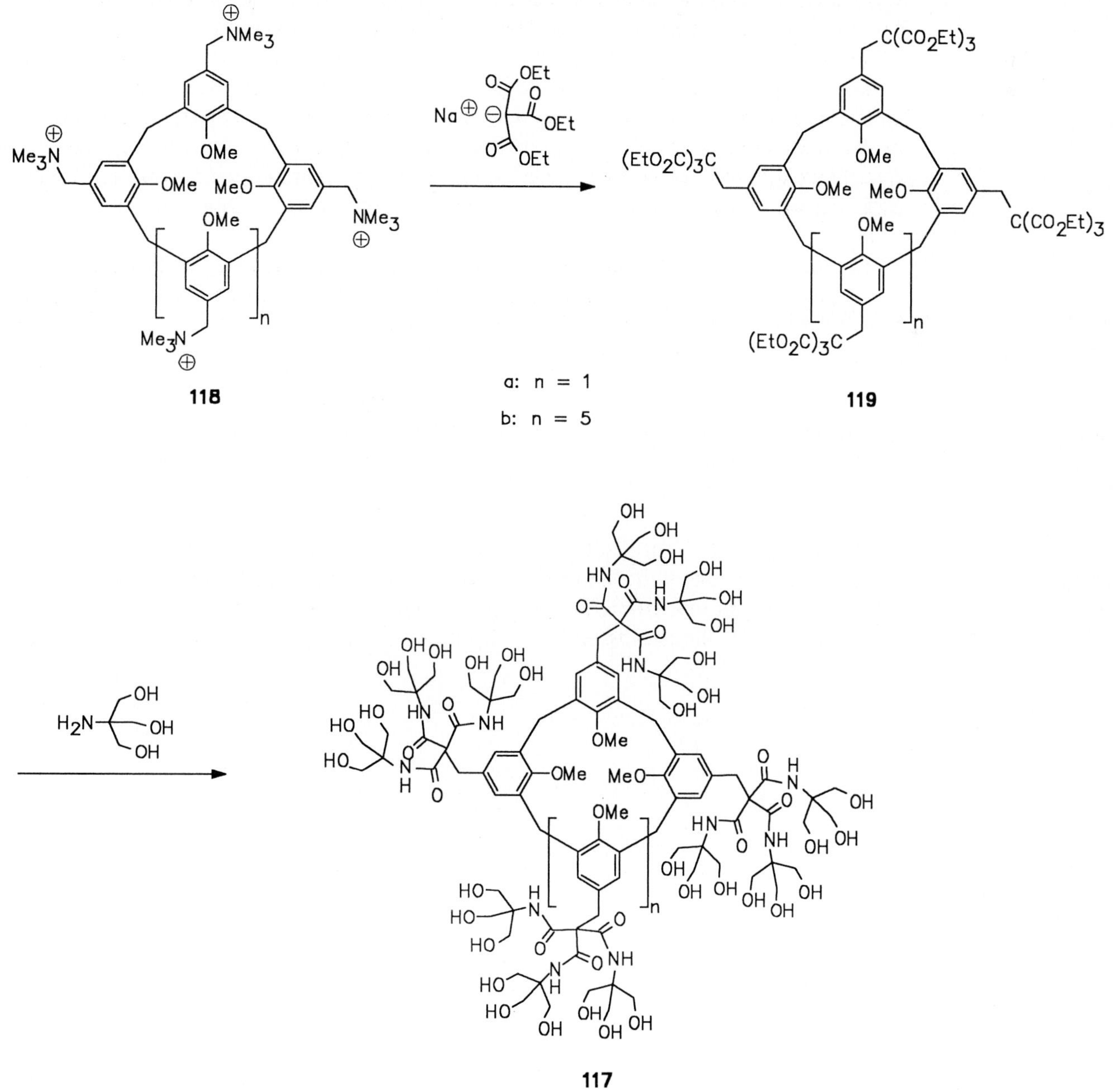

a: n = 1
b: n = 5

Scheme 4.32. Sylvanol construction of molecular forests atop a calixarene plateau.

subsequent light scattering data suggested that **116** aggregated by the stacking of its hydrophilic triad of three small spheres into a spherical array of ca. 20 nm (diameter) reminiscent of globular micelles.

Results of dynamic light scattering experiments with aqueous solutions of **116** have been reported.[136] This benzene [9]³-arborol in water forms aggregates that have dynamic properties very similar to those of single polymer chains in solvents in the crossover region (qr_h) ≈ 1, where q is the absolute value of the scattering vector, and r_h is the hydrodynamic radius of the scattering particles. The size of these particles appears to be concentration independent in the concentration range ($3.5 \times 10^{-3} c^+ \leq c \leq 13.37 \times 10^{-3} c^+$; where $c^+ = 1$ mol dm^{-3}) studied. From the ratio of the scattered light intensity to the square of the absolute value of the scattering vector ($\Gamma_{max}/q^2)_0$ at the limit $q \to 0$, a hydrodynamic radius of $r_h = 0$ (100 nm) was calculated.

Expansion of this technology to other more complex systems was demonstrated[137] via the synthesis of water-soluble calixarenes (**117**; Scheme 4.32), termed "silvanols", possessing dendritic polyhydroxy spheres on the upper rims of [1$_n$]metacyclophanes. The initial polytrimethylammonium calixarene (i.e., **118a**)[138] was converted by established procedures to the crystalline dodecaester [1$_4$]metacyclophane **119a**, X-ray structure of which confirms the assigned structure. Treatment of polyester **119a** with tris generated the [36]-silvanol (**117a**). Electron micrographs of **117a** showed it to possess a diameter of 57 Å relating to six aggregating macromolecules, as predicted by molecular modeling. The use of [1$_8$]metacyclophane (**118b**), as the starting material, gave rise to the corresponding [72]-silvanol (**117b**) via the same series of steps.

In 1985, application of this simple dendritic construction to a polymer core, specifically chloromethyl functionalized polystyrene was reported.[139] Another example of the identical procedure was reported[140] except utilizing another polymeric core backbone, derived from a functionalized vinyl ether monomer. These are very early examples of dendritic "comb" macromolecules.

4.12.1.2   1→3 (1→2) *C*-Branched, *Amide* ('Tris') Connectivity

Although Newkome et al. reported[127] a series of two-directional arborols, this original process was a 1 → 3 *C*-branching scheme; under more drastic amidation conditions, the triester, especially the methyl ester, readily decomposed to the 1 → 2 *C*-branched products – the same as those derived from the monoalkylation of malonates. Subsequent treatment of the malonate esters with tris gave rise to the [6]–(X)$_n$–[6] arborol series (e.g., **114**).

Jørgensen et al.[141] utilized this molecular organization process (dumbbell-like stacking) to incorporate tetrathiafulvalene (TTF; a substrate currently of interest in such areas as molecular electronics and organomagnetism) within the central lypophilic region (Scheme 4.33) of the self-assembled, supramolecular structures. A multistage synthesis

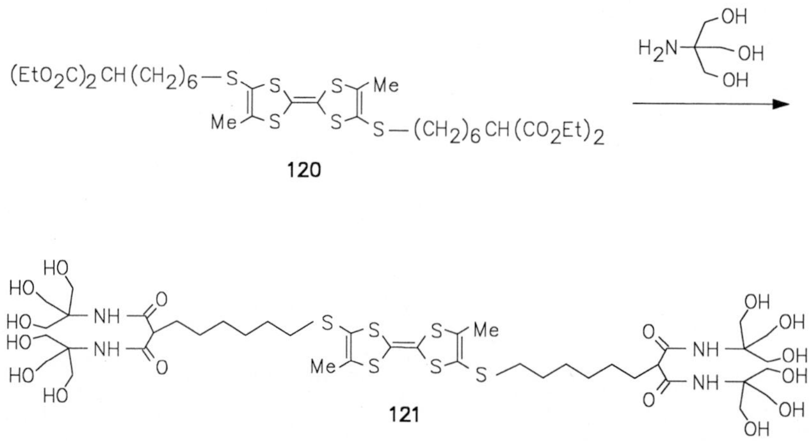

Scheme 4.33. Introduction of tetrathiafulvalene units into the lipophilic backbone of two-directional arborols for the construction of "molecular wires".

was undertaken in which the derivatized TTF core **120** was assembled. Treatment of tetra-ester **120** with tris generated the desired TTF-bis-arborol (**121**). Calculations based on an orthogonal stacking with the TTF core possessing the trans conformations indicated that the diameter of the stack of molecules should be ca. 3.5 nm. Aggregates derived from dodecaol **121** clearly reveal (microscopy) thin string-like assemblages with lengths in the order of tens of microns and diameters in the order of ca. 100 nm. These structures are therefore superstructures of the single strands, an observation analogous to that previously reported.[126, 127]

4.12.1.3 1 → 3 *C*-Branched, *Amide* ('Bis*homo*tris') Connectivity

In order to circumvent unfavored S_N2-type substitution at neopentyl positions preventing continued iteration with 'triester–tris' methodology,[142] a new monomer 'bis*homo*-tris' (**122**) was prepared[143–146] (Scheme 4.34). Thus, Michael-type addition of acrylonitrile to nitromethane afforded **123**, which was quantitatively hydrolyzed (conc. HCl) to the corresponding nitro*tris*acid **124**. Reduction of triacid **124** with a borane–THF solution gave the nitro*tris*alcohol (**125**), which was reduced with T1 Raney nickel[47] to yield the doubly homologated bis*homo*tris[148] (**122**). It should be noted that alcohol **125** is an excellent precursor to surface derivatization of cascade polymers.[149]

The use of 'bis*homo*tris' (**122**) to replace tris (**107**) in the original alkylation–amidation sequence gave rise to transesterification products. This suggested that the amidation procedure using tris proceeded via a five-membered intermediate ester **126** to give amide **127** via an intramolecular rearrangement (Scheme 4.35). It was therefore postulated[127] that an unfavorable seven-membered transition state (**128**) precluded amide formation. Treat-

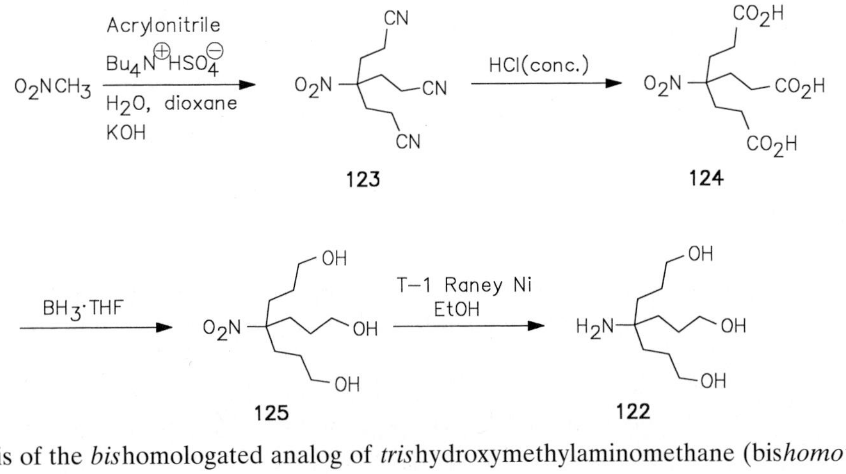

Scheme 4.34. Synthesis of the *bis*homologated analog of *tris*hydroxymethylaminomethane (bis*homo*tris).

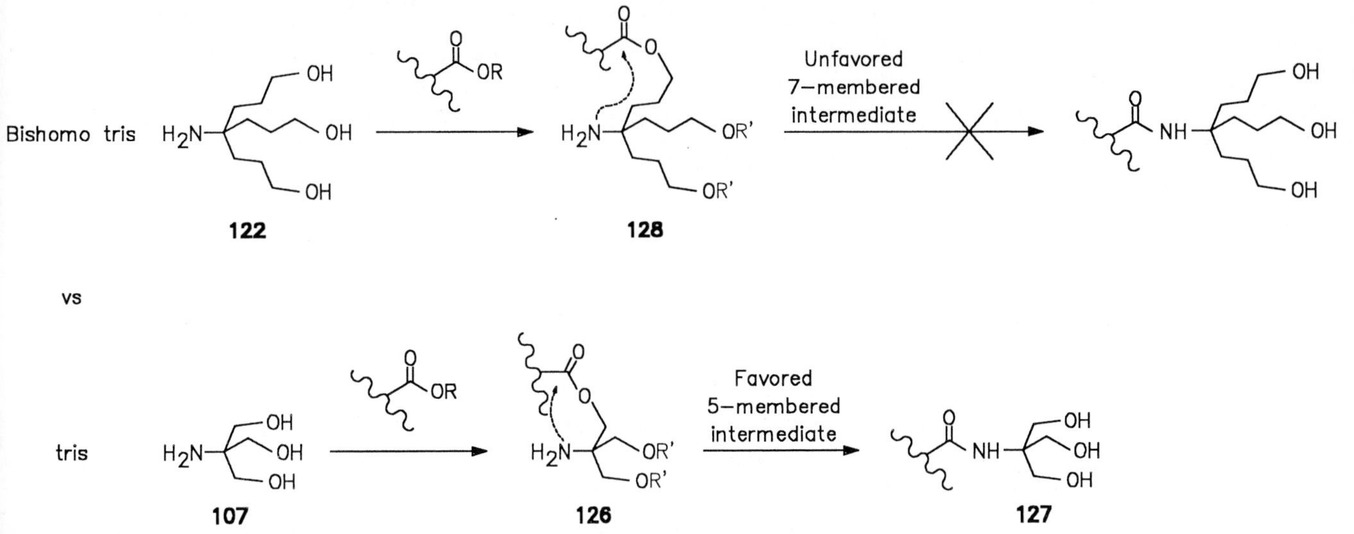

Scheme 4.35. Rationale for the difficulties encountered with the amino acylation of bis*homo*tris.

ment of this intermediate ester **128** with base (KOH) in DMSO forced the amidation to completion, albeit in extremely poor (<10%) yields.

Whitesell and Chang[150] reported the preparation of a unique $H_2NC(CH_2CH_2CH_2SH)_3$ as a building block from bis*homo*tris in four steps. This aminotrithiol was used to directionally align helical peptide polymerization on gold and indium–tin oxide glass surfaces. Unidirectional alignment of macromolecules and their polarizability are of interest in the area of supramolecular chemistry and molecular electronic devices.[151] Hence, dendritic branching combined with "anchoring" units (e.g., sulfur affinity for gold) are logical choices to assist in non-covalent as well as covalent molecular organization.

4.12.1.4 $1 \rightarrow 3$ *C*-Branched, *Amide* ("Behera's Amine") Connectivity

Addition of the anion of nitromethane to α,β-unsaturated carbonyls and nitriles followed by reduction of the nitro group to an amine (Scheme 4.36) provided the basis for the preparation[152] of "Behera's amine"[153] (**129**). Treatment of nitromethane with *tert*-butyl acrylate in the presence of base gave di-*tert*-butyl 4-[2-*tert*-butoxycarbonyl)ethyl]-4-nitroheptanedioate (**130**) via modification of a literature procedure.[154] Catalytic reduction of nitrotriester **130** quantitatively gives rise to Behera's amine (**129**). Uniquely, Behera's amine does not undergo facile intramolecular lactam formation that is predominant in the other related but less branched esters[155, 156] during the hydrogenation process. The X-ray structure of Behera's amine confirms the extended conformation with 15 of the 16 torsion angles in the *anti* orientation (mean value of 176.6°).[157]

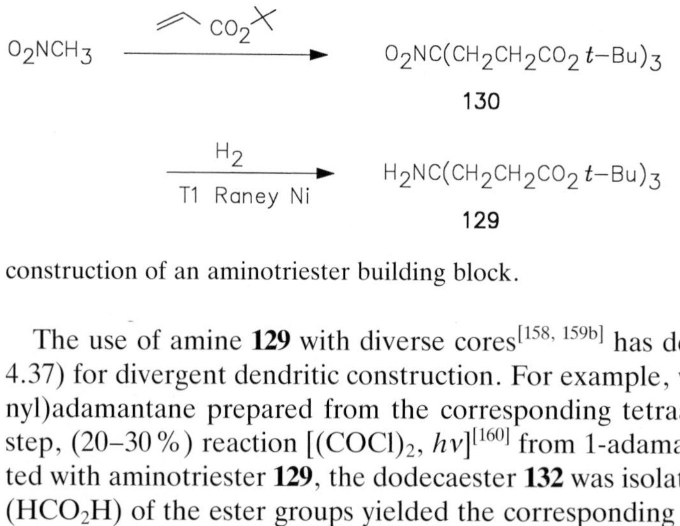

Scheme 4.36. Two-step construction of an aminotriester building block.

The use of amine **129** with diverse cores[158, 159b] has demonstrated its utility (Scheme 4.37) for divergent dendritic construction. For example, when 1,3,5,7-tetra(chlorocarbonyl)adamantane prepared from the corresponding tetraacid **131** (synthesized in a one-step, (20–30%) reaction [(COCl)$_2$, $h\nu$][160] from 1-adamantanecarboxylic acid) was treated with aminotriester **129**, the dodecaester **132** was isolated. The quantitative hydrolysis (HCO$_2$H) of the ester groups yielded the corresponding acid **133**. Amidation using peptide coupling conditions[161] of dodecaacid **133** with Behera's amine (**129**) afforded the second tier 36-ester (**134**), which, when treated with formic acid, gave 36-Cascade:tricyclo[3.3.1.1[3,7]]decane[4-1,3,5,7]:(3-oxo-2-azapropylidyne):(3-oxo-2-azapentylidyne):propanoic acid (**135**).

Newkome et al.[162] reported the use of Behera's amine (**129**) in the synthesis of the [12]- (**137**), [36]- (**138**), [108]- (**139**), [324]- (**140**), and [972]- (**141**) polyamido cascade series by an iterative, divergent procedure (Scheme 4.38) based on the ethereal core **136**, constructed via Bruson's method,[163] by exhaustive 1,4-addition of pentaerythritol to acrylonitrile, followed by hydrolysis. Repetition of the amidation[161]–deprotection sequence (DCC coupling–HCO$_2$H hydrolysis) allowed the construction of the fifth generation dendrimers with purported molecular weights of 165,909 amu [972]-ester) and 111,373 amu for 972-Cascade:methane[4]:(3-oxo-6-oxa-2-azaheptylidene):(3-oxo-2-azapentylidyne)^4:propanoic acid (**141**). Structural support for these amide-based dendrimers included typical spectroscopy procedures as well as 2-dimensional diffusion ordered spectroscopy (DOSY) NMR,[164] whereby diffusion coefficients were ascertained via pulse field gradient NMR for each generation of the water-soluble, carboxylic acid dendrimer. Application of the Stokes–Einstein equation gave dendritic hydrodynamic radii (tabulated at acid, neutral, and basic pH) that correlate well with those obtained by SEC

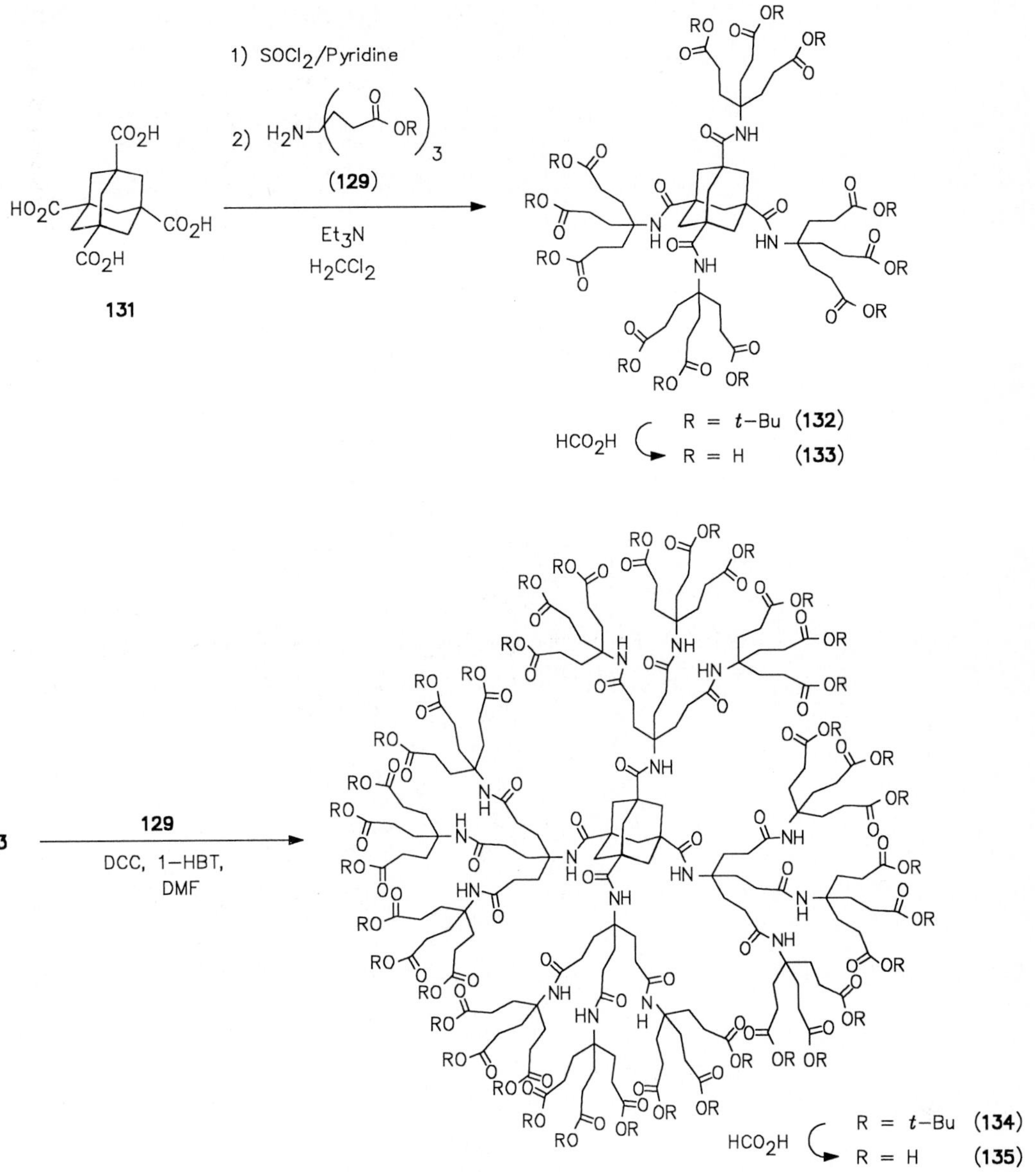

Scheme 4.37. Construction of dendrimers using the tetravalent core-adamantane tetracarboxylic acid.

measurements and computer generated molecular modeling.[162] Hydrodynamic radii changes in the acid terminated series at various solution pH values are notable; in general, the size of the [108]-acid **139** increases 35 % from pH 3.64 to pH 7.04, corresponding to a 264 % increase in overall dendritic volume. This dendritic macromolecule series (**137–141**) was examined[165a] as micellar substitutes in electrokinetic capillary chromatography employing aqueous mobile phase conditions; separations of a series of alkyl parabens using these dendrimers yielded significantly enhanced efficiency and resolution, when compared to more traditional methods using surfactants such as SDS.

Monning and Kuzdzal[165b] have extended these experiments to estimate analyte dendrimer/solvent distribution coefficients (K). Thermodynamic parameters (i.e., H, S, G) pertaining to analyte solubilization within dendritic structures can be obtained via examination of K with respect to temperature.

Three generations of the related alcohol- and amine-terminated polyamide cascades were prepared by coupling the appropriate polyacid with either the aminotris(*tert*-butyl carbamate) **142** or aminotris(acetate) **143** building block. Scheme 4.39 shows the preparation of tricarbamate **142** and triacetate **143**. Trinitrile **123** was reduced with borane to give the triamine **144**, which, when treated with di-*tert*-butyl dicarbonate[166, 167] followed by catalytic reduction with T1 Raney nickel,[168] gave the desired amine **142** in excellent overall yield. The precursor to bis*homo*tris (Scheme 4.34), triol **125**, prepared in two steps from nitrile **123**, was acylated with acetic anhydride, and reduced catalytically with T1 Raney nickel to give the corresponding amine **143** in high yield.

+
1) DCC, 1-HBT,
DMF
2) HCO_2H
136
129
1) 129
2) HCO_2H
1), 2)
1), 2)
1), 2)
36 acid = 138
108 acid = 139
324 acid = 140
972 acid = 141
137

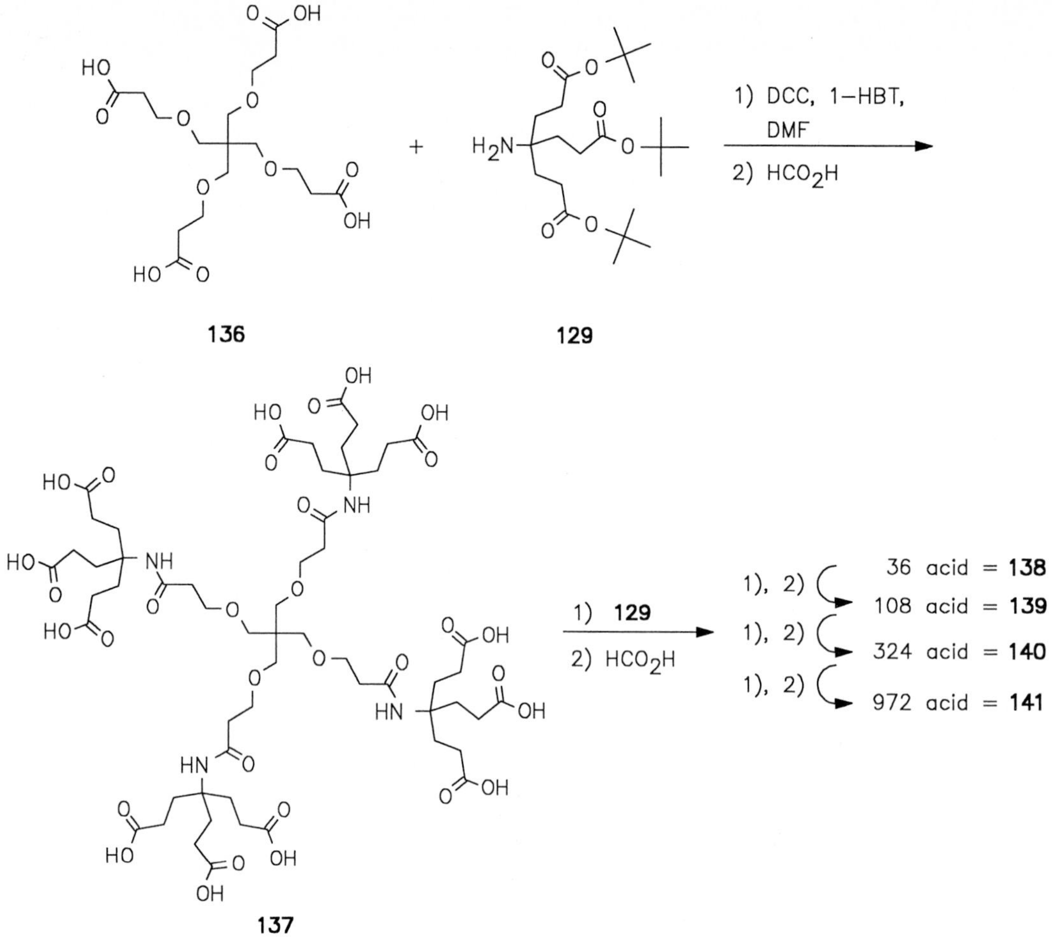

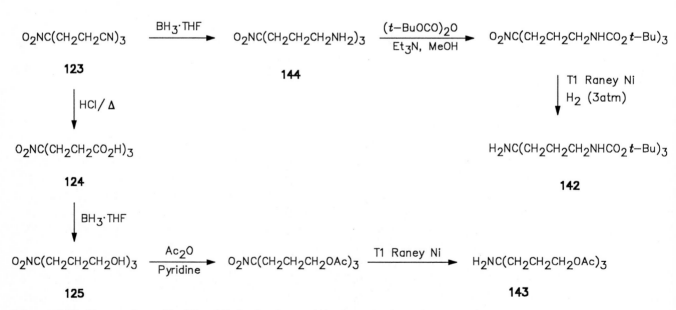

Scheme 4.38. Iterative procedure for the preparation of dendrimers with flexible pentaerythritol-based cores.

$O_2NC(CH_2CH_2CN)_3$
123
$BH_3 \cdot THF$
$O_2NC(CH_2CH_2CH_2NH_2)_3$
144
$(t\text{-}BuOCO)_2O$
Et_3N, MeOH
$O_2NC(CH_2CH_2CH_2NHCO_2\,t\text{-}Bu)_3$
T1 Raney Ni
H_2 (3atm)
$H_2NC(CH_2CH_2CH_2NHCO_2\,t\text{-}Bu)_3$
142
HCl/ Δ
$O_2NC(CH_2CH_2CO_2H)_3$
124
$BH_3 \cdot THF$
$O_2NC(CH_2CH_2CH_2OH)_3$
125
Ac_2O
Pyridine
$O_2NC(CH_2CH_2CH_2OAc)_3$
T1 Raney Ni
$H_2NC(CH_2CH_2CH_2OAc)_3$
143

Scheme 4.39. Preparation of building blocks for the modular introduction of terminal amines and alcohols.

The 'modular' syntheses[169] of the three related cascade families from the corresponding polyacids and the appropriately designed monomer are shown in Scheme 4.40. In each case, building block coupling utilized standard dicyclohexylcarbodiimide[161, 170]/1-hydroxybenzotriazole[171] (DCC/^{1}-HBT) peptide coupling conditions. Removal (HCO_2H) of the protecting *tert*-butyl groups afforded the corresponding polyacids or polyamines, and reaction of the acetate coated cascades with potassium carbonate in ethanol (transesterification) liberated the hydroxy-terminated dendrimers.

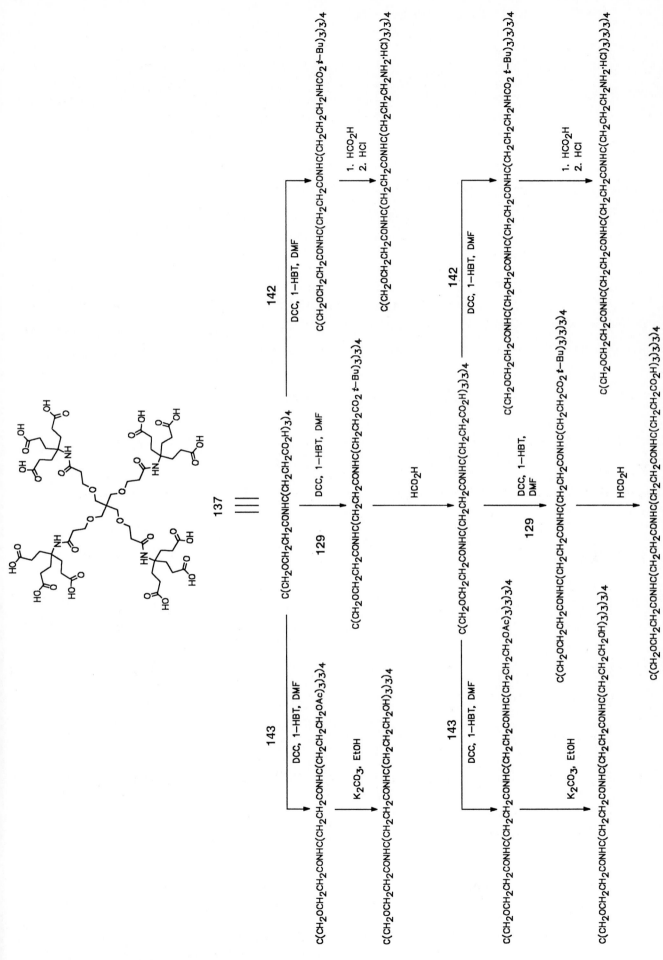

Scheme 4.40. Synthesis of a complimentary series of hydroxyl, amine, and carboxylic acid terminated dendrimers.

4.12.2 1 →3 *C*-Branched and Connectivity

Although initial efforts utilizing bis*homo*tris as a dendritic building block were disappointing due to the inability to effect complete amine acylation, the nitro intermediates proved to be excellent precursors to a diverse series of *C*-based, alkyl building blocks. The construction of a cascade polymer possessing an all-saturated, symmetrical, tetrahedral branched hydrocarbon interior framework was designed based on the bis*homo*tris nitro precursor **125**. The preparation of Micellanol™ cascades **145** and **155** was subsequently accomplished by Newkome et al.[172–174] Scheme 4.41 shows the synthesis of the core **146** and the key building block **147** for their preparation. The nitrotriol **125** was protected with benzyl chloride to give triether **148**, which underwent denitration–cyanoethylation (Ono Reaction)[175] on treatment with tri-*n*-butyltin hydride in the presence of acrylonitrile to give the key nitriletriether intermediate **149**. Nitrile **149** played a critical role in the synthesis, since it could be uniquely converted to both the core **146** previously prepared from tetrakis(2-bromoethyl)methane via citric acid[176] in 17 steps (ca. 1 % overall yield), from tetrakis(β-carbethoxyethyl)methane[177] (ca. 70 %), or from γ-pyrone[178] in 8 steps (24 % overall yield)], as well as the alkyne building block **147**. Hydrolysis of nitrile **149** gave the corresponding acid **150**, which was quantitatively reduced with

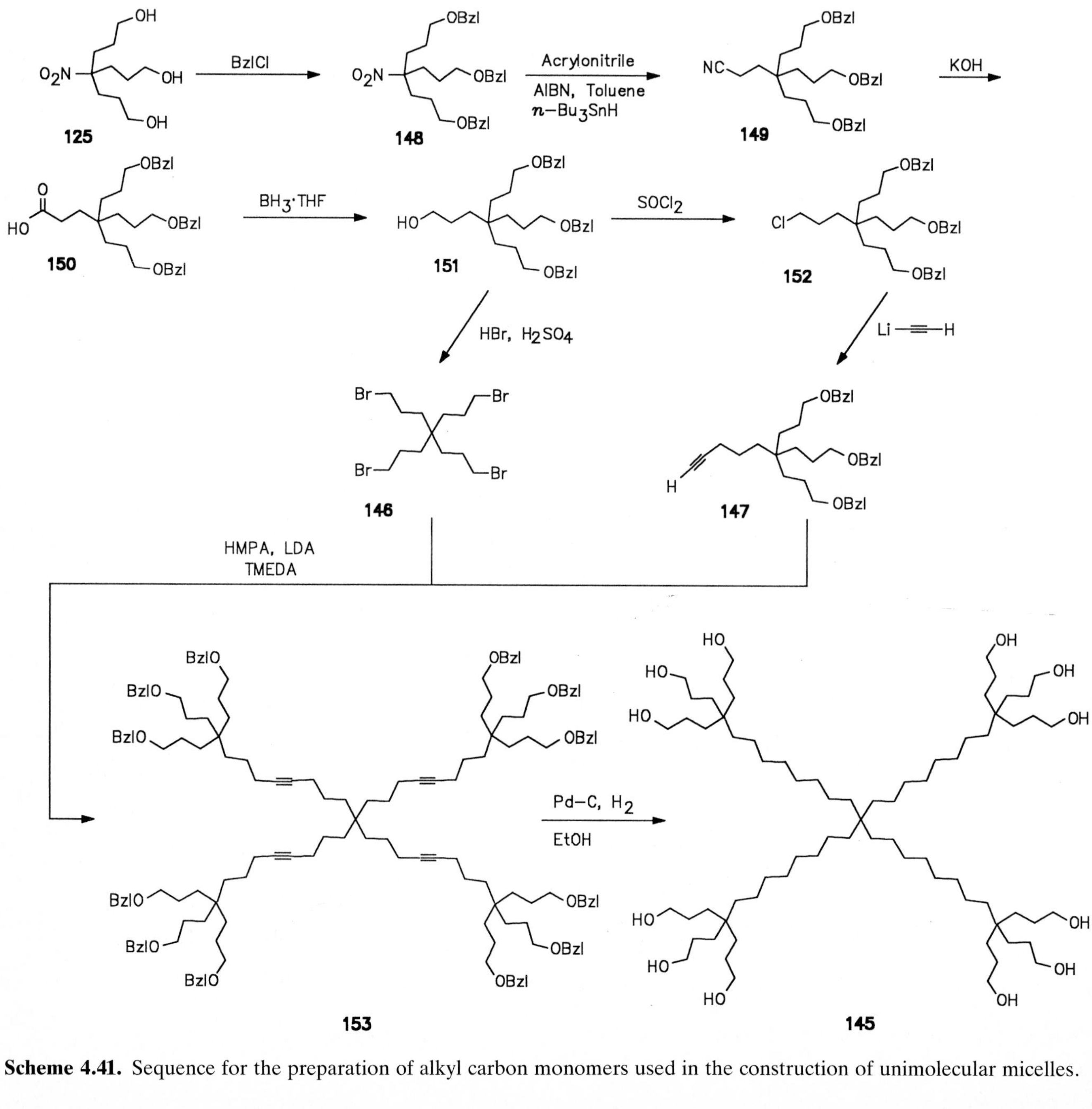

Scheme 4.41. Sequence for the preparation of alkyl carbon monomers used in the construction of unimolecular micelles.

borane to give alcohol **151**. This was transformed by treatment with HBr via concomitant deprotection and dehydroxylation–bromination to give core **146** in excellent overall yield. On the other hand treatment of alcohol **151** with thionyl chloride gave the corresponding monochloride **152**, which gave the desired functionally differentiated alkyne building block **147** with lithium acetylide.

Treatment of tetrabromide **146** with four equivalents of the lithium salt of alkyne **147** afforded the desired tetraalkyne **153**, which was concomitantly reduced and deprotected to give the saturated dodecaalcohol **145** (Scheme 4.42). Polyol **145** was then converted to the corresponding polychloride, which was treated with slightly over 12 equivalents of alkyne building block **147** to yield the 36-benzyl ether **154**. Reduction and hydrogenolysis of tetraalkyne **154** gave rise to the 36-Micellanol™ (**155**; 36-Cascade:methane[4]:(nonylidyne)²:propanol). Its water solubility was further enhanced by oxidation with ruthenium tetraoxide and conversion with tetramethylammonium hydroxide to the corresponding polytetramethylammonium 36-Micellanoate™ (**156**).

The "unimolecular" micellar characteristics of this poly(ammonium carboxylate) **156** were demonstrated[179] by UV analysis of guest molecules, such as: pinacyanol chloride, phenol blue, and naphthalene; combined with fluorescence lifetime decay experiments employing diphenylhexatriene as a molecular probe. The monodispersity, or absence of intermolecular aggregation, and molecular size were determined by electron microscopy.

The polyalkyne precursors to the hydrocarbon-based unimolecular micelles[179] allowed the testing of chemical modification at specific sites within the interior of a cascade infrastructure.[180] Thus, treatment of the alkynes **153** or **154** with decaborane afforded excellent yields of the 1,2-dicarba-*closo*-dodecaboranes[181] (*o*-carboranes) or with Co₂(CO)₈ afforded the desired poly(dicobalt carbonyl) clusters.[182] Details are given in Chapter 8.

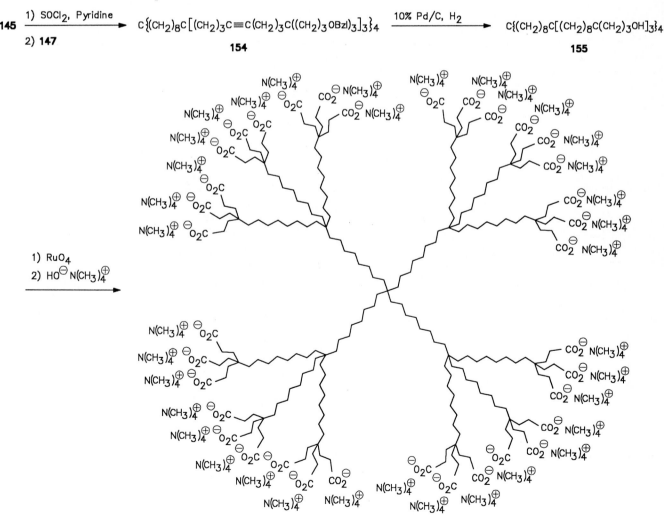

Scheme 4.42. Preparation of poly(tetramethylammonium salt) of Micellanoic™ Acid.

4.12.3   **1 → 3** *C*-Branched, *Ether* **Connectivity**

Hall et al.[183, 184] reported the synthesis of starburst polyether and polythioether dendrimers, as shown in Scheme 4.43. These cascades possess the shortest distance between branching centers yet reported. The use of pentaerythrityl tetrabromide (**157**)[185] as the core, with the potassium anion of the corresponding orthoester of pentaerythritol (**158**) as the building block, afforded the protected dodecaol **159**. Deprotection with acid and subsequent two-step conversion of the hydroxy groups to the dodecabromide via the dodecatosylate provided the precursor for the construction of the next tier. The 36-polyol (**160**) was prepared and subsequently transformed by this simple procedure to the 108-polyol **161**, which is the most densely packed cascade polymer yet reported, as evidenced by the branching defects encountered after the formation of the second and third tiers[14, 186] presumably resulting from an increasing number of neopentyl displacements[142] required for tier construction.

Ford et al.[187, 188] have reported an improved synthesis of this series of ethereal dendrimers. This polyol series was converted to the corresponding homologated polyamines, which were then alkylated (excess CH$_3$I) to generate the polyammonium salts, for example, the 36-tetraalkyl ammonium salt **162** [PETMAI(36); Scheme 4.44]. Moore also noted[189] that the use of the orthoacetate of pentaerythritol (**158c**), instead of the orthoformate (**158a**), adds to the stability of this building block and thus enhances its versatility. Use of these 'tied-back' building blocks aids in facile nucleophilic substitution, even at hindered neopentyl centers.

A series of related "cascadols", prepared from pentaerythritol, has been reported,[190] but little supportive data are available.

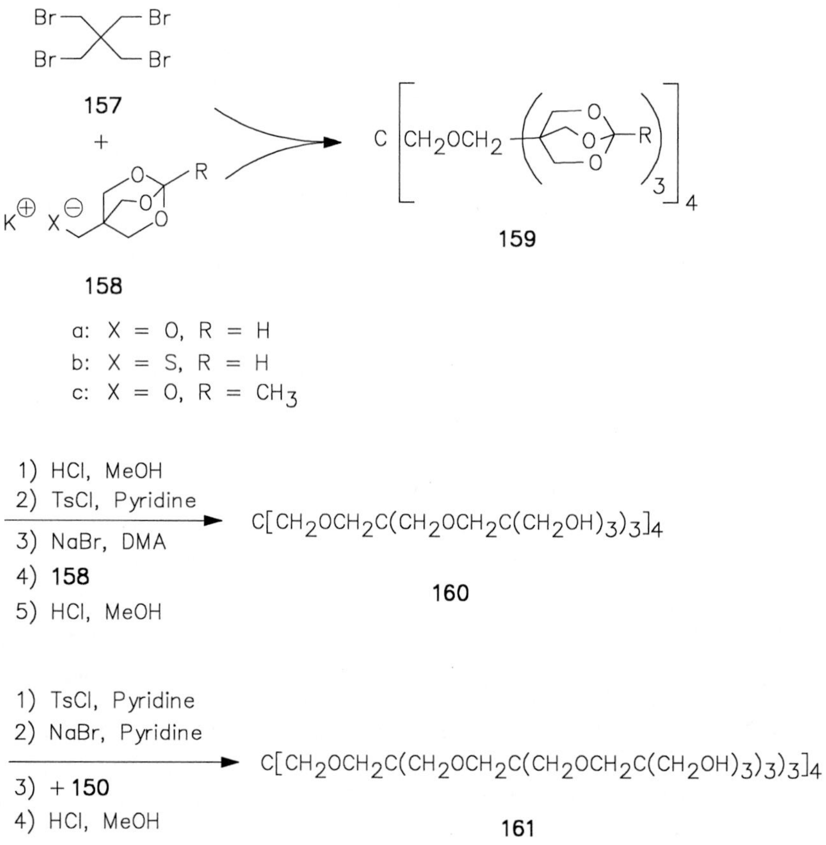

Scheme 4.43. Construction of highly compact pentaerythrityl-based dendrimers.

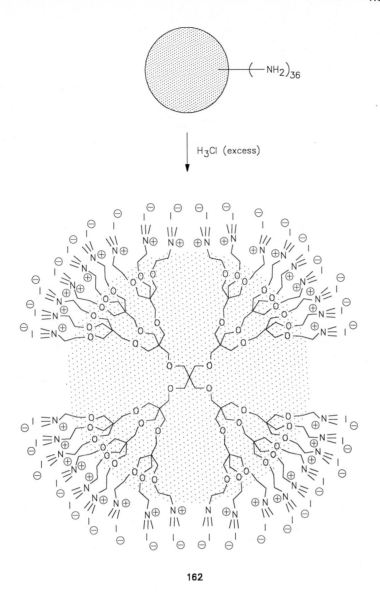

162

Scheme 4.44. Poly(ammonium iodide) dendrimers prepared for catalytic ester hydrolysis.

4.12.4 1 → 3 *C*-Branched, *Ether & Amide* Connectivity

Using a series of simple monomers,[125] a diverse collection of dendritic macromolecules can be devised and easily prepared. The four-directional ethereal amine building block (**163**) is but one example, since it is readily prepared[191] in two-steps from tris and Michael-type addition of acrylonitrile, followed by ethanolysis. There is a small amount of *N*-addition but the major product of the reaction resulted from exclusive *O*-addition. Treatment of the tetraacyl chloride of acid **136** with the ethereal amine **163** gives the dodecaester in high yields; subsequent saponification affords the corresponding dodecaacid **164** (Scheme 4.45). Additional tiers, such as the 36-Cascade:methane[4]:(3-oxo-6-oxa-2-azaheptylidyne)^2:4-oxapentanoic acid (**165**), were prepared by the use of standard peptide coupling conditions (DCC/1-HBT/DMF) giving rise to the poly(ethereal–amido) cascade series.[191]

Diederich et al.[192] reported the divergent synthesis of dendrimers possessing porphyrin cores with the aim of modeling redox potentials of electroactive chromophores via environmental polarity modification. The dendrimers thus can be considered as electron-transfer protein mimics for such proteins as cytochrome c; oxidation potentials for cytochrome c in aqueous solution are known to be 300–400 mV more positive than those reported for similarly ligated heme mimics lacking hydrophobic peptide encapsulation.[193a]

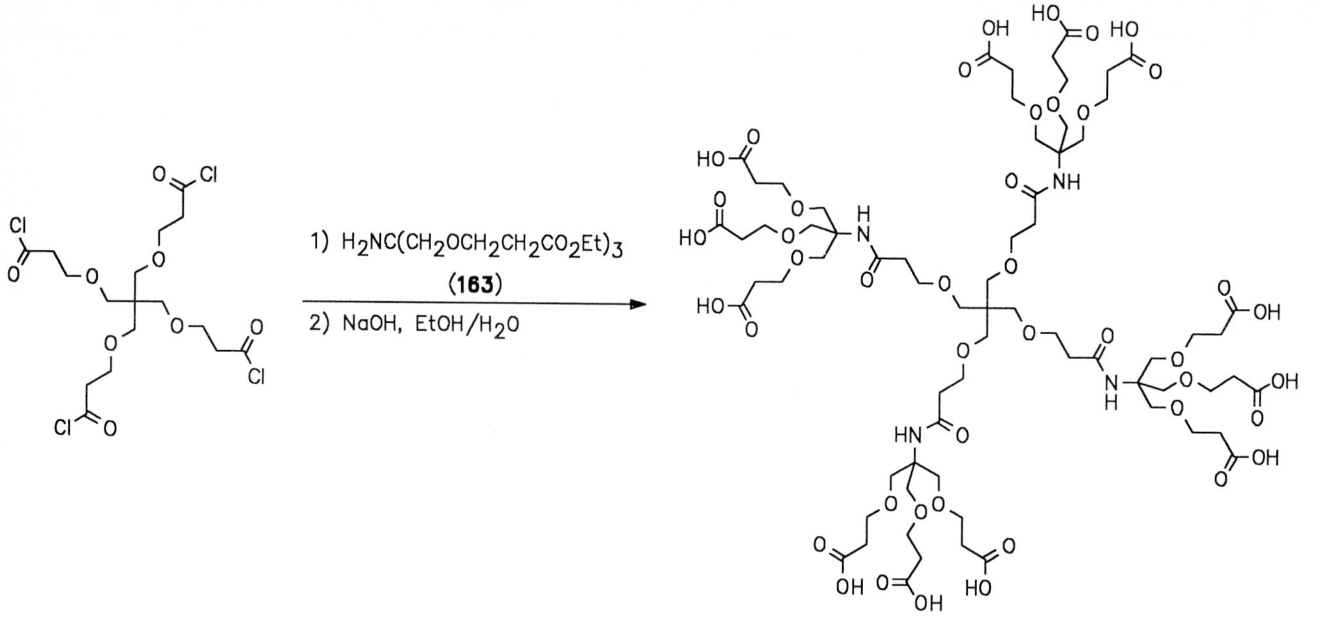

Scheme 4.45. Preparation of poly(ether-amido) cascades employing a "tris"-based aminotriester monomer.

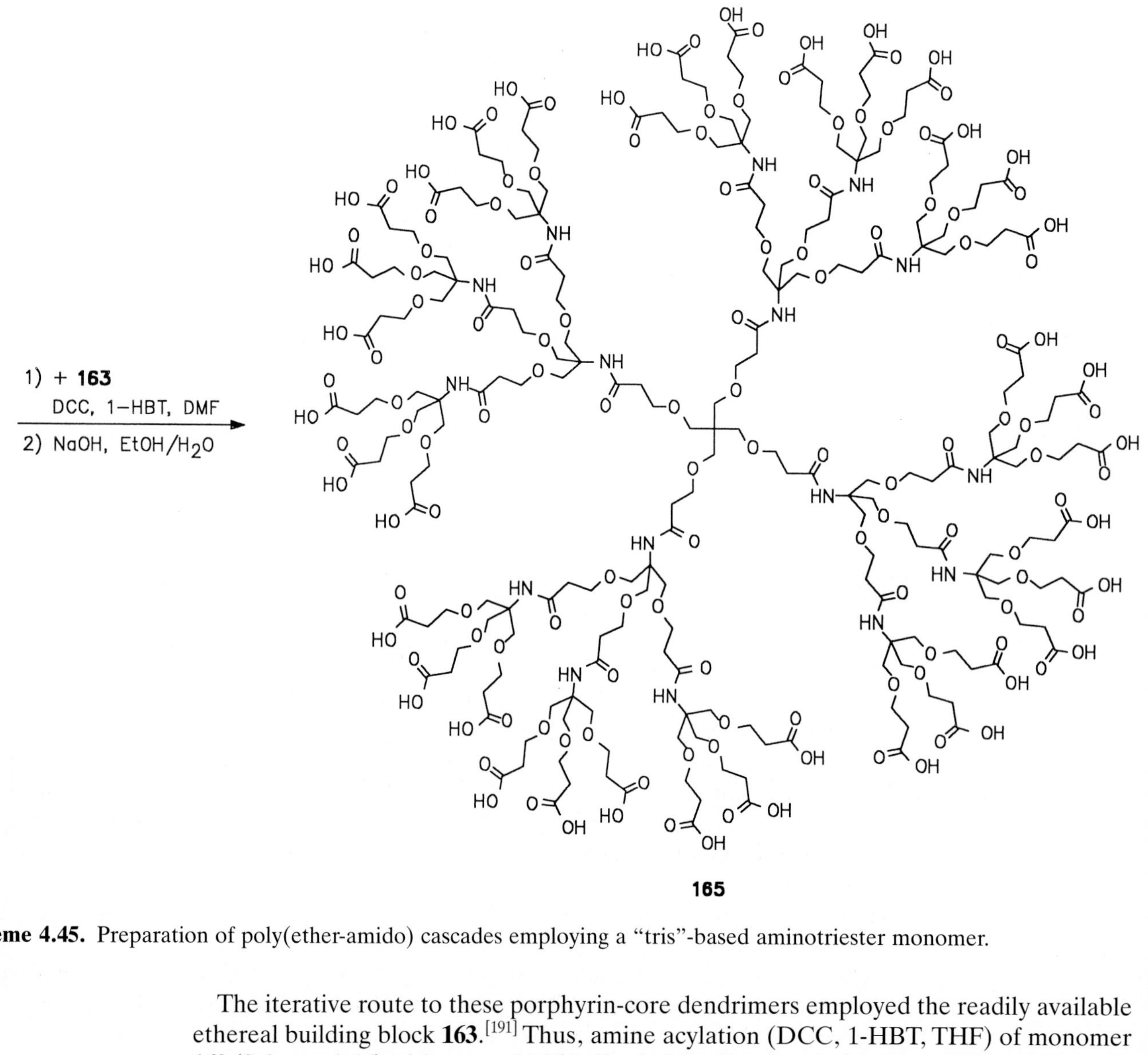

The iterative route to these porphyrin-core dendrimers employed the readily available ethereal building block **163**.[191] Thus, amine acylation (DCC, 1-HBT, THF) of monomer **163** (Scheme 4.46) with tetraacid **166** afforded the first tier dodecaester **167 a**, which upon termini transformation (LiOH, MeOH/H₂O (1:1)) of the polyester gave the polyacid **167 b**. Repetition of this sequence allowed the construction of two additional tiers, e. g., ester **168**.

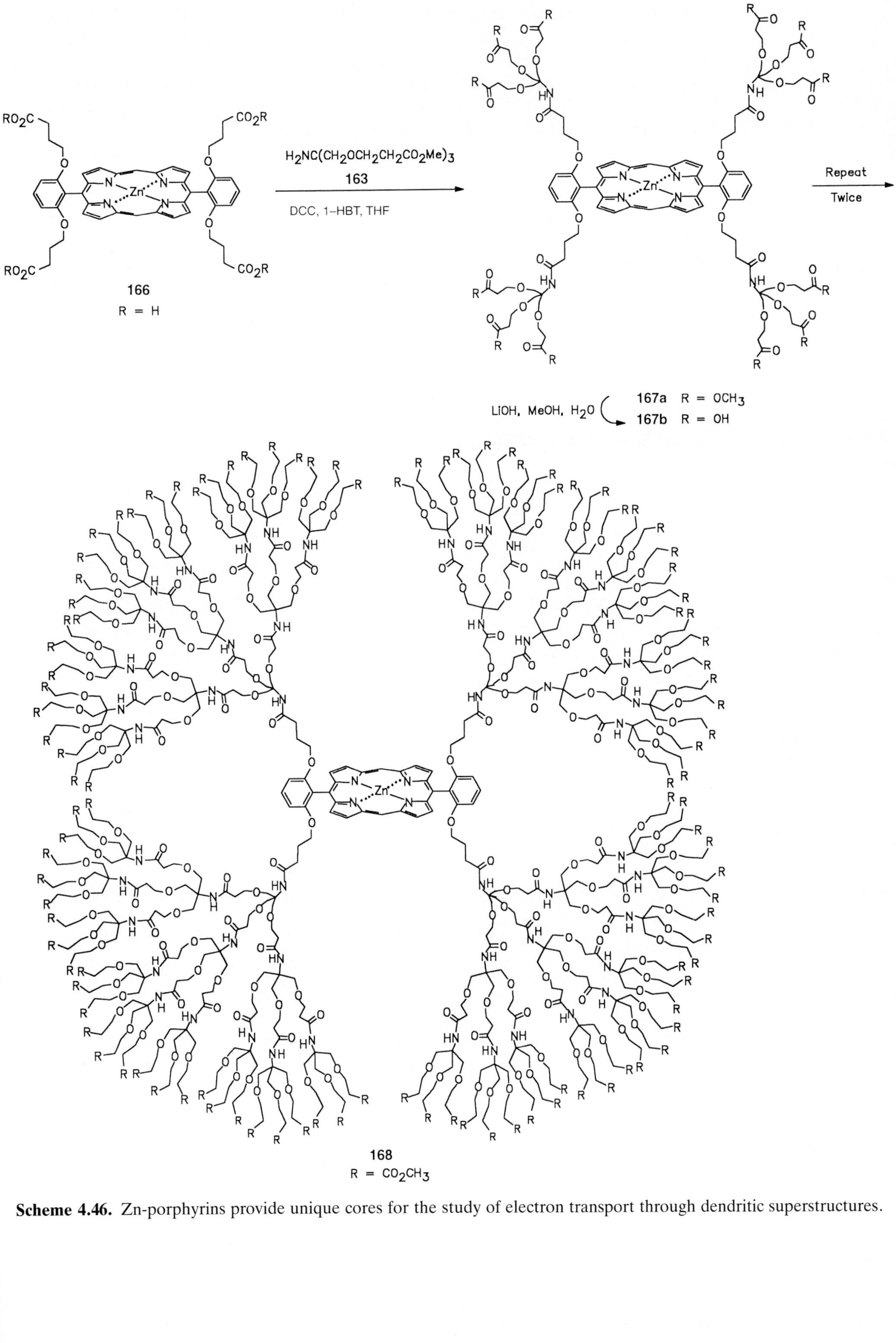

168
R = CO_2CH_3

Scheme 4.46. Zn-porphyrins provide unique cores for the study of electron transport through dendritic superstructures.

Dendrimer characterization was accomplished by ^{13}C, ^{1}H NMR and FT-IR spectroscopy as well as mass spectrometry using fast atom bombardment (FAB) and matrix-assisted laser desorption ionization time-of-flight (MALDI-TOF) techniques. Molecular ion base peaks were observed in the MALDI-TOF MS of polyester **168** (m/z 18 900 (calcd. 19 044)) along with minor peaks at $\approx$ 37 000 and $\approx$ 54 000 m/z corresponding to ionic gasphase dimer and trimer complexes.

Examination of the cyclic voltammetry (CV) of the Zn-porphyrin dendrimers in THF and CH$_2$Cl$_2$ with Bu$_4$N$^+$PF$_6^-$ (0.1 M) electrolyte revealed first oxidation potentials up to 300 mV (THF) less positive than the corresponding values obtained for the "unshielded" tetraester, Zn-porphyrin core. These preliminary electrochemical experiments suggested dendritic encapsulation of redox-active chromophores can effectively influence the electrophoric environment; controlled and well-conceived cascade architecture can lead to new avenues of selective redox catalyst design.

Similar dendritic building blocks have been used to prepare „dendrophanes"[193b] (*dendri*mer + cyclo*phane*), which possess cyclophane cores. Internal Complexation of naphthalene derivatives was subsequently examined by ^{1}H NMR and fluorescence titration.

4.13   1 → 3 *N*-Branched and Connectivity

Rengan and Engel[194, 195] reported and reviewed[196] the preparation of polyammonium cascade polymers (Scheme 4.47) starting with the quaternization of triethanolamine with an alkyl halide or 2-chloroethanol to give a three- (**169**) or four- (**170**) directional core. Tosylation (TsCl/pyr) of the terminal alcohol groups of ammonium chloride **170** and treatment with excess triethanolamine afforded the first generation pentaammonium dendrimer. Following two iterations, the second tier, four-directional dendrimer **171** possessing 17 ammonium branching centers and 36 terminal hydroxy groups was prepared.[197] Attaching these polyammonium polyols to a polymeric backbone has generated a new, high capacity ion exchange substrate.[198]

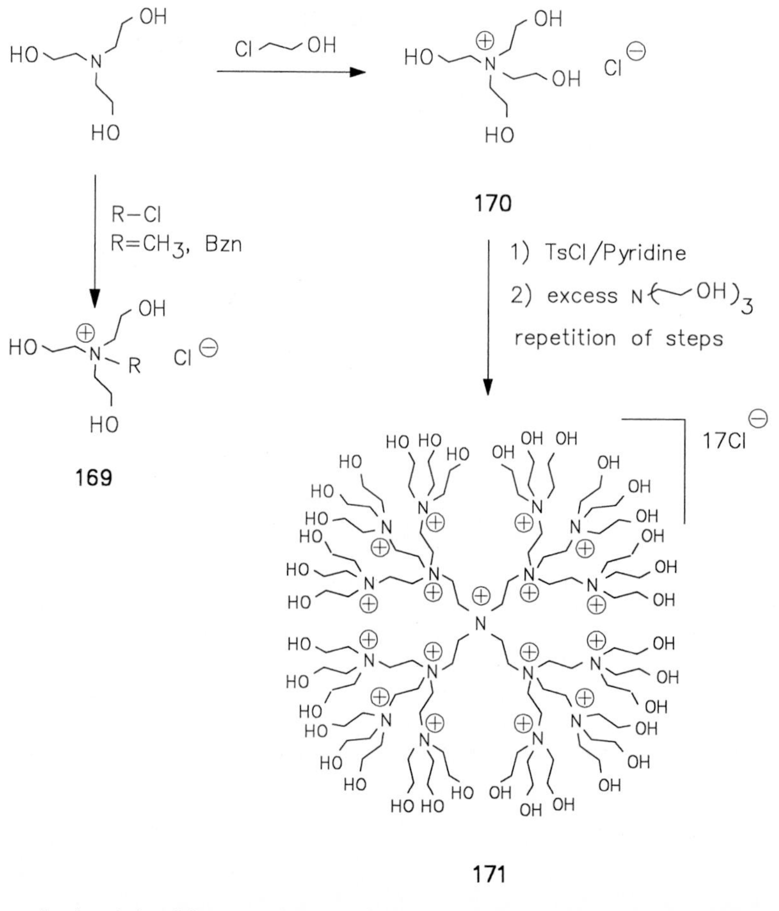

Scheme 4.47. Iterative synthesis of dendrimers using tetrahedral alkyl ammonium moieties as branching centers.

4.14 1 → 3 *P*-Branched and Connectivity

Engel and Rengan[199, 200] also reported and reviewed[196] the preparation of polyphosphonium cascade polymers (Scheme 4.48). A tetradirectional phosphonium core **172** was synthesized via treatment of phosphine **173** with 4-methoxymethylbromobenzene in the presence of NiBr$_2$. The tetramethoxy core **172** was then transformed to the tetraiodide (Me$_3$SiI) and subsequently reacted with phosphine monomer **173** to generate the first tier pentaphosphonium dendrimer **174**. The second tier cascade **175** possessing 17 phosphonium moieties was generated by iteration. Good solubility of dendrimer **175** in common organic solvents (e.g., MeCN, CHCl$_3$) was reported. Related three-directional cascades were similarly prepared using phosphonium core **176**.

Another series of phosphonium dendrimers were constructed by Engel and Rengan[201] also starting with tris(*p*-methoxymethyl)phenylphosphine (**173**). Their goal was to construct dendrimers possessing (trivalent) phosphine and (pentavalent) phosphorane core moieties. These *P*-based cascades were prepared by oxidation (H$_2$O$_2$, AcOH) of building block **173** to the corresponding phosphine oxide **177**. Divergent elaboration was then accomplished by repetitive benzyl ether transformation to the benzyl iodide (Me$_3$SiI/MeCN), followed by phosphine monomer **173** addition. After two iterations, the central phosphine oxide of dendrimer **178** was reduced (Cl$_3$SiH) to afford the trivalent phosphine core dendrimer **179** (Scheme 4.49). Treatment of phosphine **179** with NaAuCl$_4$ gave (97 %) the mono gold chloride–phosphorus dendrimer complex.

A neutral phosphorane core (**181**) was generated from the treatment of tetra(*p*-methoxymethyl)phenylphosphonium bromide (**180**)[201] with 4-lithiobenzylmethyl ether. This pentavalent core was subjected to the previously described procedures to generate the first five-directional dendritic macromolecule **182** (Scheme 4.50).

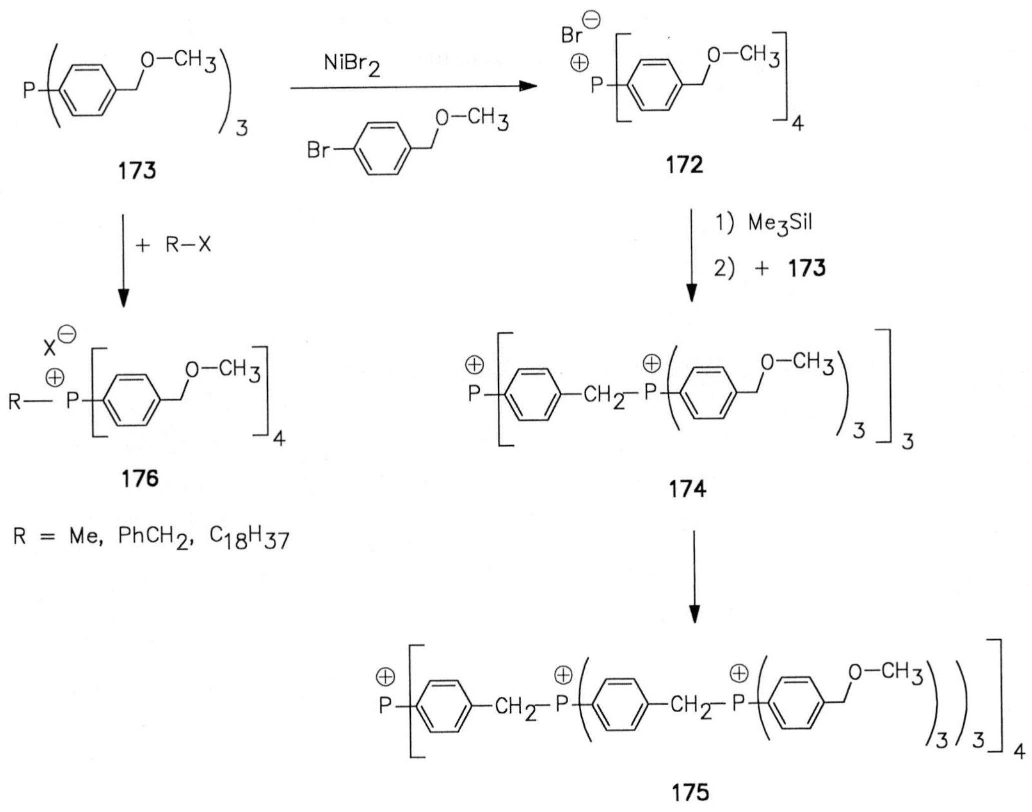

Scheme 4.48. Construction of novel charged *P*-based dendrimers.

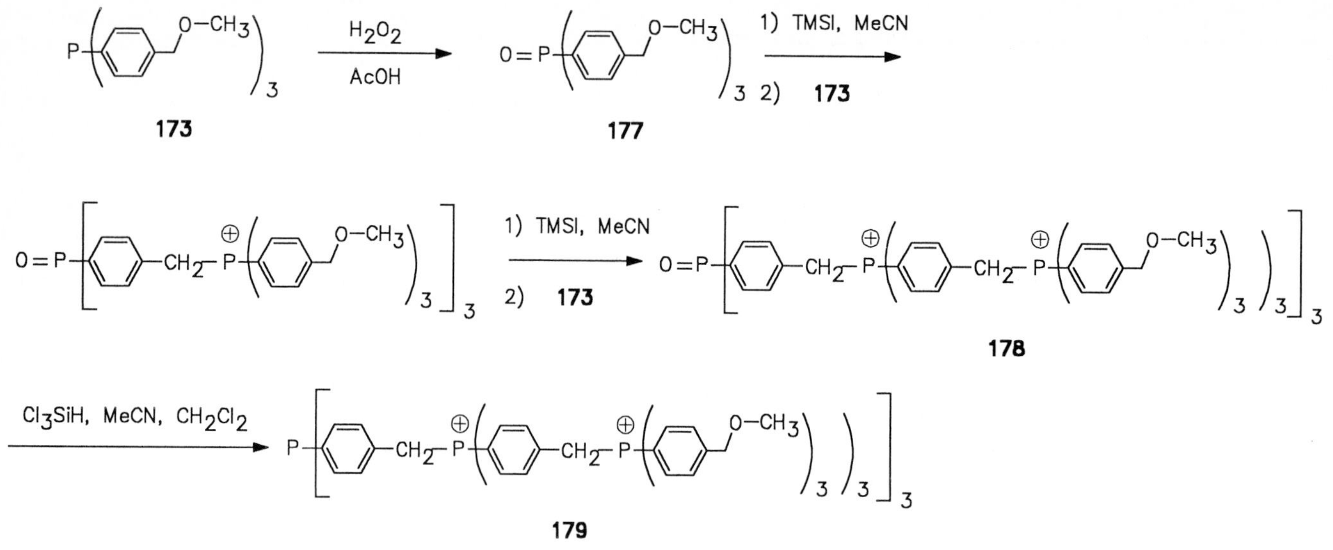

Scheme 4.49. Construction of *P*-based dendrimers with neutral trivalent phosphorus cores.

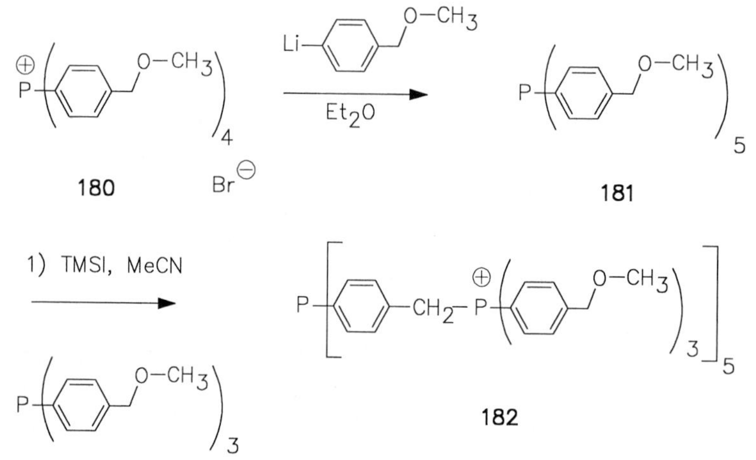

Scheme 4.50. Cascade synthesis employing a pentavalent, neutral phosphorus core.

4.15 1 → 3 *Si*-Branched and Connectivity

Van der Made and van Leeuwen reported[202, 203] the synthesis of *Si*-based dendrimers via a repetitive hydrosilylation and alkenylation sequence (Scheme 4.51). Hence, the 0th tier, tetraalkene **82** (see Scheme 4.23), prepared from tetrachlorosilane and allylmagnesium bromide, was treated with trichlorosilane in the presence of a Pt-catalyst to give the first generation tetrakis(trichlorosilane) **183**. Subsequent exhaustive alkylation with a 10 % excess of allylmagnesium bromide afforded dodecaalkene **184**. Up to the fifth generation dendrimer was prepared by this simple iterative procedure. Branching (1 → 3) employing tetravalent Si gave the fifth tier cascade a molecular weight of 73,912 amu with 972 peripheral groups.

Seyferth and coworkers[104] prepared a series of *Si*-based dendrimers up to the fourth generation. Employing the tetravalent nature of silicon, these macromolecules possessed a tetrahedral, four-directional core, as well as 1 → 3 branching centers. The divergent strategy (Scheme 4.52) utilized two repetitive transformations: displacement of halide at silicon with vinylmagnesium bromide (vinylation) and Pt-catalyzed alkyl trichlorosilane formation (hydrosilylation).

Reaction of tetravinylsilane (**78**) with $H_2PtCl_6 \cdot 6\ H_2O$ and trichlorosilane afforded the first generation tetrakis(trichlorosilane) (**185**), which was treated with vinylmagnesium bromide to produce the corresponding dodecavinylsilane (**186**). Hydrosilylation of silane

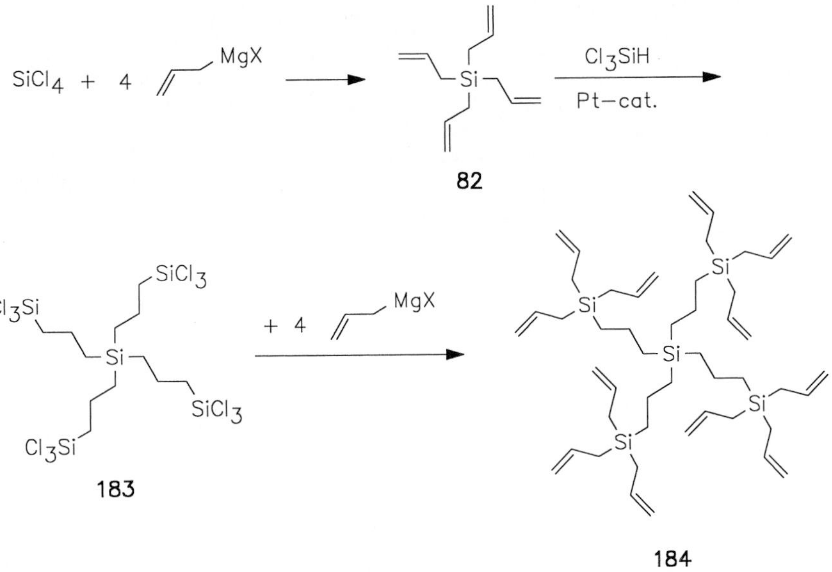

Scheme 4.51. Dendrimers prepared using tetraalkyl substituted silicon, as branching centers.

186 using similar catalytic conditions was unsuccessful resulting in the formation of impure products. Acceptable yields of the second generation dodeca(trichlorosilane) (**187**) were obtained by the employment of the Karstedt catalyst,[204a] which is the product of $[(CH_2=CH)(CH_3)_2Si]_2$ and $H_2PtCl_6 \cdot 6 H_2O$; further vinylation gave good yields of the second generation dodeca(trivinylsilane) (**188**).

Attempted transformation of vinylsilane **188** to the corresponding 36-trichlorosilane (**189**) using the Karstedt catalyzed hydrosilylation procedure led to "grossly impure" products, as indicated by ^1H NMR spectroscopy. However, a simple solvent change from THF to diethyl ether gave the desired third tier dendrimer (**189**) and suppressed impurity formation. Conversion of the third tier trichlorosilane to the corresponding trivinylsilane (not shown) proceeded smoothly, while the construction of the fourth tier, 108-(trichlorosilane) (**190**) (Scheme 4.53) required forcing reaction conditions (excess $HSiCl_3$, Karstedt catalyst, Et_2O, 140 °C, 45 h, Pyrex sealed glass vessel).

Reduction of the fourth tier poly(trichlorosilane) (**190**) with $LiAlH_4$ gave the terminal poly(trihydridosilane) (**191**) as a clear, hard solid. Similar reduction of the first through third tier trichlorosilanes afforded the related hydrido-terminated silanes (e.g., **192** and **193**; see Scheme 4.52). Interest in the utility of these materials for ceramic applications provided the impetus for cross-linking experiments with the hydrido-terminated series, in particular, the first generation dendrimer **192** was examined due in part to its availability. Use, however, of a Zr-cross-linking agent resulted in the formation of products, that were insoluble in most common organic solvents. X-Ray crystallographic data and the corresponding structure (Figure 4.6) were reported for the second generation polyhydrido **193**. The molecular structure for **193** has been published and the pertinent bond distances are given in Table 4.1. Frey et al. have prepared analogues carbosilane dendrimers bearing alcohols[204b] and rigid, cyanobiphenyl mesogenic termini.[204c]

Lamert, Pflug, and Stern[205] reported the preparation of the first dendritic polylisane consisting of an all silicon framework. The small dendrimer can be visualized by considering the following formula: $MeSi[SiMe_2Si(SiMe_3)_3]_3$ {methyl[tris(permethyl-neopentasilyl)]silane} (**194**). Impetus for the construction of this material stems from the electronic, optical, and chemical properties of oligo- and polymeric silanes (i.e., $(-SiR_2-)_n$). However, Si–Si bond lability can, under the proper conditions, adversely affect these properties. Branched silane structures might inhibit internal Si–Si bond scission and thereby maintain bulk properties.

Preparation of the branched silane began with the reaction of tris(trimethylsilyl)silane with $CHCl_3$ (CCl_4) and MeLi to afford the peralkylated methyl[tris(trimethylsilyl)]silane. Subsequent treatment with $AlCl_3$ and $ClSiMe_3$ gave the trichlorosilane, which was then reacted with tris(trimethylsilyl)silyllithium yielding the final silane dendrimer.

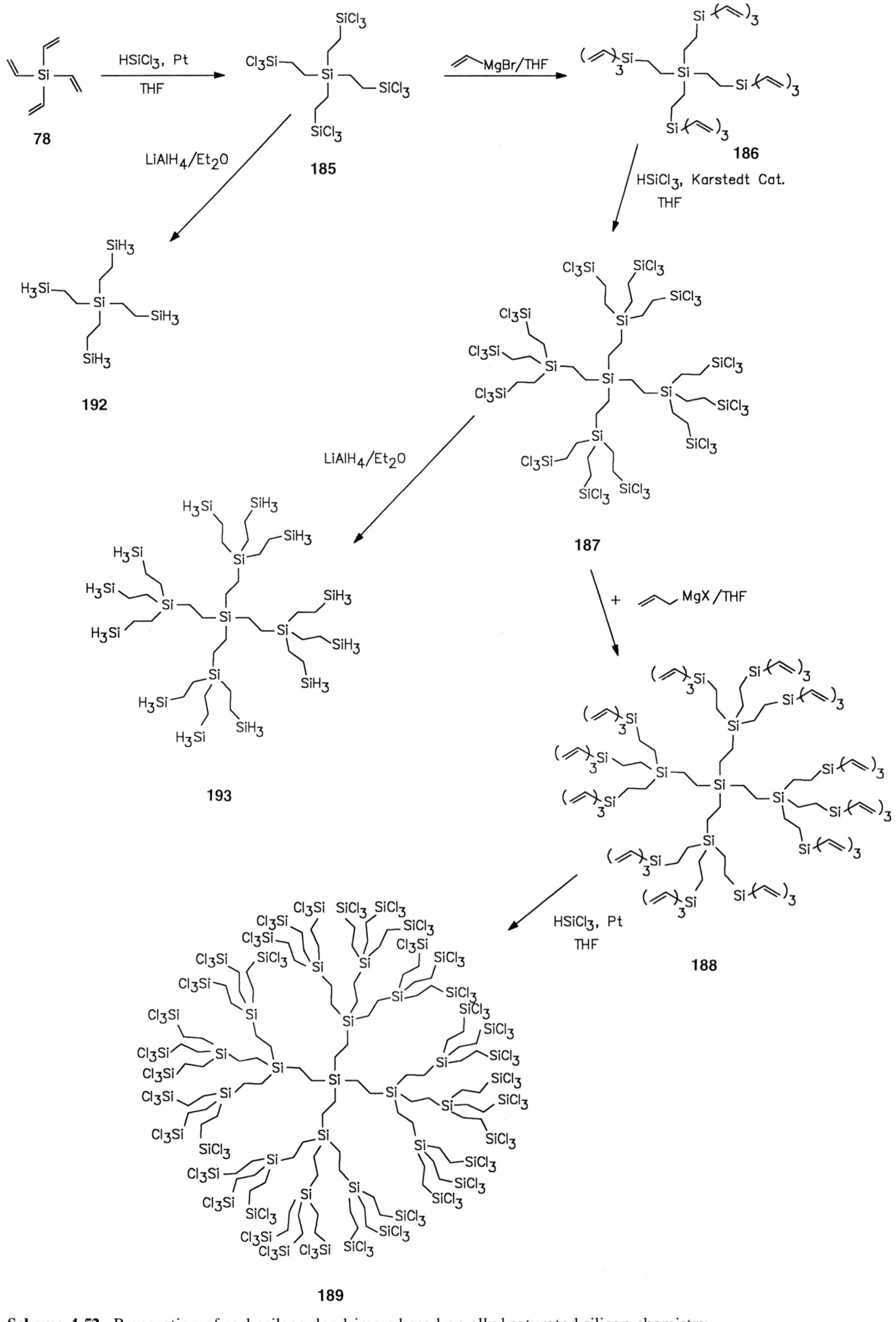

Scheme 4.52. Preparation of carbosilane dendrimers based on alkyl saturated silicon chemistry.

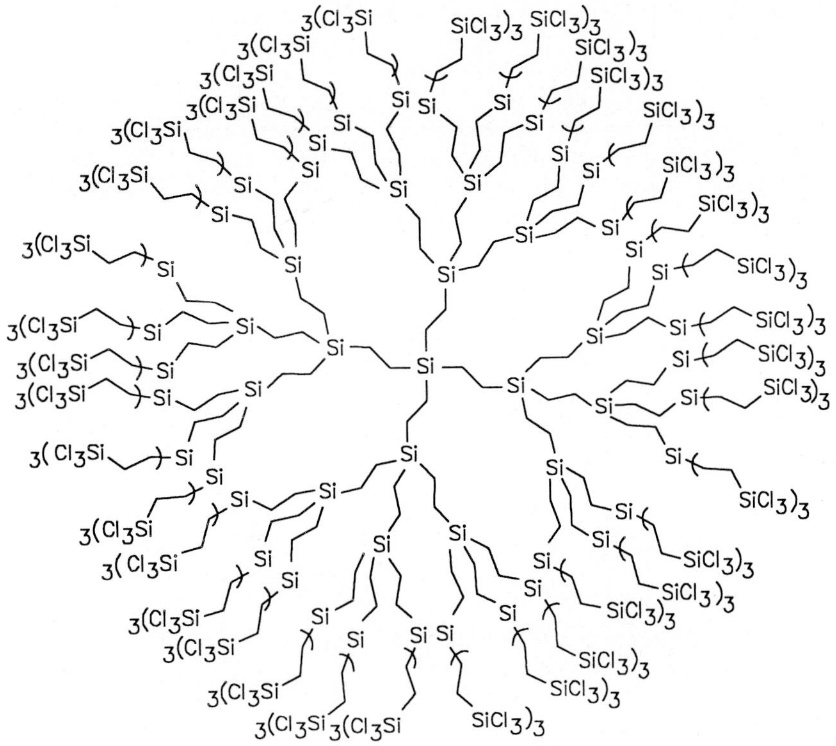

190

LiAlH$_4$/THF

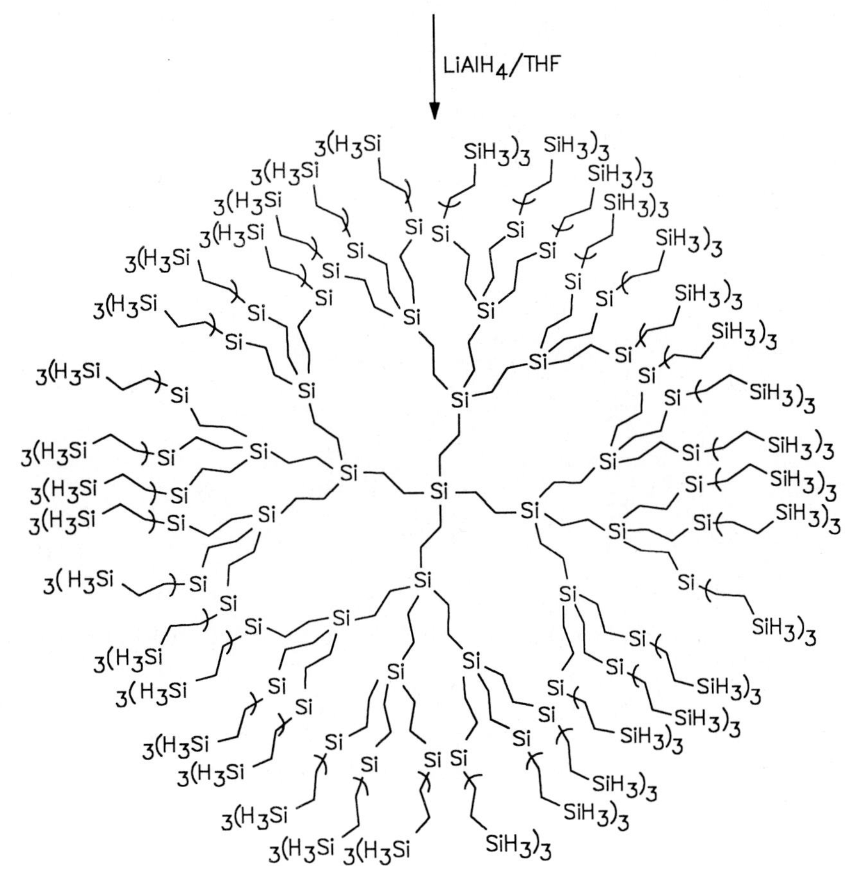

191

Scheme 4.53. Reduction of terminal trichlorosilane moieties to trihydridosilane units.

Table 4.1 Pertinent bond distances for **193**.

Bond[a)	Bond distances, Å averaged	shortest	longest
Si^0-C^1	1.862(14)	1.841	1.872
C^1-C^2	1.543(18)	1.517	1.555
C^2-Si^3	1.854(19)	1.838	1.877
Si^3-C^4	1.866(41)	1.822	1.977
C^4-C^5	1.40(16)	1.094	1.660
C^5-Si^6	1.825(75)	1.667	1.946

a) The superscripts denote the relative relationship of the atom from the *Si*-core.

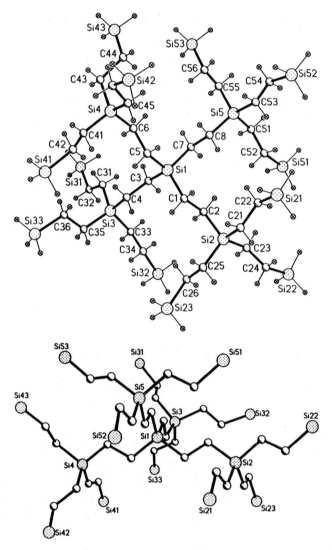

Figure 4.6. X-ray structure of a dendrimer possessing an all silicon superstructure (**193**). [Reproduced by permission of the American Chemical Society.][104]

Seven silicon nuclei comprise the longest silane chain which is repeated 27 times. X-ray crystallography confirmed a threefold axis of symmetry with respect to the core Si–C bond.

4.16   1 → 3 *Adamantane*-Branched, *Ester* Connectivity

Chapman et al.[206] recently reported the construction of "polycules" (**197**) which were generated from "*quasi*-atoms", in this case substituted 1,3,5,7-tetraphenyladamantanes, and the related diadamantanes. This assembly process can be demonstrated (Scheme 4.54) by the treatment of core **195**[207–209] with building block **196**.

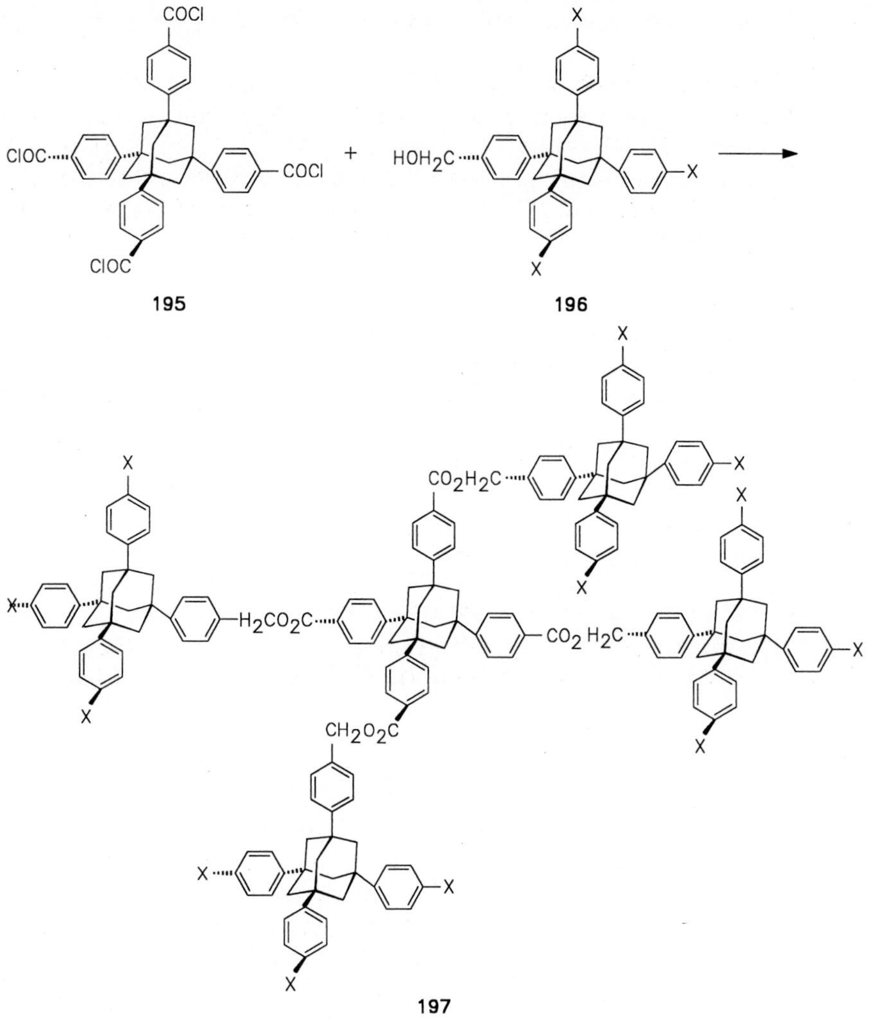

Scheme 4.54. Use of rigid building blocks for the construction of adamantane-based dendrimers.

4.17 References

[1] E. Buhleier, W. Wehner, F. Vögtle, *Synthesis* **1978**, 155.

[2] R. Moors, F. Vögtle, *Chem. Ber.* **1993**, *126*, 2133.

[3] G. R. Newkome, C. N. Moorefield, in *Mesomolecules From Molecules to Materials*, (Eds.: G. D. Mendenhall, A. Greenberg, J. F. Lieberman), Chapman & Hall, New York, **1995**, pp. 27–67.

[4] R. G. Denkewalter, J. Kolc, W. J. Lukasavage, U. S. Pat. 4,289,872 (Sept. 15, **1981**).

[5] R. G. Denkewalter, J. Kolc, W. J. Lukasavage, U. S. Pat. 4,360,646 (Nov. 23, **1982**).

[6] R. G. Denkewalter, J. F. Kolc, W. J. Lukasavage, U. S. Pat. 4,410,688 (Oct. 18, **1983**).

[7] G. R. Newkome, Z.-Q. Yao, G. R. Baker, V. K. Gupta, J. Org. Chem. **1985**, *50*, 2003.

[8] D. A. Tomalia, H. Baker, J. R. Dewald, M. Hall, G. Kallos, S. Martin, J. Roeck, J. Ryder, P. Smith, *Polym. J.* **1985**, *17*, 117.

[9] C. Wörner, R. Mülhaupt, *Angew. Chem.* **1993**, *105*, 1367; *Angew. Chem., Int. Ed. Engl.* **1993**, *32*, 1306.

[10] R. J. Bergeron, J. R. Garlich, *Synthesis* **1984**, 782.

[11] E. M. M. de Brabander-van den Berg, E. W. Meijer, *Angew. Chem.* **1993**, *105*, 1370; *Angew. Chem., Int. Ed. Engl.* **1993**, *32*, 1308.

[12] M. H. P. van Genderen, M. W. P. L. Baars, J. C. M. van Hest, E. M. M. de Brabander-van den Berg, E. W. Meijer, *Recl. Trav. Chim. Pays-Bas.* **1994**, *113*, 573.

[13] E. M. M. de Brabander-van den Berg, A. Nijenhuis, M. Mure, J. Keulen, R. Reintjens, F. Vandenbooren, B. Bosman, R. de Raat, T. Frijns, S. van der Wal, M. Castelijns, J. Put, E. W. Meijer, *Macromol. Symp.* **1994**, *77*, 51.

[14] D. A. Tomalia, A. M. Naylor, W. A. Goddard III, *Angew. Chem.* **1990**, *102*, 119; *Angew. Chem., Int. Ed. Engl.* **1990**, *29*, 138.

[15] J. F. G. A. Jansen, E. M. M. de Brabander-van den Berg, E. W. Meijer, *Science* **1994**, *266*, 1226.

[16] M. Maciejewski, *J. Macromol. Sci.-Chem.* **1982**, *A17*, 689.

[17] a) J. F. G. A. Jansen, E. W. Meijer, E. M. M. de Brabander-van den Berg, *J. Am. Chem. Soc.* **1995**, *117*, 4417. b) J. C. M. van Hest, M. W. P. L. Baars, C. Elissen-Román, M. H. P. van Genderen, E. W. Meijer, *Macromolecules* **1995**, *28*, 6689. c) J. C. M. van Hest, D. A. P. Delnoye, M. W. P. L. Baars, M. H. P. van Genderen, E. W. Meijer, *Science* **1995**, *268*, 1592.

[18] Y. Shirota, T. Kobata, N. Noma, *Chem. Lett.* **1989**, 1145.

[19] K. Yoshizawa, A. Chano, A. Ito, K. Tanaka, T. Yamabe, H. Fujita, J. Yamauchi, M. Shiro, *J. Am. Chem. Soc.* **1992**, *114*, 5994.

[20] D. A. Tomalia, H. Baker, J. R. Dewald, M. Hall, G. Kallos, S. Martin, J. Roeck, J. Ryder, P. Smith, *Macromolecules* **1986**, *19*, 2466.

[21] D. A. Tomalia, D. R. Swanson, J. W. Klimash, H. M. Brothers III, *Polym. Preprints* **1993**, *34* (1), 52.

[22] D. A. Tomalia, J. R. Dewald, U. S. Pat. 4,507,466 (**1985**).

[23] D. A. Tomalia, J. R. Dewald, U. S. Pat. 4,558,120 (**1985**).

[24] J. March, *Advanced Organic Chemistry, 4th ed.*, Wiley, New York, **1992**, pp. 795 & 768.

[25] A. M. Naylor, W. A. Goddard III, *Polym. Preprints* **1988**, *29* (1), 215.

[26] D. A. Tomalia, V. Berry, M. Hall, D. M. Hedstrand, *Macromolecules* **1987**, *20*, 1164.

[27] P. B. Smith, S. J. Martin, M. J. Hall, D. A. Tomalia, in *Applied Polymer Analysis and Characterization*, (Ed.: J. Mitchell, Jr.), Hansen, New York, **1987**, p. 357.

[28] A. M. Naylor, W. A. Goddard III, G. E. Kiefer, D. A. Tomalia, *J. Am. Chem. Soc.* **1989**, *111*, 2339.

[29] A. D. Meltzer, D. A. Tirrell, A. A. Jones, P. T. Inglefield, *Macromolecules* **1992**, *25*, 4549.

[30] A. D. Meltzer, D. A. Tirrell, A. A. Jones, P. T. Inglefield, D. M. Hedstrand, D. A. Tomalia, *Macromolecules* **1992**, *25*, 4541.

[31] A. D. Meltzer, D. A. Tirrell, A. A. Jones, P. T. Inglefield, D. M. Downing, D. A. Tomalia, *Polym. Preprints* **1989**, *30* (1), 121.

[32] G. J. Kallos, D. A. Tomalia, D. M. Hedstrand, S. Lewis, J. Zhou, *Rapid Commun. Mass Spectrom.* **1991**, *5*, 383. P. R. Dvornic, D. A. Tomalia, *Macromol. Symp.* **1995**, *98*, 403.

[33] M. L. Mansfield, L. I. Klushin, *J. Phys. Chem.* **1992**, *96*, 3994.

[34] Y. Gao, K. H. Langley, F. E. Karasz, *Macromolecules* **1992**, *25*, 4902.

[35] D. J. Evans, A. Kanagasooriam, A. Williams, R. J. Pryce, *J. Mol. Catal.* **1993**, *85*, 21.

[36] G. Caminati, N. J. Turro, D. A. Tomalia, *J. Am. Chem. Soc.* **1990**, *112*, 8515.

[37] G. Caminati, D. A. Tomalia, N. J. Turro, *Prog. Colloid Polym. Sci.* **1991**, *84*, 219.

[38] M. C. Moreno-Bondi, G. Orellana, N. J. Turro, D. A. Tomalia, *Macromolecules* **1990**, *23*, 910.

[39] K. R. Gopidas, A. R. Leheny, G. Caminati, N. J. Turro, D. A. Tomalia, *J. Am. Chem. Soc.* **1991**, *113*, 7335.

[40] N. J. Turro, J. K. Barton. D. Tomalia, in *Photochemical Conversion and Storage of Solar Energy*, (Eds.: E. Pelizzetti, M. Schiavello), Kluwer, The Netherlands, **1991**, p. 121.

[41] a) N. J. Turro, J. K. Barton, D. A. Tomalia, *Acc. Chem. Res.* **1991**, *24*, 332. b) Y. Li, P. L. Dubin, R. Spindler, D. A. Tomalia, *Macromolecules* **1995**, *28*, 8426: Dubin et al. have studied complex formation between carboxylated, PAMAM-type dendrimers and poly(dimethyldiallylammonium chloride) to gain insight into colloid diameter, or surface curvature, with polyelectrolyte binding. Based on turbidity experiments, observations were consistent with those of Turro et al.;[38–41] higher generation dendrimers behave as "hindered Stern layer structures" whereas early generations exhibit "simple electrolyte characteristics."

[42] M. F. Ottaviani, E. Cossu, N. J. Turro, D. A. Tomalia, *J. Am. Chem. Soc.* **1995**, *117*, 4387.

[43] L. L. Miller, T. Hashimoto, I. Tabakovic, D. R. Swanson, D. A. Tomalia, *Chem. Mater.* **1995**, *7*, 9.

[44] J. F. Penneau, B. J. Stallman, P. H. Kasai, L. L. Miller, *Chem. Mater.* **1991**, *3*, 791.

[45] a) C. J. Zhong, W. S. V. Kwan, L. L. Miller, *Chem. Mater.* **1992**, *4*, 1423. b) R. G. Duan, L. L. Miller, D. A. Tomalia, *J. Am. Chem. Soc.* **1995**, *117*, 10783.

[46] P.-G. de Gennes, H. Hervet, *J. Phys. Lett.* **1983**, *44*, 351.

[47] D. Farin, D. Avnir, *Angew. Chem.* **1990**, *103*, 1409; *Angew. Chem., Int. Ed. Engl.* **1991**, *30*, 1379.

[48] P. Pfeifer, D. Avnir, *J. Chem. Phys.* **1983**, *79*, 3558.

[49] P. Pfeifer, D. Avnir, *J. Chem. Phys.* **1984**, *80*, 4573.

[50] S. Watanabe, S. L. Regen, *J. Am. Chem. Soc.* **1994**, *116*, 8855.

[51] R. K. Iler, *Colloid Interface Sci.* **1966**, *21*, 569.

[52] a) K. Yu, P. S. Russo, *Polym. Preprints* **1994**, *35*(2), 773. b) T. D. James, H. Shinmori, M. Takeuchi, S. Shinkai, *J. Chem. Soc., Chem. Commun.* **1996**, 705. c) M. Wells, R. M. Crooks, *J. Am. Chem. Soc.* **1996**, *118*, 3988.

[53] H.-B. Mekelburger, K. Rissanen, F. Vögtle, *Chem. Ber.* **1993**, *126*, 1161.

[54] H.-B. Mekelburger, F. Vögtle, *Supramolecular Chem.* **1993**, *1*, 187.

[55] R. Moors, F. Vögtle, in *Advances in Dendritic Macromolecules*, (Ed.: G. R. Newkome) JAI Press, Greenwich, Connecticut, Vol. 2, Chapter 2, p. 41.

[56] K. Kadei, R. Moors, F. Vögtle, *Chem. Ber.* **1994**, *127*, 897.

[57] F. Vögtle, M. Zuber, R. G. Lichtenthaler, *Chem. Ber.* **1973**, *106*, 717.

[58] H. S. Sahota, P. M. Lloyd, S. G. Yeates, P. J. Derrick, P. C. Taylor, D. M. Haddleton, *J. Chem. Soc., Chem. Commun.* **1994**, 2445.

[59] S. M. Aharoni, C. R. Crosby III, E. K. Walsh, *Macromolecules* **1982**, *15*, 1093.

[60] J. P. Tam, *Proc. Natl. Acad. Sci. USA* **1988**, *85*, 5409.

[61] K. J. Chang, W. Pugh, S. G. Blanchard, J. McDermed, J. P. Tam, *Proc. Natl. Acad. Sci. USA* **1988**, *85*, 4929.

[62] D. N. Posnett, H. McGrath, J. P. Tam, *J. Biol. Chem.* **1988**, *263*, 1719.

[63] C. Rao, J. P. Tam, *J. Am. Chem. Soc.* **1994**, *116*, 6975.

[64] J. P. Tam, Y.-A. Lu, *Proc. Natl. Acad. Sci. USA* **1989**, *86*, 9084.

[65] J.-P. Defoort, B. Nardelli, W. Huang, D. D. Ho, J. P. Tam, *Proc. Natl. Acad. Sci. USA* **1992**, *89*, 3879.

[66] J. Shao, J. P. Tam, *J. Am. Chem. Soc.* **1995**, *117*, 3893.

[67] C.-F. Liu, J. P. Tam, *J. Am. Chem. Soc.* **1994**, *116*, 4149.

[68] A. Pessi, E. Bianchi, F. Bonelli, L. Chiappinelli, *J. Chem. Soc., Chem. Commun.* **1990**, 8.

[69] A. Dryland, R. C. Sheppard, *J. Chem. Soc., Perkin Trans. 1* **1986**, 125.

[70] R. Roy, D. Zanini, S. J. Maunier, A. Romanowska, in *Synthetic Oligosaccharides*, (Ed.: P. Kovác), American Chemical Society, Washington, D.C., **1994**, Chapter 7, p. 104.

[71] R. Roy, D. Zanini, S. J. Meunier, A. Romanowska, *J. Chem. Soc., Chem. Commun.* **1993**, 1869.

[72] a) T. M. Chapman, G. L. Hillyer, E. J. Mahan, K. A. Shaffer, *J. Am. Chem. Soc.* **1994**, *116*, 11195. b) P. Scrimin, A. Veronese, P. Tecilla, U. Tonellato, V. Monaco, F. Formaggio, M. Crima, C. Toniolo, *J. Am. Chem. Soc.* **1996**, *118*, 2505.

[73] H. Hart, A. Bashir-Hashemi, J. Luo, M. A. Meador, *Tetrahedron* **1986**, *42*, 1641.

[74] K. Shahlai, H. Hart, *J. Am. Chem. Soc.* **1990**, *112*, 3687.

[75] K. Shahlai, H. Hart, *J. Org. Chem.* **1991**, *56*, 6905.

[76] Review: H. Hart, *Pure Appl. Chem.* **1993**, *65*, 27.

[77] S. B. Singh, H. Hart, *J. Org. Chem.* **1990**, *55*, 3412.

[78] K. Shahlai, H. Hart, A. Bashir-Hashemi, *J. Org. Chem.* **1991**, *56*, 6912.

[79] H. Hart, S. Shamouilian, T. Takehira, *J. Org. Chem.* **1981**, *46*, 4427.

[80] C. F. Huebner, R. T. Puckett, M. Brzechffa, S. L. Schwartz, *Tetrahedon Lett.* **1970**, 359.

[81] P. Venugopalan, H.-B. Bürgi, N. L. Frank, K. K. Baldridge, J. S. Seigel, *Tetrahedron Lett.* **1995**, *36*, 2419.

[82] O. W. Webster, *Polym. Preprints* **1993**, *34* (1), 98.

[83] N. S. Zefirov, S. I. Kozhushkov, B. I. Ugrak, K. A. Lukin, O. V. Kokoreva, D. S. Yufit, Y. T. Struchkov, S. Zoellner, R. Boese, A. de Meijere, *J. Org. Chem.* **1992**, *57*, 701.

[84] A. de Meijere, S. I. Kozhushkov, T. Spaeth, N. S. Zefirov, *J. Org. Chem.* **1993**, *58*, 502.

[85] S. I. Kozhushkov, T. Haumann, R. Boese, A. de Meijere, *Angew. Chem.* **1993**, *105*, 426; *Angew. Chem., Int. Ed. Engl.* **1993**, *32*, 401.

[86] G. W. Griffin, A. P. Marchand, *Chem. Rev.* **1989**, *89*, 997.

[87] a) P. E. Eaton, *Angew. Chem.* **1992**, *104*, 1447; *Angew. Chem., Int. Ed. Engl.* **1992**, *31*, 1421. b) H.-D. Beckhaus, C. Rüchardt, S. T. Kozhushkov, V. N. Belov, S. P. Verevkin, A. de Meijere, *J. Am. Chem. Soc.* **1995**, *117*, 11854.

[88] J. Veciana, C. Rovira, M. I. Crespo, O. Armet, V. M. Domingo, F. Palacio, *J. Am. Chem. Soc.* **1991**, *113*, 2552.

[89] J. Veciana, C. Rovira, E. Hernández, N. Ventosa, *An. Quim.* **1993**, *89*, 73.

[90] J. Veciana, C. Rovira, N. Ventosa, M. I. Crespo, F. Palacio, *J. Am. Chem. Soc.* **1993**, *115*, 57.

[91] N. Ventosa, D. Ruiz, C. Rovira, J. Veciana, *Mol. Cryst. Liq. Cryst. Sect. A* **1993**, *232*, 333.

[92] J. Veciana, C. Rovira, in *Magnetic Molecular Materials*, (Eds.: D. Gatteschi, et al.), Kluwer, The Netherlands, **1991**, p. 121.

[93] N. Nakamura, K. Inoue, H. Iwamura, *Angew. Chem.* **1993**, *105*, 900; *Angew. Chem., Int. Ed. Engl.* **993**, *32*, 872.

[94] H. Bock, A. John, Z. Havlas, J. W. Bats, *Angew. Chem.* **1993**, *105*, 416; *Angew. Chem., Int. Ed. Engl.* **1993**, *32*, 416.

[95] B. Kirste, M. Grimm, H. Kurreck, *J. Am. Chem. Soc.* **1989**, *111*, 108.

[96] R. Chaler, J. Carilla, E. Brillas, A. Labarta, L. Fajar, J. Riera, L. Guliá, *J. Org. Chem.* **1994**, *59*, 4107.

[97] H. K. Hall, Jr., D. W. Polis, *Polym. Bull.* **1987**, *17*, 409.

[98] M. Schuster, P. Wang, J. C. Paulson, C.-H. Wong, *J. Am. Chem. Soc.* **1994**, *116*, 1135.

[99] Sialyl Lewis X-type glycoproteins are ligands for *P*- and *L*-selectins; for an excellent review, see: Lasky, L. A. *Science* **1992**, *258*, 964.

[100] H. Uchida, Y. Kabe, K. Yoshino, A. Kawamata, T. Tsumuraya, S. Masamune, *J. Am. Chem. Soc.* **1990**, *112*, 7077.

[101] A. Morikawa, M.-a. Kakimoto, Y. Imai, *Polym. J.* **1992**, *24*, 573.

[102] J. Roovers, P. M. Toporowski, L.-L. Zhou, *Polym. Preprints* **1992**, *33*, 182.

[103] L.-L. Zhou, J. Roovers, *Macromolecules* **1993**, *26*, 963.

[104] D. Seyferth, D. Y. Son, A. L. Rheingold, R. L. Ostrander, *Organometallics* **1994**, *13*, 2682.

[105] P. S. Chang, T. S. Hughes, Y. C. Zhang, G. R. Webster Jr., D. Poczynok, M. A. Buese, *J. Polym. Sci. Part A: Polym. Chem.* **1993**, *31*, 891.

[106] J. Roovers, L.-L. Zhou, P. M. Toporowski, M. van der Zwan, H. Iatrou, N. Hadjichristidis, *Macromolecules* **1993**, *26*, 4324.

[107] Also see: L.-L. Zhou, N. Hadjichristidis, P. M. Toporowski, J. Roovers, *Rubber Chem. Technol.* **1992**, *65*, 303.

[108] B. Alonso, I. Cuadrado, M. Morán, J. Losada, *J. Chem. Soc., Chem. Commun.* **1994**, 2575.

[109] I. Cuadrado, M. Morán, J. Losada, *J. Chem. Soc., Chem. Commun.* **1994**, 2575.

[110] A. Sekiguchi, M. Nanjo, C. Kabuto, H. Sakurai, *J. Am. Chem. Soc.* **1995**, *117*, 4195.

[111] A. Miedaner, C. J. Curtis, R. M. Barkley, D. L. DuBois, *Inorg. Chem.* **1994**, *33*, 5482.

[112] R. B. King, J. C. Cloyd, Jr., *J. Am. Chem. Soc.* **1975**, *97*, 46.

[113] R. B. King, J. C. Cloyd, Jr., P. N. Kapoor, *J. Chem. Soc., Perkin Trans. 1* **1973**, 2226.

[114] N. Launay, A.-M. Caminade, R. Lahana, J.-P. Majoral, *Angew. Chem.* **1994**, *106*, 1682; *Angew. Chem., Int. Ed. Engl.* **1994**, *33*, 1589.

[115] N. Launay, A.-M. Caminade, J.-P. Majoral, *J. Am. Chem. Soc.* **1995**, *117*, 3282.

[116] J. Mitjaville, A.-M. Caminade, J.-P. Majoral, *Tetrahedron Lett.* **1994**, *35*, 6865.

[117] J. Y. Chang, H. J. Ji, M. J. Han, S. B. Rhee, S. Cheong, M. Yoon, *Macromolecules* **1994**, *27*, 1376.

[118] C. Galliot, D. Prévoté, A.-M. Caminade, J.-P. Majoral, *J. Am. Chem. Soc.* **1995**, *117*, 5470.

[119] F. Hallé, R. A. A. Oldeman, in *Essai sur l'architecture et la dynamique de croissance des arbes tropicaux*, Masson, Paris, **1970**.

[120] F. Hallé, R. A. A. Oldeman, P. B. Tomlinson, in *Tropical Trees and Forests: An Architectural Analysis*, Springer, Berlin, **1982**.

[121] P. B. Tomlinson, *Am. Sci.* **1983**, *71*, 141.

[122] G. R. Newkome, G. R. Baker, *Org. Prep. Proced. Int.* **1986**, *18*, 117.

[123] J. Skarzewski, *Tetrahedron* **1989**, *45*, 4593.

[124] J. Skarzewski, *Synthesis* **1990**, 1125.

[125] G. R. Newkome, C. N. Moorefield, G. R. Baker, *Aldrichimica Acta* **1992**, *25* (2), 31.

[126] G. R. Newkome, G. R. Baker, M. J. Saunders, P. S. Russo, V. K. Gupta, Z.-q. Yao, J. E. Miller, K. Bouillon, *J. Chem. Soc., Chem. Commun.* **1986**, 752.

[127] G. R. Newkome, G. R. Baker, S. Arai, M. J. Saunders, P. S. Russo, K. J. Theriot, C. N. Moorefield, L. E. Rogers, J. E. Miller, T. R. Lieux, M. E. Murray, B. Phillips, L. Pascal, *J. Am. Chem. Soc.* **1990**, *112*, 8458.

[128] R. M. Fuoss, D. J. Edleson, *J. Am. Chem. Soc.* **1951**, *73*, 269.

[129] J.-H. Fuhrhop, J. Mathieu, *Angew. Chem.* **1984**, *96*, 124; *Angew. Chem., Int. Ed. Engl.* **1984**, *23*, 100.

[130] G. H. Escamilla, G. R. Newkome, *Angew. Chem.* **1994**, *106*, 2013; *Angew. Chem., Int. Ed. Engl.* **1994**, *33*, 1937.

[131] G. H. Escamilla, in *Advances in Dendritic Macromolecules* (Ed.: G. R. Newkome), JAI Press, Greenwich, Connecticut, **1995**, Vol. 2, Chapter 6, p. 157.

[132] G. R. Newkome, C. N. Moorefield, G. R. Baker, R. K. Behera, G. H. Escamilla, M. J. Saunders, *Angew. Chem.* **1992**, *104*, 901; *Angew. Chem., Int. Ed. Engl.* **1992**, *31*, 917.

[133] J.-M. Lehn, *Angew. Chem.* **1990**, *102*, 1347; *Angew. Chem., Int. Ed. Engl.* **1990**, *29*, 1304.

[134] G. R. Newkome, X. Lin, Y.-X. Chen, G. H. Escamilla, *J. Org. Chem.* **1993**, *58*, 3123.

[135] G. R. Newkome, Z.-q. Yao, G. R. Baker, V. K. Gupta, P. S. Russo, M. J. Saunders, *J. Am. Chem. Soc.* **1986**, *108*, 849.

[136] Th.P. Engelhardt, L. Belkoura, D. Woermann, W. Grimme, *Ber. Bunsenges. Phys. Chem.* **1993**, *97*, 33.

[137] G. R. Newkome, Y. Hu, M. J. Saunders, F. R. Fronczek, *Tetrahedron Lett.* **1991**, *32*, 1133.

[138] C. D. Gutsche, *Monographs in Supramolecular Chemistry*, (Ed.: J. F. Stoddart) Royal Society of Chemistry, London, **1989**.

[139] G. R. Newkome, V. K. Gupta, G. R. Baker, et al., National Meeting of the American Chemical Society, Miami, FL, April, **1985**, ORGN-166.

[140] Ref. 6 in M. Sawamoto, *Kagaku (Kyoto)* **1990**, *45*, 537.

[141] M. Jørgensen, K. Bechgaard, T. Bjørnholm, P. Sommer-Larsen, L. G. Hansen, K. Schaumburg, *J. Org. Chem.* **1994**, *59*, 5877.

[142] J. March, *Advanced Organic Chemistry*, 4th ed., Wiley, New York, **1992**, p. 339.

[143] G. R. Newkome, C. N. Moorefield, K. J. Theriot, *J. Org. Chem.* **1988**, *53*, 5552.

[144] G. R. Newkome, C. N. Moorefield, U. S. Pat. 5,136,096 (Aug. 4, **1992**).

[145] G. R. Newkome, C. N. Moorefield, U. S. Pat. 5,206,410 (Apr. 27, **1993**).

[146] G. R. Newkome, C. N. Moorefield, U. S. Pat. 5,210,309 (May 11, **1993**).

[147] C. D. Weis, G. R. Newkome, *Synthesis* **1995**, 1053.

[148] M. Broussard, B. Juma, F. R. Fronczek, S. F. Watkins, G. R. Newkome, C. N. Moorefield, *Acta Cryst. C* **1991**, *47*, 1245.

[149] G. R. Newkome, G. R. Baker, R. K. Behera, A. L. Johnson, C. N. Moorefield, C. D. Weis, W.-J. Cao, J. K. Young, *Synthesis* **1991**, 839.

[150] J. K. Whitesell, H. K. Chang, *Science* **1993**, *261*, 73.

[151] For an interesting overview see: J. G. Tirrell, M. J. Fournier, T. L. Mason, D. A. Tirrell, *Chem. & Eng. News* **1994**, Dec. 19th, 40.

[152] G. R. Newkome, C. D. Weis, *Org. Prep. Proced. Int.* **1996**, in press.

[153] This name was derived from Dr. Rajani K. Behera, who was the first to prepare this aminotriester.

[154] H. A. Bruson, T. W. Riener, *J. Am. Chem. Soc.* **1943**, *65*, 23.

[155] S. J. Allen, J. G. N. Dywitt, U. S. Pat. 2,502,548, **1950**.

[156] D. E. Butler, U. S. Pat. 4,454,327, **1984**.

[157] G. R. Newkome, R. K. Behera, G. R. Baker, *Acta Cryst.* **1994**, *C50*, 120.

[158] G. R. Newkome, A. Nayak, R. K. Behera, C. N. Moorefield, G. R. Baker, *J. Org. Chem.* **1992**, *57*, 358.

[159] a) G. R. Newkome, R. J. Behera, C. N. Moorefield, G. R. Baker, *J. Org. Chem.* **1991**, *56*, 7162. b) G. R. Newkome, V. V. Narayanan, A. K. Patri, J. Groß, C.N. Moorefield, G. R. Baker, *Polym. Mater. Sci. Eng.* **1995**, *73*, 222.

[160] A. Bashir-Hashemi, J. Li, *Tetrahedron Lett.* **1995**, *36*, 1233.

[161] J. Klausner, B. Bodansky, *Synthesis* **1972**, 453.

[162] G. R. Newkome, J. K. Young, G. R. Baker, R. L. Potter, L. Audoly, D. Cooper, C. D. Weis, K. F. Morris, C. S. Johnson, Jr., *Macromolecules* **1993**, *26*, 2394.

[163] H. A. Bruson, U. S. Pat. 2,401,607; *Chem. Abstr.* **1946**, *40*, 5450.

[164] K. F. Morris, C. S. Johnson, Jr., *J. Am. Chem. Soc.* **1993**, *115*, 4291.

[165] a) S. A. Kuzdzal, C. A. Monnig, G. R. Newkome, C. N. Moorefield, *J. Chem. Soc., Chem. Commun.* **1994**, 2139. b) C. Monnig, S. Kuzdzal, University of California at Riverside, personal communication, **1996**.

[166] D. S. Tarbell, Y. Yamamoto, B. M. Pope, *Proc. Natl. Acad. Sci., USA* **1972**, *69*, 730.

[167] E. Ponnusamy, U. Fotadar, A. Spisni, D. Fiat, *Synthesis* **1986**, 48.

[168] X. A. Dominguez, I. C. Lopez, R. Franco, *J. Org. Chem.* **1961**, *26*, 1625.

[169] J. K. Young, G. R. Baker, G. R. Newkome, K. F. Morris, C. S. Johnson, Jr., *Macromolecules* **1994**, *27*, 3464.

[170] D. F. DeTar, R. Silverstein, F. F. Rogers, Jr., *J. Am. Chem. Soc.* **1966**, *88*, 1024.

[171] W. König, R. Geiger, *Chem. Ber.* **1970**, *103*, 788.

[172] G. R. Newkome, G. R. Baker, C. N. Moorefield, M. J. Saunders, *Polym. Preprints* **1991**, *32*, 625.

[173] G. R. Newkome, C. N. Moorefield, G. R. Baker, A. L. Johnson, R. K. Behera, *Angew. Chem.* **1991**, *103*, 1205; *Angew. Chem., Int. Ed. Engl.* **1991**, *30*, 1176.

[174] G. R. Newkome, C. N. Moorefield, U. S. Pat. 5,154,853 (Oct. 13, **1992**).

[175] N. Ono, H. Miyake, A. Kamimura, I. Hamamoto, R. Tamura, A. Kaji, *Tetrahedron* **1985**, *41*, 4013.

[176] C. K. Ingold, L. C. Nickolls, *J. Chem. Soc.* **1922**, *121*, 1646.

[177] L. M. Rice, B. S. Sheth, T. B. Zalucky, *J. Pharm. Sci.* **1971**, 1760.

[178] G. R. Newkome, V. K. Gupta, R. W. Griffin, S. Arai, *J. Org. Chem.* **1987**, *52*, 5480.

[179] G. R. Newkome, C. N. Moorefield, G. R. Baker, M. J. Saunders, S. H. Grossman, *Angew. Chem.* **1991**, *103*, 1207; *Angew. Chem. Int. Ed. Engl.* **1991**, *30*, 1178.

[180] G. R. Newkome, C. N. Moorefield, *Polym. Preprints* **1993**, *34* (1), 75.

[181] G. R. Newkome, C. N. Moorefield, J. M. Keith, G. R. Baker, G. H. Escamilla, *Angew. Chem.* **1994**, *106*, 701; *Angew. Chem., Int. Ed. Engl.* **1994**, *33*, 666.

[182] G. R. Newkome, C. N. Moorefield, *Macromol. Symp.* **1994**, *77*, 63.

[183] A. B. Padias, H. K. Hall, Jr., D. A. Tomalia, *Polym. Preprints 30* (1) **1989**, 119.

[184] A. B. Padias, H. K. Hall, Jr., D. A. Tomalia, J. R. McConnell, *J. Org. Chem.* **1987**, *52*, 5305.

[185] S. Rustad, R. Stølevik, *Acta Chim. Scand., Ser. A.* **1976**, *30*, 209.

[186] D. A. Tomalia, D. M. Hedstrand, L. R. Wilson, in *Encyclopedia of Polymer Science and Engineering*, Wiley, New York, **1990**, p. 46.

[187] J.-J. Lee, W. T. Ford, J. A. Moore, Y. Li, *Macromolecules* **1994**, *27*, 4632.

[188] Ref. 6 in reference 187.

[189] J. A. Moore, personal communication, **1994**.

[190] A. F. Bochkov, B. E. Kalganov, V. N. Chernetskii, *Izv. Akad. Nauk SSSR, Ser. Khim.* **1989**, 2394 (*Chem. Abstr.* **1990**, *112*, 216174p).

[191] G. R. Newkome, X. Lin, *Macromolecules* **1991**, *24*, 1443.

[192] P. J. Dandliker, F. Diederich, M. Gross, C. B. Knobler, A. Louati, E. M. Sanford, *Angew. Chem.* **1994**, *106*, 1821; *Angew. Chem., Int. Ed. Engl.* **1994**, *33*, 1739.

[193] a) See refs. 5 and 6 in Dandliker, et al., ref. 192. b) S. Mattei, P. Seiler, F. Diederich, V. Gramlich, *Helv. Chim. Acta* **1995**, *78*, 1904.

[194] K. Rengan, R. Engel, *J. Chem. Soc., Chem. Commun.* **1990**, 1084.

[195] K. Rengan, R. Engel, *J. Chem. Soc., Chem. Commun.* **1992**, 757.

[196] R. Engel, in *Advances in Dendritic Macromolecules,* Vol. 2, (Ed.: G. R. Newkome), JAI Press, Greenwich, Connecticut, **1995**, Chapter 3, pp. 73–100.

[197] R. Engel, *Polym. News* **1992**, *17*, 301.

[198] A. Cherestes, R. Engel, *Polymer* **1994**, *35*, 3343.

[199] R. Engel, K. Rengan, C. Milne, *Polym. Preprints* **1991**, *32* (3), 601.

[200] K. Rengan, R. Engel, *J. Chem. Soc., Perkin Trans. 1* **1991**, 987.

[201] R. Engel, K. Rengan, C.-S. Chan, *Heteroatom Chem.* **1993**, *4*, 181.

[202] A. W. van der Made, P. W. N. M. van Leeuwen, *J. Chem. Soc., Chem. Commun.* **1992**, 1400.

[203] A. W. van der Made, P. W. N. M. van Leeuwen, J. C. de Wilde, R. A. C. Brandes, *Adv. Mater.* **1993**, *5*, 466.

[204] a) B. D. Karstedt, U. S. Pat. 3,775,452, **1973**. b) K. Lorenz, R. Mülhaupt, H. Frey, U. Rapp, F. J. Mayer-Poser, *Macromolecules* **1995**, *28*, 6657. c) H. Frey, K. Lorenz, P. Hölter, R. Mülhaupt, *Polym. Preprints* **1996**, *37* (1), 758.

[205] J. B. Lambert, J. L. Pflug, C. L. Stern, *Angew. Chem.* **1995**, *107*, 106; *Angew. Chem., Int. Ed. Engl.* **1995**, *34*, 98. J. B. Lambert, J. L. Pflug, J. M. Denari, *Organometallics* **1996**, *15*, 615.

[206] O. L. Chapman, J. Magner, R. Ortiz, *Polym. Preprints* **1995**, *36* (1), 739.

[207] V. R. Reichert, L. J. Mathias, *Macromolecules* **1994**, *27*, 7015.

[208] V. R. Reichert, L. J. Mathias, *Polym. Preprints* **1993**, *34* (1), 495.

[209] M. Dotrong, M. H. Dotrong, G. J. Moore, R. C. Evars, *Polym. Preprints* **1994**, *35*(2), 673.

5 Synthetic Methodologies: Convergent Procedures

5.1 General Concepts

The "convergent" mode of dendritic construction is another strategy whereby branched polymeric arms (dendrons) are synthesized from the "outside-in". This concept (Scheme 5.1), initially described by Fréchet and his coworkers[1,2] and shortly thereafter by Miller and Neenan,[3] can best be illustrated and reviewed[4,5] by envisioning the attachment of two terminal units (1) containing one reactive group X to one monomer (2) possessing protected functionality Z, resulting in the preparation of the first generation or tier (e.g., 3). Transformation of the active, or focal site (Z → X), followed by treatment with 0.5 equivalent of the masked monomer 2 affords the next higher generation (e.g., 4). A notable advantage of this procedure is the requirement of a minimum number of transformations for tier construction. Thus, using a three-directional building block, only two collisions (reactions of Y at X) need to be effected in order to attach each consecutive generation. In general, the active site of attachment of the dendron or dendritic wedge is crowded further into the infrastructure with each successive tier attachment; at some point of development, therefore, chemical connectivity can become more difficult as a result of the juxtaposed steric interference.

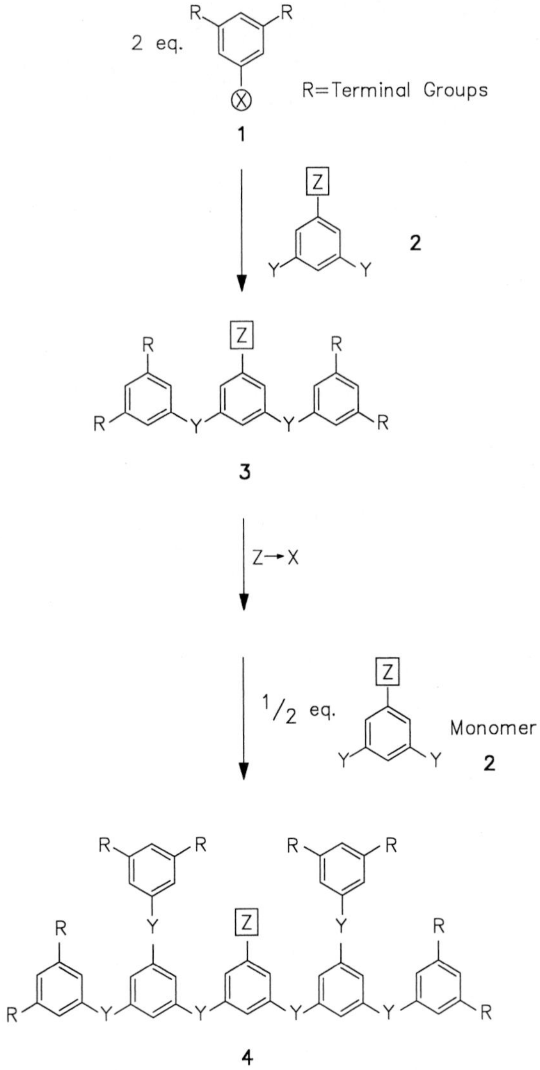

Scheme 5.1. Illustration of the general concept of convergent synthesis.

5.2   1 → 2 *C*-Branched

5.2.1   1 → 2 *C*-Branched and Connectivity

Rajca[6–8] reported a novel route to the construction of dendritic triphenylmethyl poly-radicals and polyanions; his elegant work has been reviewed.[9–12] Construction of these macromolecules was aimed toward examination of high spin, organomagnetic materials.[13–16] An example of this 1 → 2 *C*-branched iterative methodology (Scheme 5.2) starts with the monolithiation of 1,3-dibromobenzene (**5**), followed by the addition of 4,4′-di-*tert*-butylbenzophenone (**6**). Quenching with EtOCOCl in ethanol afforded the terminal unit **7**. Subsequent metal–halogen exchange and reaction with 3-bromo-4′-*tert*-butylbenzophenone (**8**), performed twice, followed by ethoxylation, afforded the desired

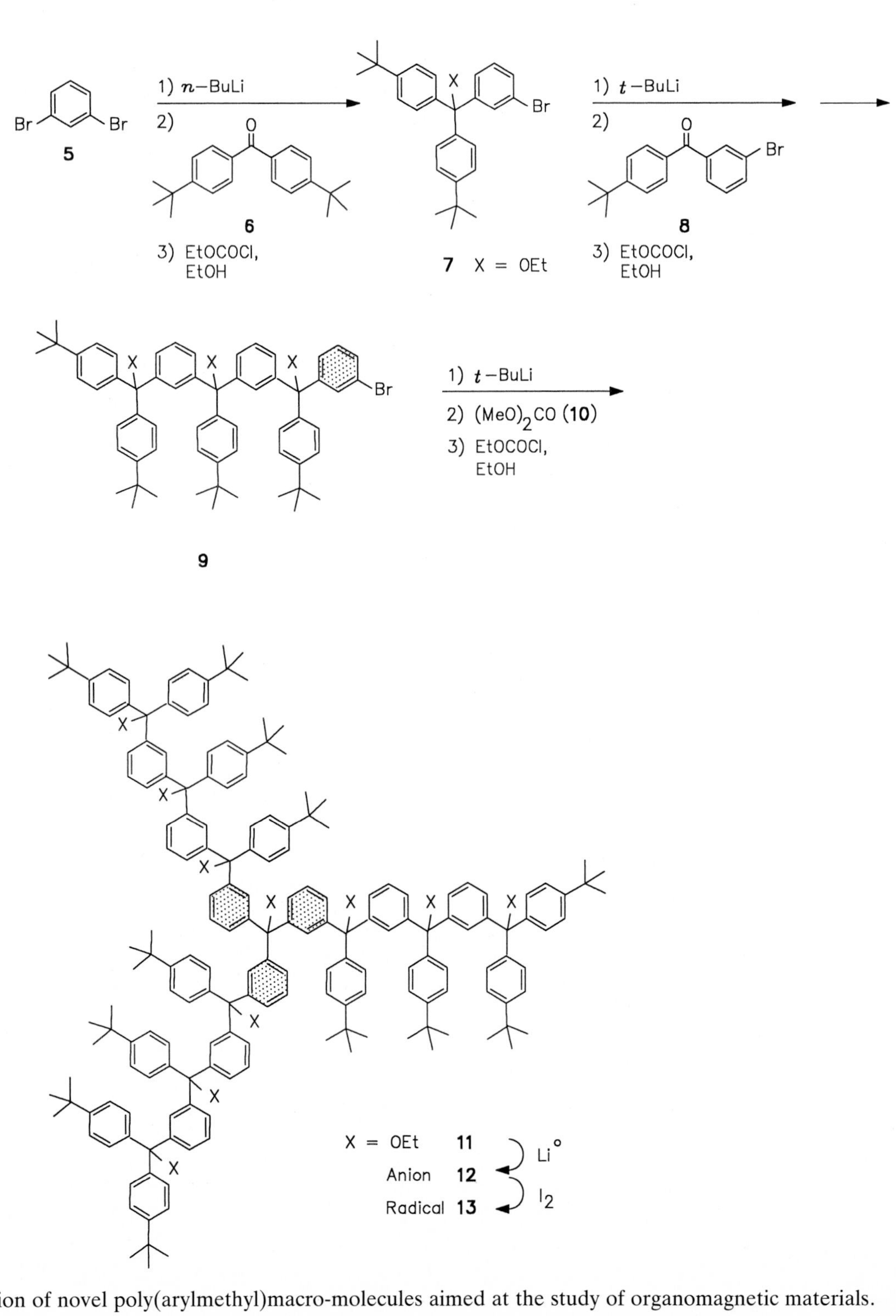

Scheme 5.2. Preparation of novel poly(arylmethyl)macro-molecules aimed at the study of organomagnetic materials.

oligomeric, linear arms. Further metallation of the heptaphenylbromide (**9**) and addition of the three-directional core **10** was followed by treatment with EtOCOCl and ethanol to give the desired poly(arylmethyl)decaether (**11**). The corresponding decaanion **12** was generated by reaction of **11** with lithium metal in THF and its subsequent oxidation with iodine afforded the related decaradical **13**.

The largest molecule yet reported in this series had an extended diameter of 30 Å and a molecular mass of ca. 2,800 Daltons. Decaether **11** and all the precursors were fully characterized; the corresponding decaradical **13** was supported[17] by ESR and magnetic susceptibility data. Further studies[18] of these unique macromolecules with respect to the "topological control of electron localization in π-conjugated carbopolyanions and radical anions" demonstrated the 'relationship between high spin in polyradicals, poly-anion charge distribution, and electron localization in the radical anions'. 1,3-Bridging benzene units result in weaker coupling between arylmethyl moieties than with the 1,4-benzenoid coupling; thus, controlled molecular construction effects electron localization. The magnetic interactions of the corresponding polyarylmethyl triplet diradicals were also studied.[19a] Inoue et al.[19b] have studied trinitroxide radicals stabilized by Mn coordination that behave as molecule-based magnets ($T_c = 46$ K).

Rajca and Utamapanya have further expanded[20] their iterative procedures to include the unprecedented synthesis (Scheme 5.3) and characterization of dendrimers possessing 15 and 31 centers of unpaired electrons. The convergent synthesis of these unique materi-

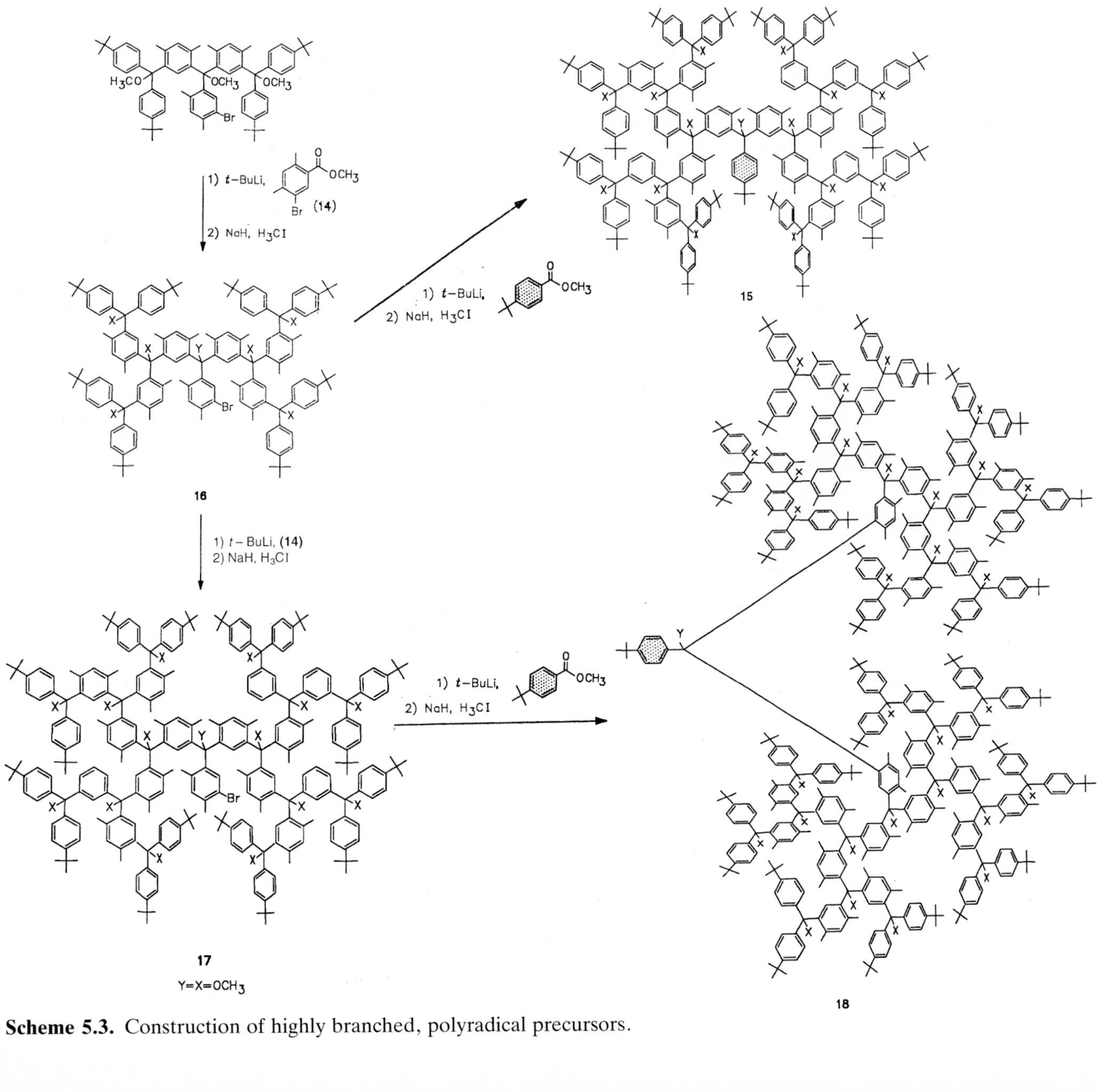

Scheme 5.3. Construction of highly branched, polyradical precursors.

als relied on the iterative application of aryl lithiation via metal−halogen exchange (*tert*-BuLi), addition of an alkyl benzoate (**14**; methyl 5-bromo-2,4-dimethylbenzoate), and etherification (MeI). Thus, pentadecaether **15** (X = Y = OMe) was prepared by the reaction of two equivalents of heptamethyl ether **16** (X = OMe) with methyl 4-*tert*-butylbenzoate and subsequent methylation of the resulting central hydroxyl moiety. Treatment of two equivalents of metallated arylbromide **16** with the benzoate **14** lead to the monobromopentadecaether **17**, which when lithiated and added to methyl *tert*-butylbenzoate afforded the 31-methyl ether **18**. The syntheses, via the repetitive addition of aryllithiums to carbonyl precursors, of a related series of sterically hindered 1,3-connected tri-, tetra-, and heptaarylmethanes, have also been reported.[21,22]

Synthesis of the closely related acyclic (**19**) and macrocyclic (**20**) polyradicals has recently been reported (Figure 5.1).[23] The π-conjugated carbanions (e.g., the calix[4]-arene-based tetraanion and the related calix[3]arene-based trianion) were synthesized and studied.[24] Oxidation of these tetra- and tri-anions gave the corresponding tetra- and tri-radicals, respectively. It has been shown in closely related systems that it is not the shape or overall geometric symmetry of the molecules, but rather it is the juxtaposition of the carbenic centers within the π-cross-conjugated structure, that is most important in determining the spin multiplicity of the alternant hydrocarbon molecule.[25]

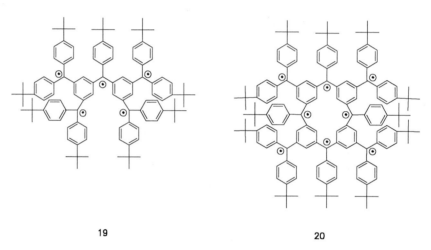

19   20

Figure 5.1. Cyclic and acyclic polyradicals were designed to minimize defect impact on conjugation.

5.2.2 1 → 2 *C*-Branched, *Ether* Connectivity

Wooley, Hawker, and Fréchet[26] described the synthesis of dendrimers via a 'double-stage convergent growth method. Essentially, this approach leads to dendrimers comprised of different monomers at different generations. Dendrons (e. g., **21**) were convergently prepared (Scheme 5.4) using 4,4-bis(4'-hydroxyphenyl)pentanol (**22**), which was obtained from the corresponding pentanoic acid (**23**) by four simple steps. Reaction of two equivalents of bis(benzyl)-protected phenol **24**, prepared from the same common acid **23**, with bis(hydroxyphenyl)pentanol **22** afforded the new 1 → 2 branching dendron **21**. Using these dendrons or wedges, 'hypercores' up to three generations were constructed. The simplest of these flexible cores, **25**, was prepared by treatment of alkyl bromide **24** with the three-directional, tris(phenolic) core **26**, as depicted in Scheme 5.5.

After debenzylation (Pd−C, cyclohexene) of the hypercores (e.g., **25**) by transfer hydrogenation, they were treated with aryl branched, benzyl ether dendrons (Scheme 5.6) that were prepared by similar iterative transformations,[27] i.e., benzylic bromination and phenolic *O*-alkylation (See: Section 5.4.2). Thus, hexaphenol core **27** was reacted with six equivalents of the benzylic bromide building block **28** to give the benzyloxy terminated dendrimer **29**. Key features of these dendrimers include cores with flexible alkyl spacers and a three-directional, quaternary carbon branching center.

Further, this was the first example of dendritic polymer syntheses that involved the use of two different monomers in a convergent synthesis leading to a species possessing a

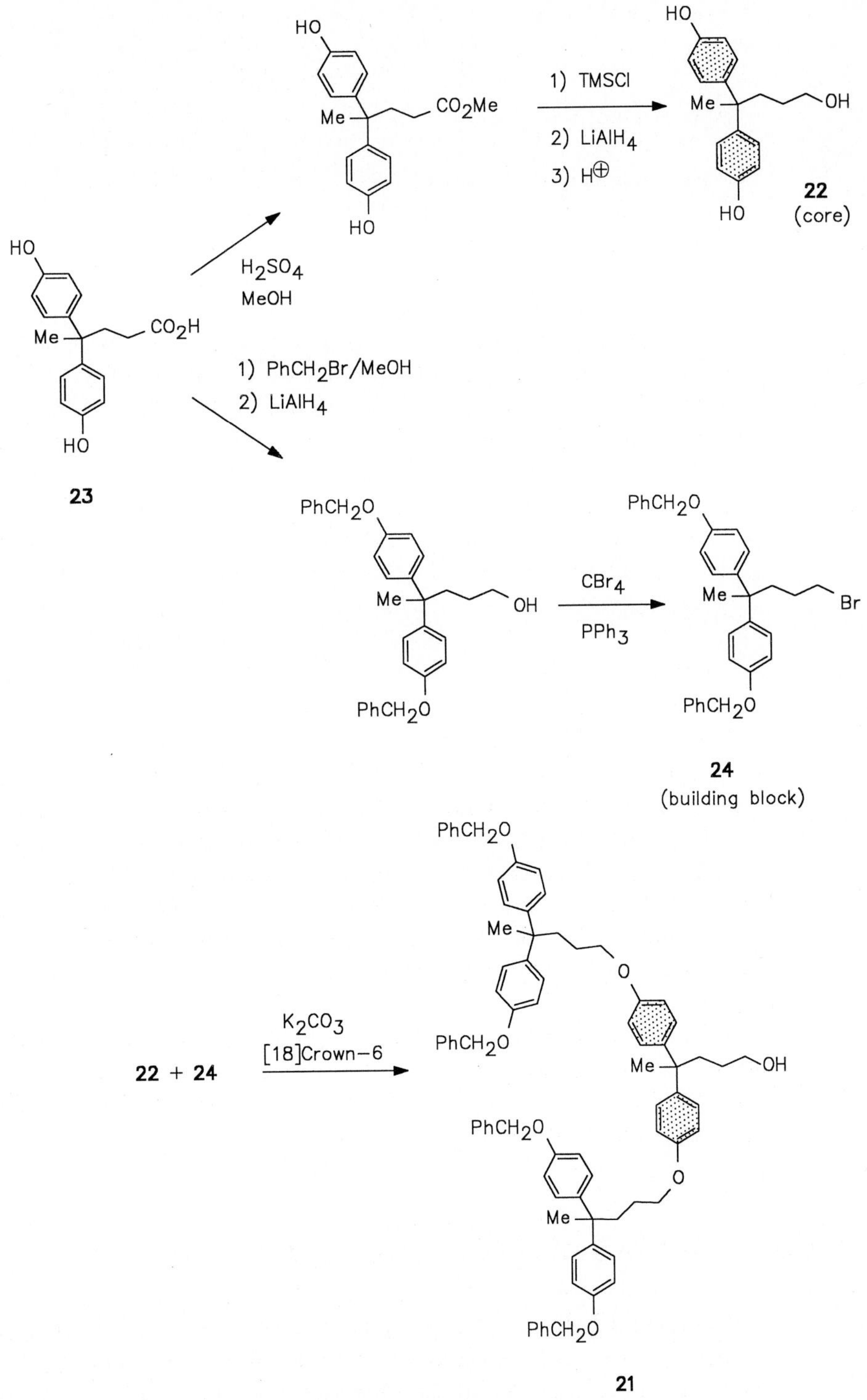

Scheme 5.4. Synthesis of dendritic "wedges" to be used in "double stage convergent growth".

nominal molecular weight of ca. 84,200 amu. Variations of glass transition temperature with molecular weight and chain-end composition for related dendritic polyethers (as well as polyesters) were also studied.[28]

Yamamoto et al.[29] reported the preparation of a simple one-directional, two-tier cascade (**30**) for the purpose of solubilizing a terminal *o*-carborane moiety (1,2-RHC$_2$B$_{10}$H$_{10}$) in an aqueous environment. Convergent cascade construction is shown in Scheme 5.7. The dibenzyl ether **31** was reacted with epichlorohydrin to afford the second tier tetrabenzyl ether (**32**) possessing 1 → 2 *C*-branching centers. Subsequent treatment with propargyl bromide gave the terminal acetylene, which was then treated with decaborane (B$_{10}$H$_{14}$/MeCN), and deprotected (Pd(OH)$_2$) to provide the water-soluble, terminal *o*-carborane cascade **30**.

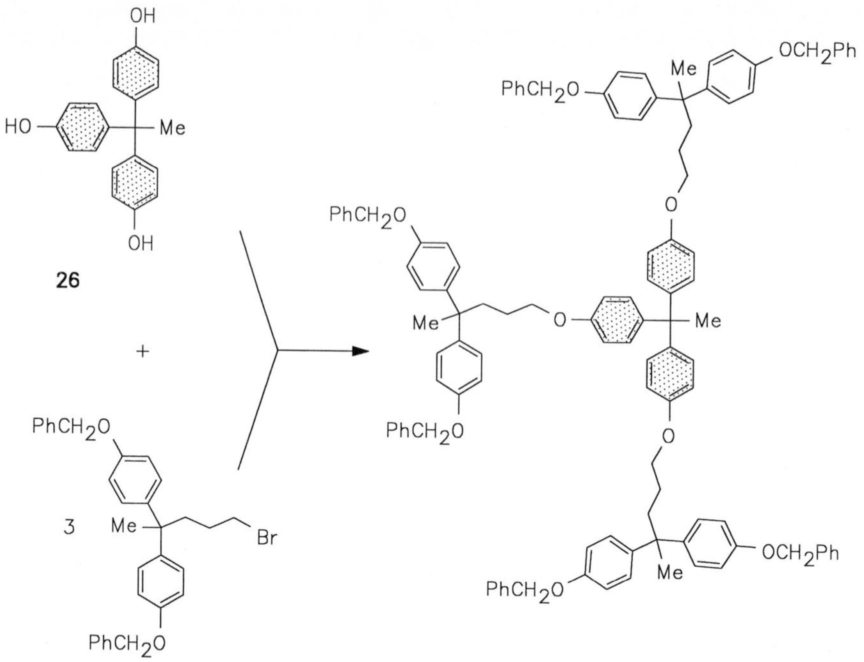

Scheme 5.5. Construction of flexible "hypercores".

Nemoto, Yamamoto, and Cai[30a] later modified the preparation of their water-soluble carborane to include the attachment of a tumor seeking uracil moiety (Scheme 5.8). Key transformations allowing the synthesis of this unique dendritic carborane (**33**) included construction of masked uracil allyl carbonate **34** and its subsequent connection to the benzyl protected *o*-carborane cascade **35**, the intermediate precursor to tetraol **30**, via palladium bis(dibenzylideneacetone) [Pd(dba)$_2$] and 1,2-bis(diphenylphosphino)ethane (dppe) mediation.

Percec et al.[30b] have employed 1 → 2 *C*-branching and ether-type connectivity for the preparation of nonspherical, thermotropic liquid crystalline dendrimers. Observed thermotropic behaviors are predicated on mesogenic monomers that are capable of conformational isomerism.

5.3   1 → 2 *Ethano*-Branched, *Phosphate* Connectivity

Damha and Zabarylo[31] reported a general method for the preparation of singly branched RNA and DNA oligonucleotides involving oligomer synthesis on a solid-phase, controlled-pore glass support. Branching of the bound, linear hydroxy-terminated oligomers was affected by treatment with a tetrazole-activated, 2′,3′-adenosine bisphosphoramidite derivative (**36**) via phosphate ester formation. Various dendrimers were constructed[32] based on thymidine and adenosine building blocks, including an 87-unit dendrimer (**37**) possessing a molecular weight of ca. 25,000 amu (Scheme 5.9). The convergent procedure was successful through the third generation; however, there was an increasing presence of incomplete structures suggesting[32] that the divergent approach might be more advantageous in this case.

5.4   1 → 2 *Aryl*-Branched

5.4.1   1 → 2 *Aryl*-Branched and Connectivity

At about the same time as Hawker and Fréchet's publication[1] of the convergent methodology, Miller and Neenan reported[3] the synthesis of monodisperse molecular

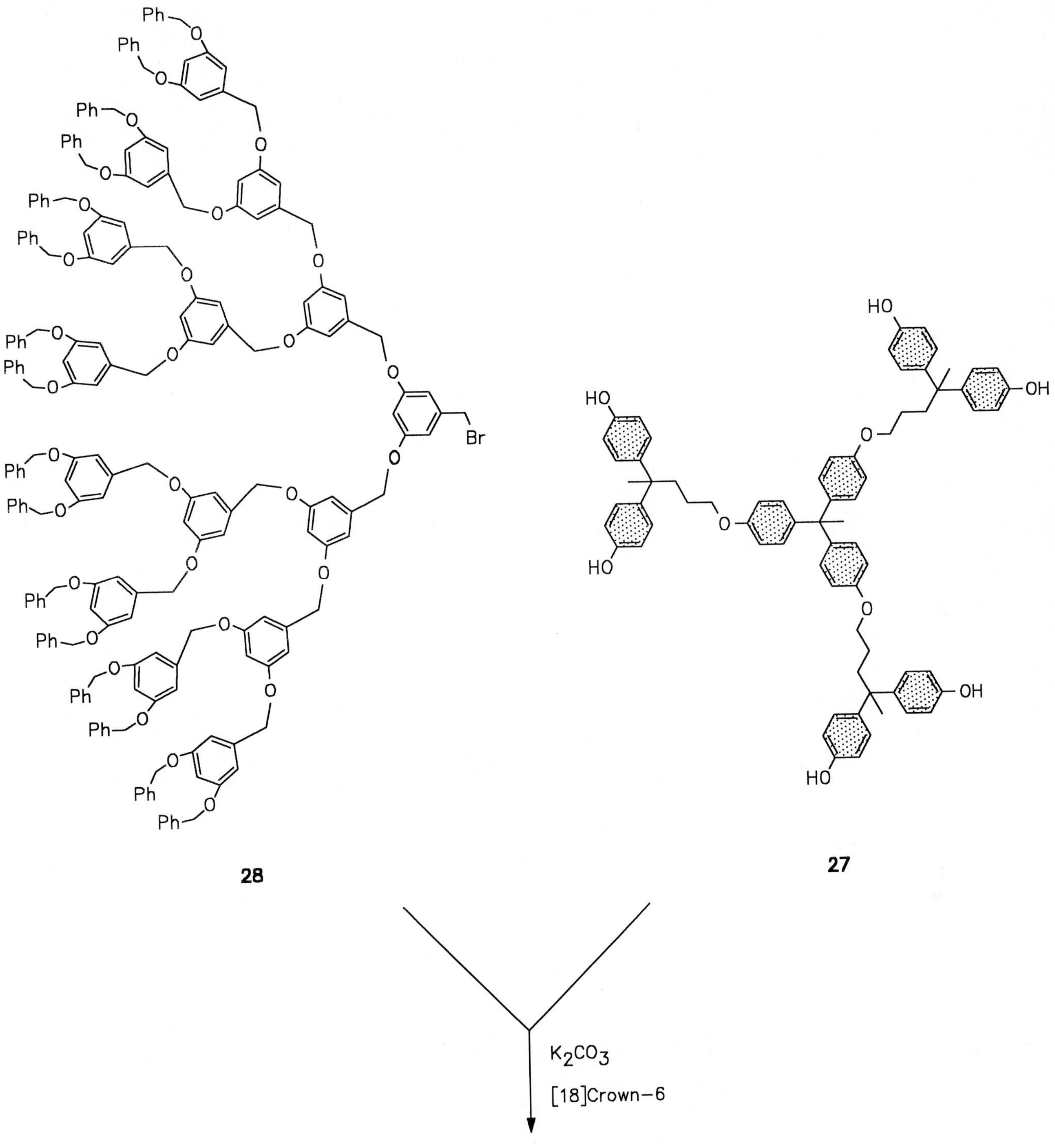

28

27

K₂CO₃

[18]Crown−6

Scheme 5.6. (Continued on page 114).

spheres based on 1,3,5-trisubstituted benzene (Scheme 5.10). The completely aromatic hydrocarbon **38** was prepared using 1,3-dibromo-5-(trimethylsilyl)benzene (**39**), as the key building block. Two equivalents of phenylboronic acid were catalytically coupled using tetrakis(triphenylphosphine)palladium(0) to the dibromobenzene **39** yielding ter-phenylsilane **40**, in which the silyl group was subsequently transformed to the corresponding boronic acid moiety and further coupled to monomer **39**. The resultant heptaphenyl trimethylsilyl dendritic arm (**41**) was then boronated and reacted to a three-directional core (e.g., 1,3,5-tribromobenzene, **42**) to afford the hydrocarbon dendrimer **38**, termed a 12-Cascade:benzene[3-1,3,5]:(1,3,5-phenylene)²:benzene.

Miller et al.[33,34] described the preparation of the related 46-phenylene dendrimer **43**, as shown in Scheme 5.11. Employment of the aryl−aryl bond forming Suzuki reaction[35] for dendron connection was again used.[3] In that the smaller second generation, 7-phenylene boronic acid monomer (**44**) was readily accessible (from the trimethylsilyl

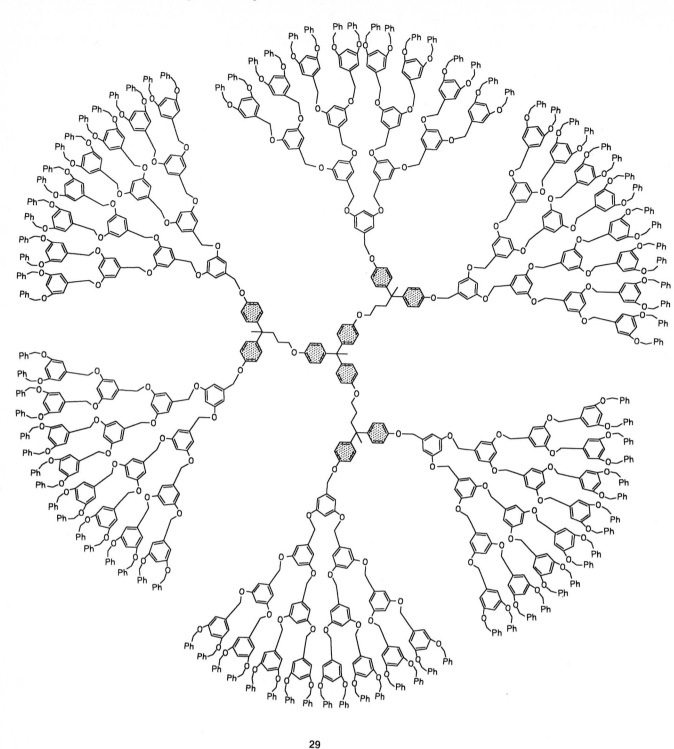

29

Scheme 5.6. Attachment of poly(benzyl ether) dendrons to flexible hypercores results in "dendritic block copolymers" possessing monomer differentiated generations.

precursor **41**), an extended hexabromotetraphenyl core **45** was prepared via the acid-catalyzed trimerization of the dibromoacetophenone **46**. Subsequent coupling of core **45** with the heptaphenylboronic acid (**44**) thus afforded the 46-phenylene hydrocarbon **43**. The fluorinated analogs were prepared[33,36] via analogous technology except that the first tier derivative was obtained, albeit in poor yield (< 5 %), from the treatment of bromopentafluorobenzene with copper at 190 °C and the tribromobenzene core **42**. Higher generations were derived in greater yields by the reaction of 3,5-bis(pentafluorophenyl)-phenylboronic acid with the tri- and hexabromoarene cores. The second generation, fluorinated dendrimer was reported to possess very limited solubility in most organic solvents; it was rationalized 'that it can adopt a pancake-like conformation in which all of the benzene rings are approximately in a plane'.

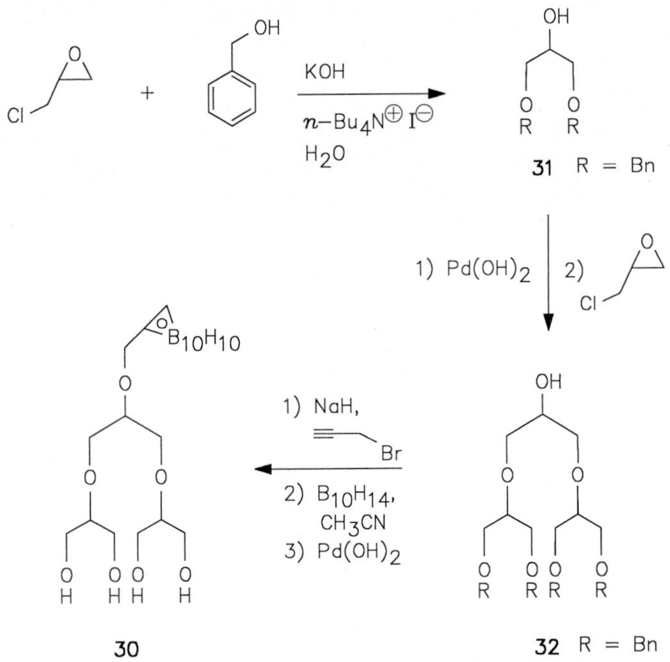

Scheme 5.7. Synthesis of water soluble *o*-carboranes for investigation as Boron Neutron Capture Therapy reagents.

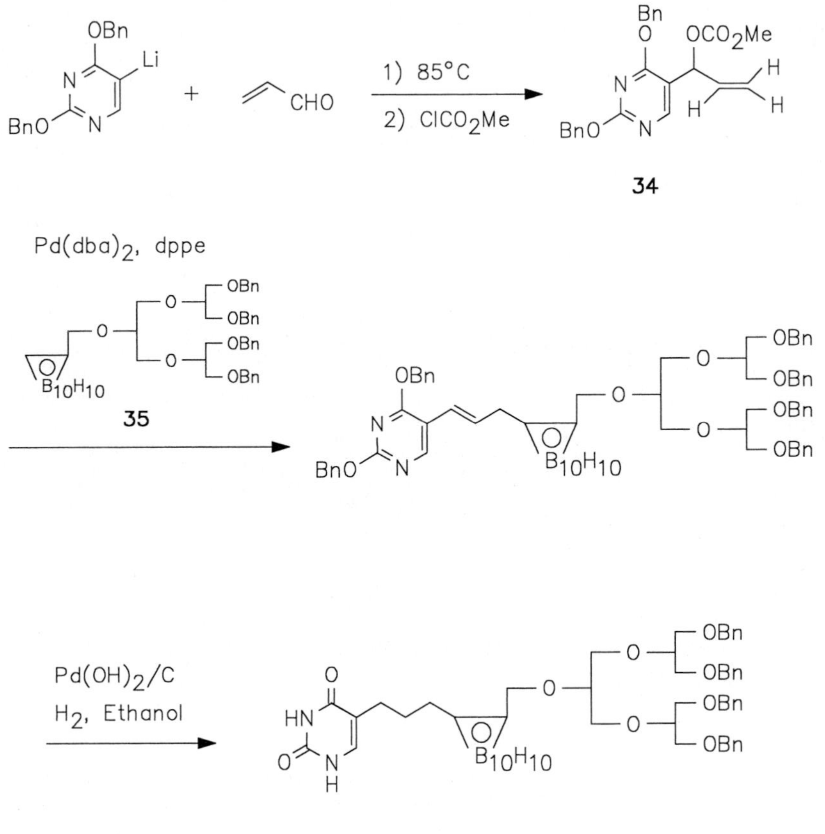

Scheme 5.8. Introduction of a tumor seeking uracil moiety onto a water-soluble *o*-carborane.

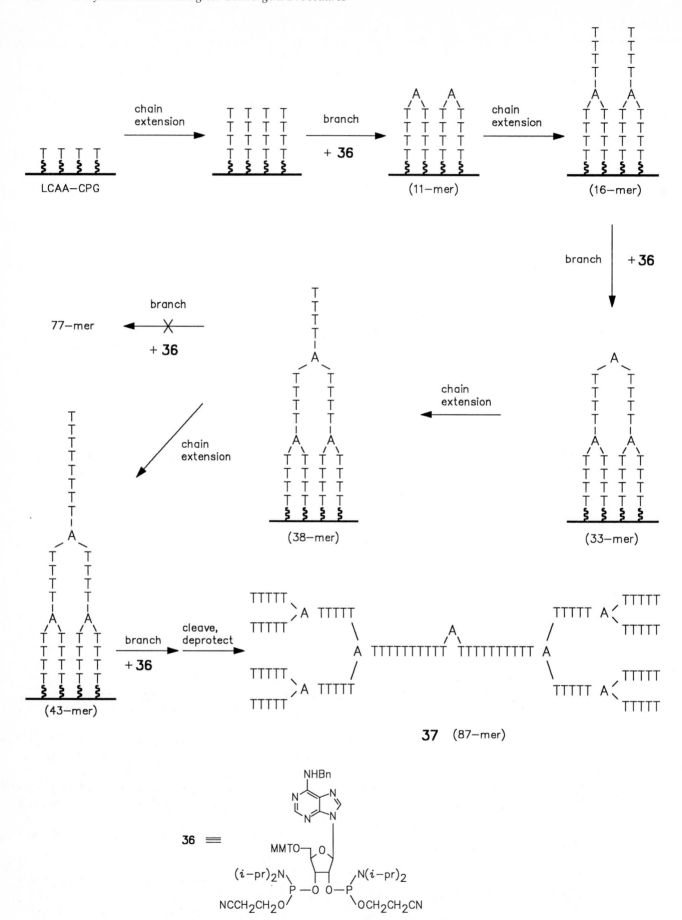

Scheme 5.9. Dendrimer synthesis employing thymidine and adenosine building blocks.

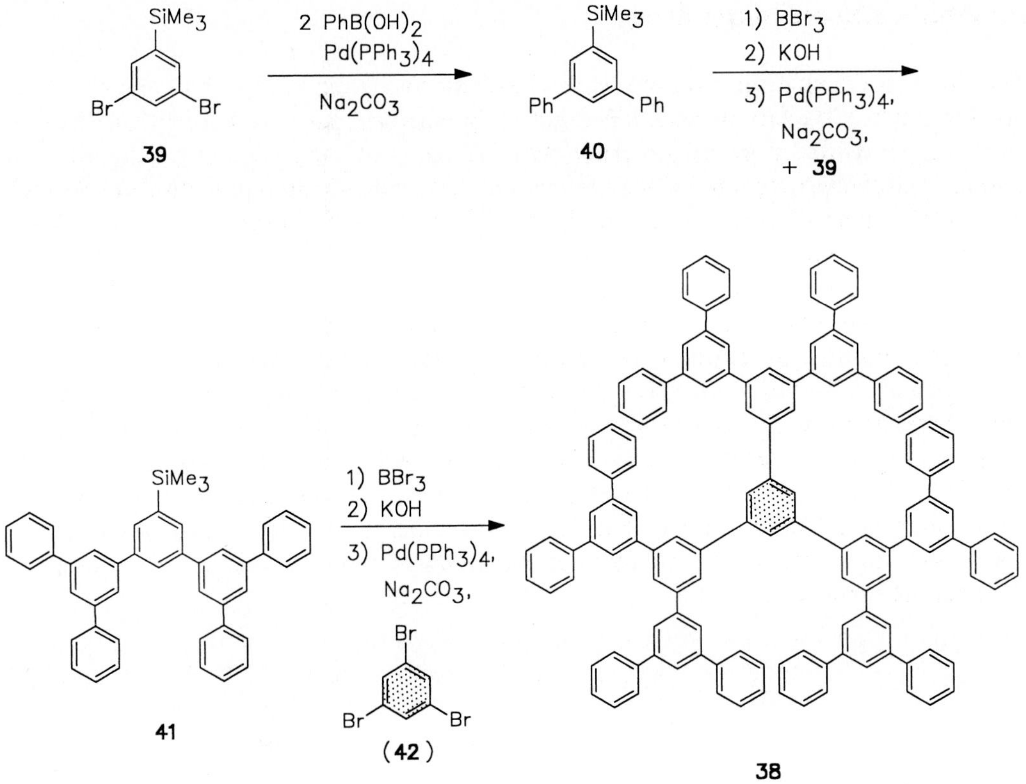

Scheme 5.10. Employment of "Suzuki coupling" leads to the construction of all aromatic hydrocarbon dendrimers.

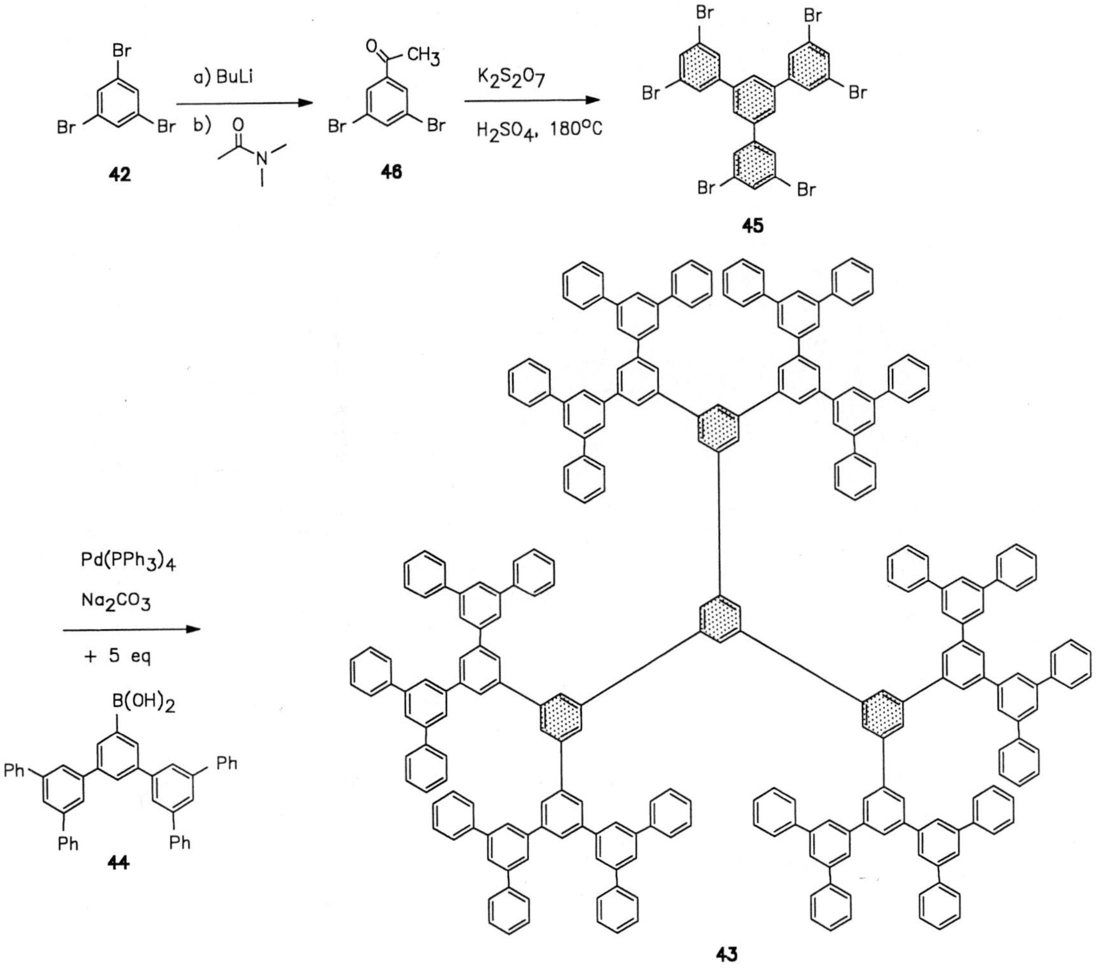

Scheme 5.11. Use of extended cores can facilitate the construction of high molecular weight dendrimers.

5.4.2 1 → 2 *Aryl*-Branched, *Ether* Connectivity

The first convergent dendritic synthesis, based on the combination of meta−aryl branching and ether connectivity, was reported[1,27] by Hawker and Fréchet in 1990. The use of two simple synthetic transformations was described: (1) the selective alkylation of phenolic hydroxyl groups; and (2) conversion of a benzylic alcohol to a benzylic bromide to generate a reactive focal moiety (Scheme 5.12). Thus, addition of two equivalents of the previously reported[37] benzylic bromide **47** to the initial monomer **48** yielded the heptaphenyl ethereal dendron, which was converted (CBr_4, $P(C_6H_5)_3$) to the corresponding bromide **49**. This sequence was successfully repeated to prepare dendritic wedges or dendrons up through the sixth generation. These one-directional cascades were attached to the three-directional core **26** to afford a benzylic ether-based dendritic series represented by **50**, the largest of which possessed 192 terminal phenyl rings, a molecular weight of 40,689 amu, and whose series could be described as: 192-Cascade: ethane[3-1,1,1]:(5-(*p*-phenyl-2-oxaethyl)phenylene):(5-(2-oxaethyl)1,3-phenylene)⁵:(2-oxaethyl)benzene. These easily accessible dendritic arms and their facile attachment to diverse cores have been used by many research groups to initiate their entrance to macromolecular chemistry.

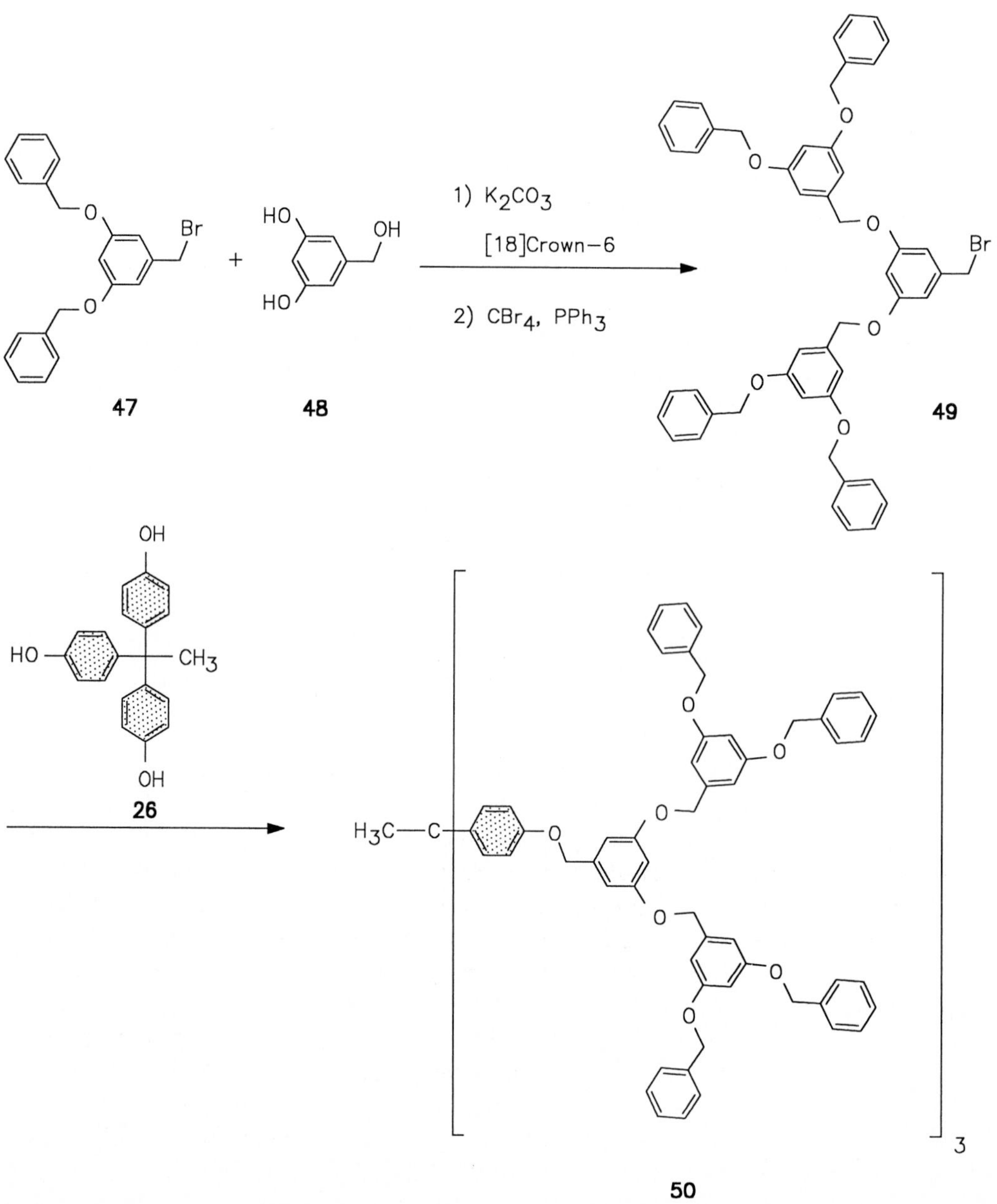

Scheme 5.12. Fréchet et al.'s[1,27] original convergent methodology.

Mourey et al.[38] evaluated the viscosity of these polybenzyl ether dendrimers[1,26] using size exclusion chromatography (SEC) coupled with differential viscometry (DV). For the 0–6th generations, the intrinsic viscosity for these three-directional cascades (e. g., **50**) was determined to pass through a maximum at the third generation; whereas, the refractive index was recorded as passing through a minimum at ca. the second generation. These data suggest a monotonic decrease in density from the center of these polyethers. These results support Lescanec and Muthukumar's[39] theoretical model of dendrimers, which indicates higher internal density, relative to surface density, but are contrary to the de Gennes and Hervet model,[40] which suggests greater surface congestion.

Saville et al.[41] studied the $\Pi - A$ (surface pressure−area) isotherms at the water−air interface at 25 °C for monomolecular films of the ethereal dendrimer series (i.e., the benzylic alcohol series represented by structures **50** in Scheme 5.12). They reported a strong dependence of the isotherms on molecular weight, which compared well with that observed for hydroxyl-terminated polystyrene.

Hawker et al.[42] have investigated the melt viscosity behavior for this family of dendrimers generated from 3,5-dihydroxybenzyl alcohol. In general, an evaluation of the melt viscosity for this series revealed a profile with no critical molecular weight for molecules as high as 85,000 amu. At high molecular weight, branching and surface congestion in this family prevent appreciable intermolecular entanglement affording "ball bearing-like" macromolecules, which are capable of only interdigitation.

Gitsov et al.[43,44,45] reported the syntheses (Scheme 5.13) of dendritic polyether co-polymers (e.g., **51**) by the treatment of the benzyl ether wedges (e.g., the fourth generation dendron **52**) with poly(ethylene glycol) (PEG) and poly(ethylene oxide) (PEO). It was determined that the rate constants for the reaction of PEG or PEO with bromo dendrons of various sizes increased as dendritic generation and linear block length increased. Synthesis of the related linear dendritic 'triblock' copolymers (i.e., **53**)[46–49] was accomplished via anionic polymerization of styrene, initiated by potassium naphthalide, modification of the resulting 'living' polystyrene with 1,1-diphenylethylene and finally quenching with a reactive dendron, such as bromide **52** (Figure 5.2).

The attachment of preformed polyethereal dendrons (e. g., **70**, see Scheme 5.17), which have been functionalized at the focal point by reaction with *p*-(chloromethyl)styrene, has been shown[50] to undergo free radical copolymerization with styrene giving rise to polystyrene with appended benzyl ether dendritic wedges.

Wooley et al.[51] have described the creation of fullerene-bound dendrimers (Scheme 5.14). Reaction of C_{60} with bis(*p*-methoxyphenyl)diazomethane **54** and subsequent cleavage of the methyl ethers afforded a 6-6 methano-bridged fullerene (**55**) possessing two phenolic moieties, as the major product.[52] Treatment of the bisphenolic fullerene **55** with 2.7 equivalents of the activated dendron (**52**) afforded the desired substituted fullerene (**56**) possessing two dendritic arms.

Similarly, treatment of C_{60} in refluxing dry chlorobenzene with dendrimer **57** possessing an azide focal point, prepared from the corresponding bromide (**52**) with sodium azide in DMSO, gave rise (69 %) to a controlled, one-step cycloaddition affording the amine substituted fullerene **58**.[53]

The use of functionalized or activated focal points or cores has been one of the most interesting attributes of the convergently generated dendritic macromolecules. Researchers can thus probe a specific or specifically created microenvironment; one such study[54] attached a solvachromatic probe (4-(*N*-methylamino)-1-nitrobenzene) at the inner focal position of a series of related ethereal dendrimers (Figure 5.3; **59**). Variations in the absorption maximum with increasing generations and different solvents demonstrated that the microenvironment near the focal point (core) had a high polarity especially with the more spherical dendrimers, i.e. the fourth or higher generation.

The coupling of two different (e.g., one coated with cyano and the other coated with benzyl ether moieties) convergently built aryl ether dendrons has been demonstrated[55,56] to generate monodisperse dendritic macromolecules possessing enhanced dipolar moments (**60–64**; Figure 5.4). Dendritic block copolymers possessing amphiphilic character such as macromolecule **65** have also been reported (Scheme 5.15).[57–59] Construction was based on the attachment of hemispherical wedges possessing lipophilic (**66**) or hydrophilic (precursor **67**) surface groups to different cores. Surfactant studies employ-

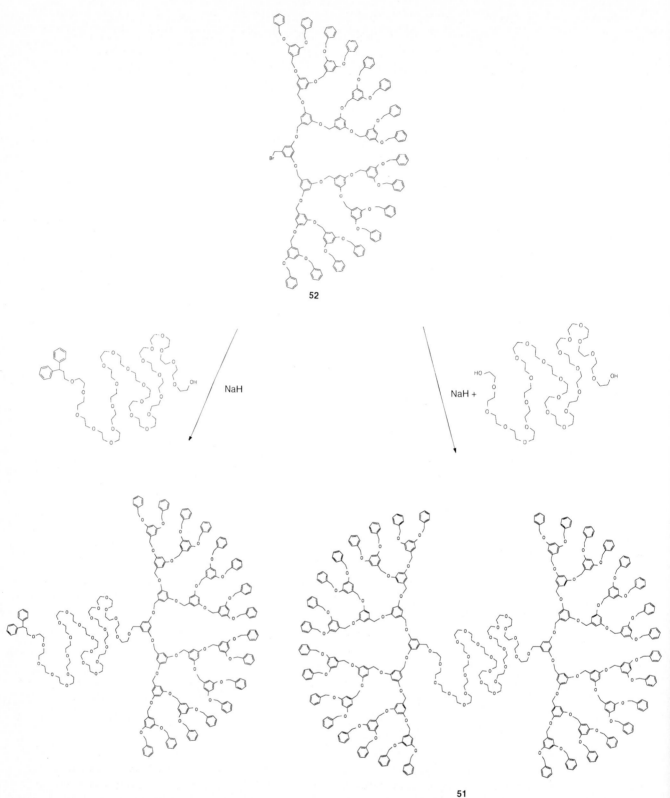

Scheme 5.13. The rate of reaction of Fréchet-type dendritic wedges with poly(ethylene oxide) [PEO] and poly(ethylene glycol) [PEG] was determined to increase dendron generation and block length increased.

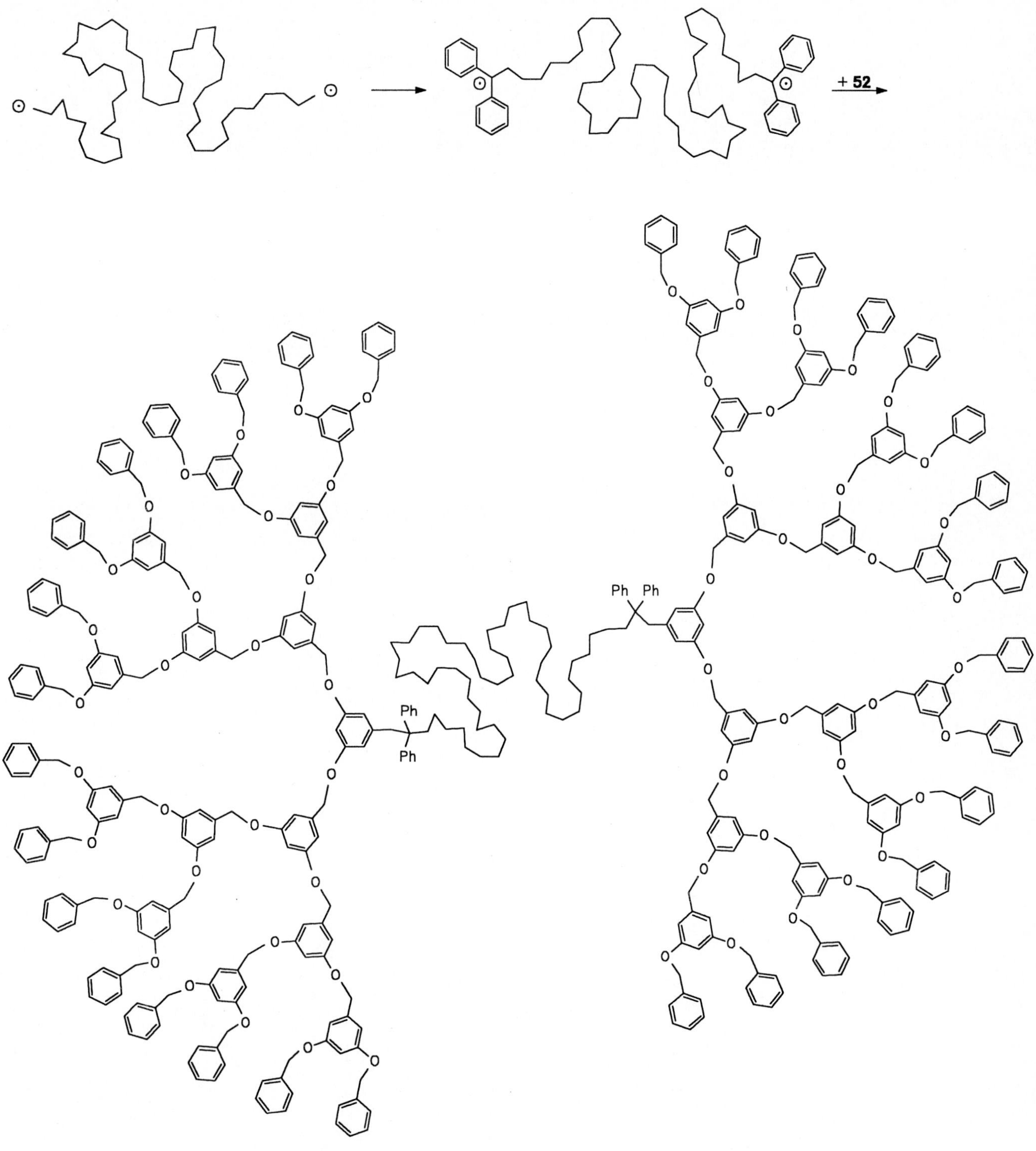

53

Figure 5.2. Linear dendritic triblock copolymers prepared via anion trapping with benzyl bromide focal groups.

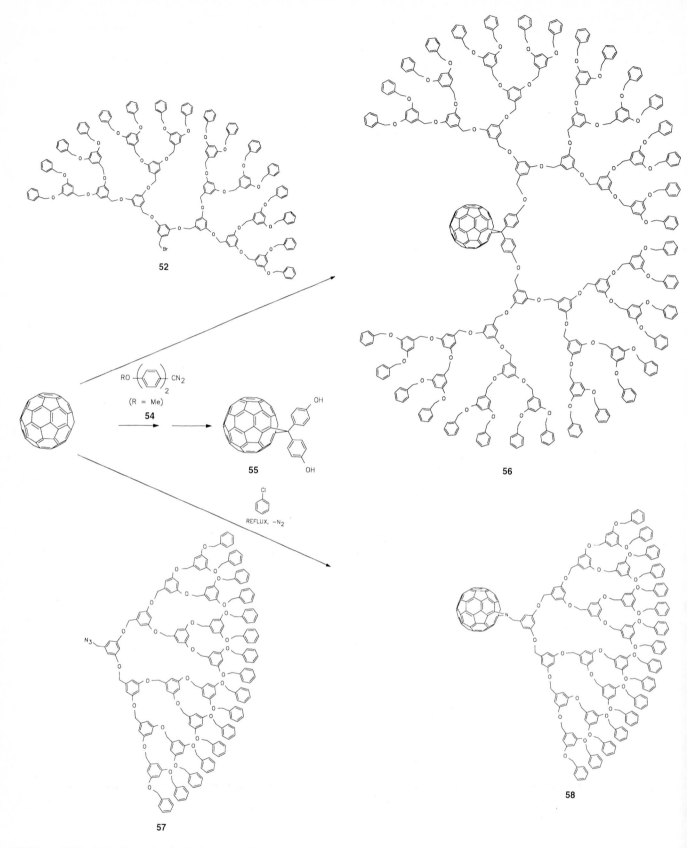

Scheme 5.14. Poly(benzyl ether) dendrons have been attached to fullerenes via carbenoid-type chemistry.

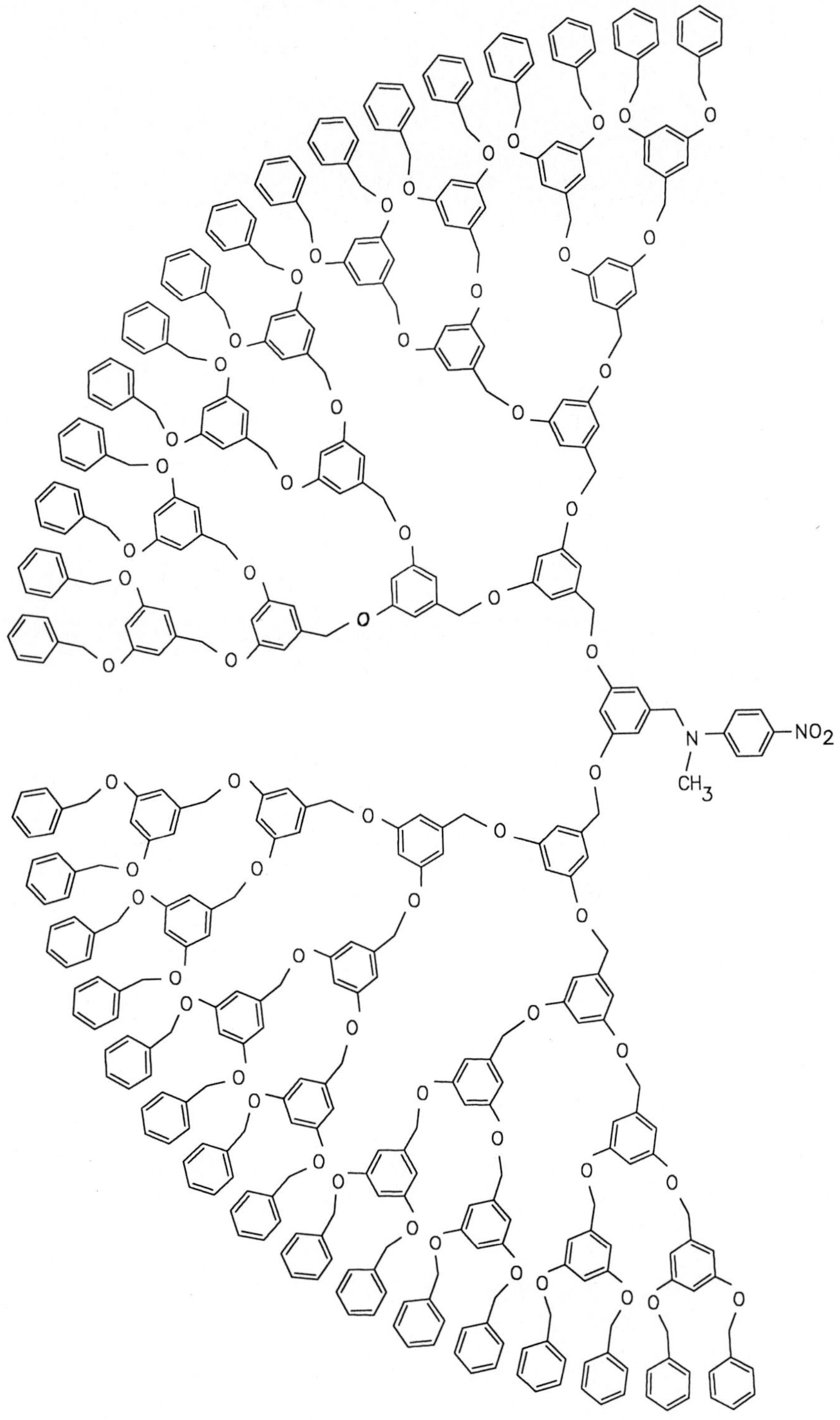

59

Figure 5.3. The microenvironment of solvachromatic probes, such as the illustrated *N*-alkylated nitroaniline, can be modified by employing dendritic superstructures.

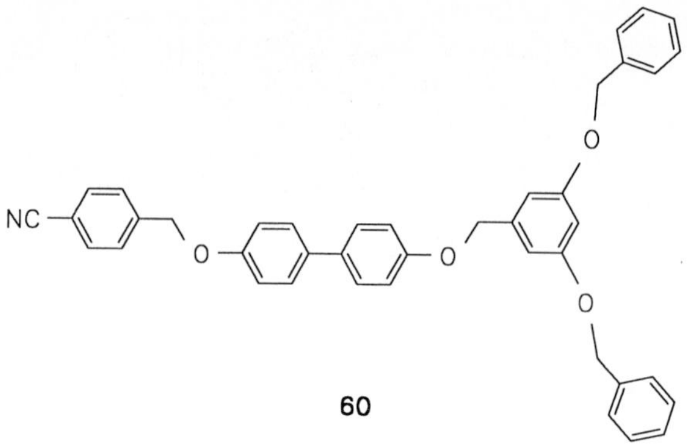

60

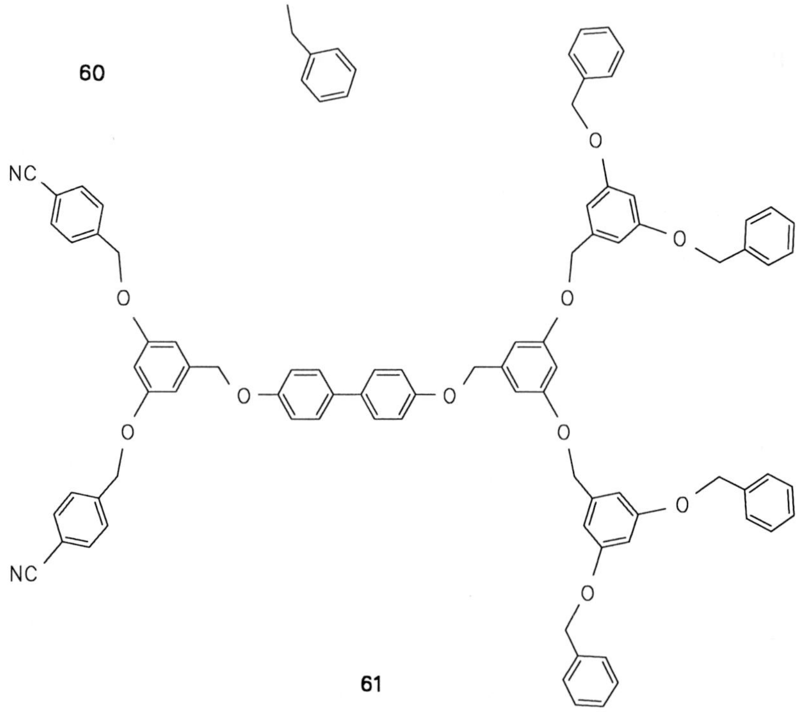

61

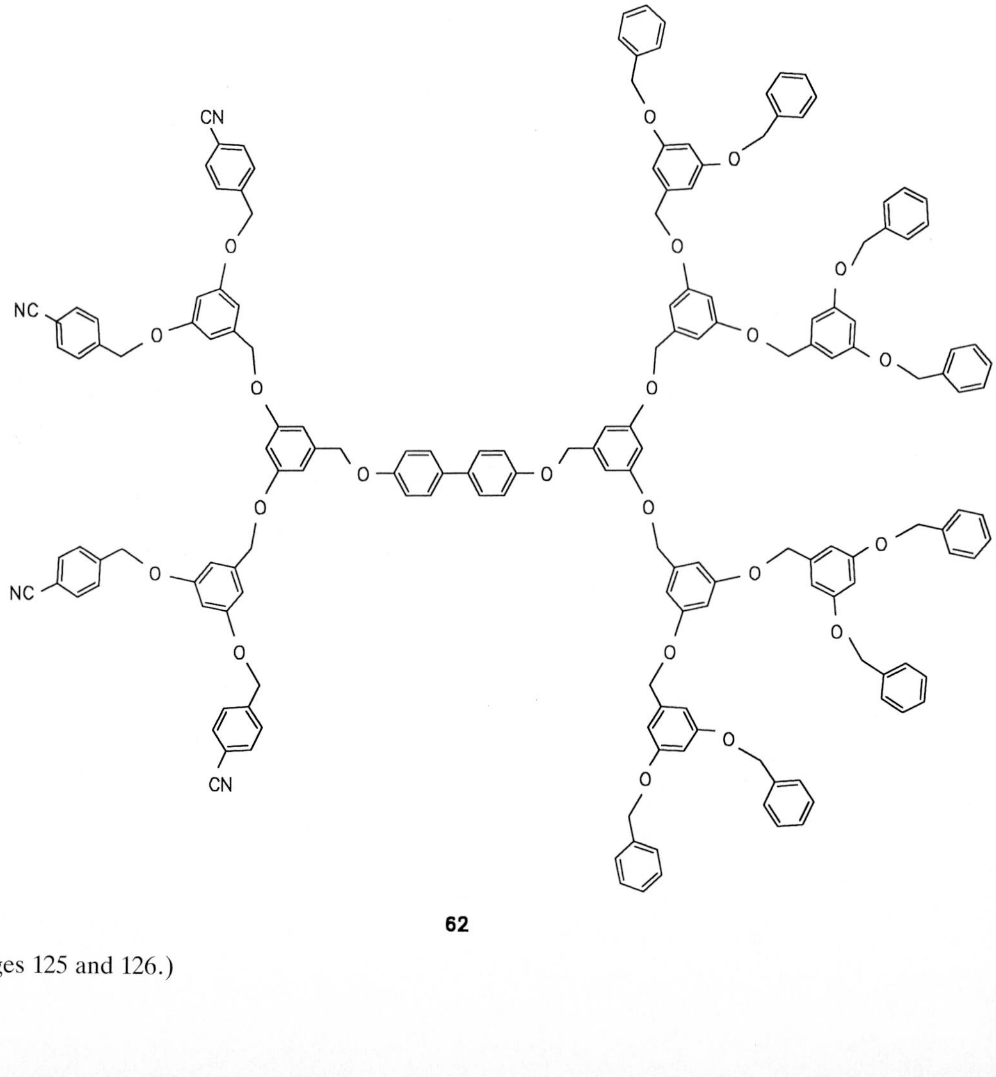

62

Figure 5.4. (Continued on pages 125 and 126.)

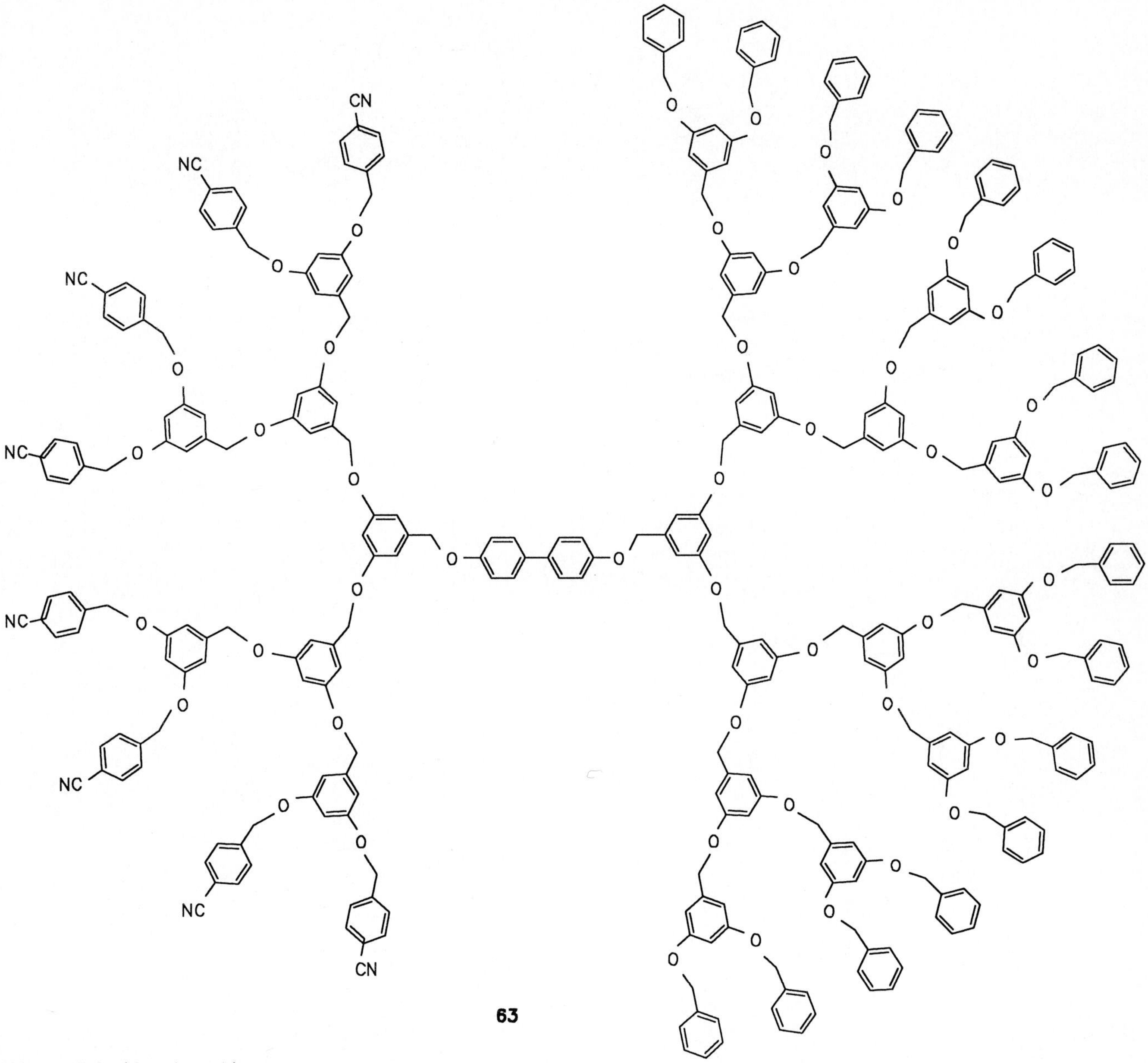

63

Figure 5.4. (Continued.)

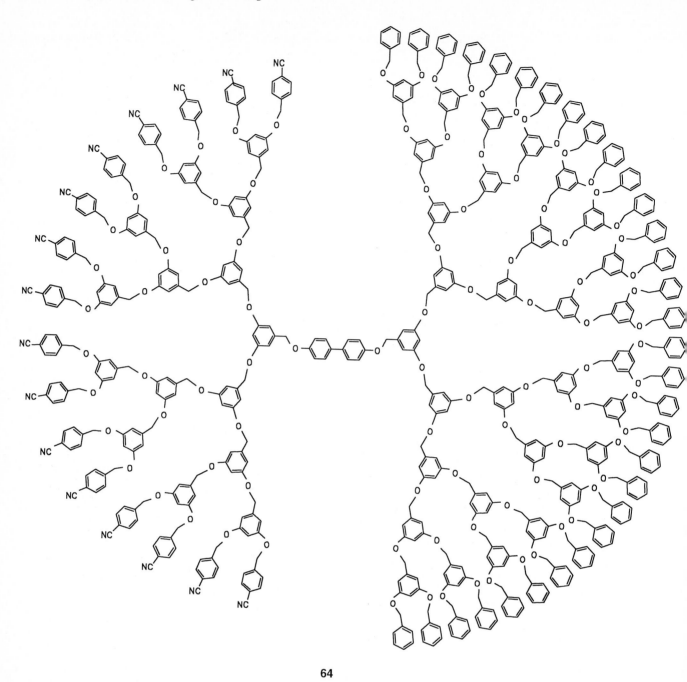

64

Figure 5.4. Surface modified dendrimers give rise to enhanced macromolecular dipole moments.

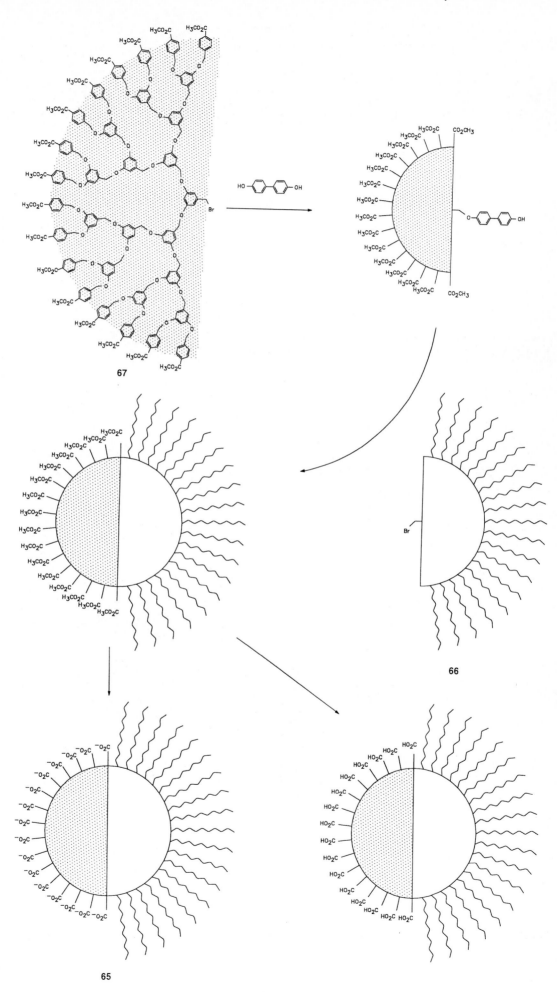

Scheme 5.15. Convergent methodology has facilitated the construction of amphiphilic micellar dendrimers.

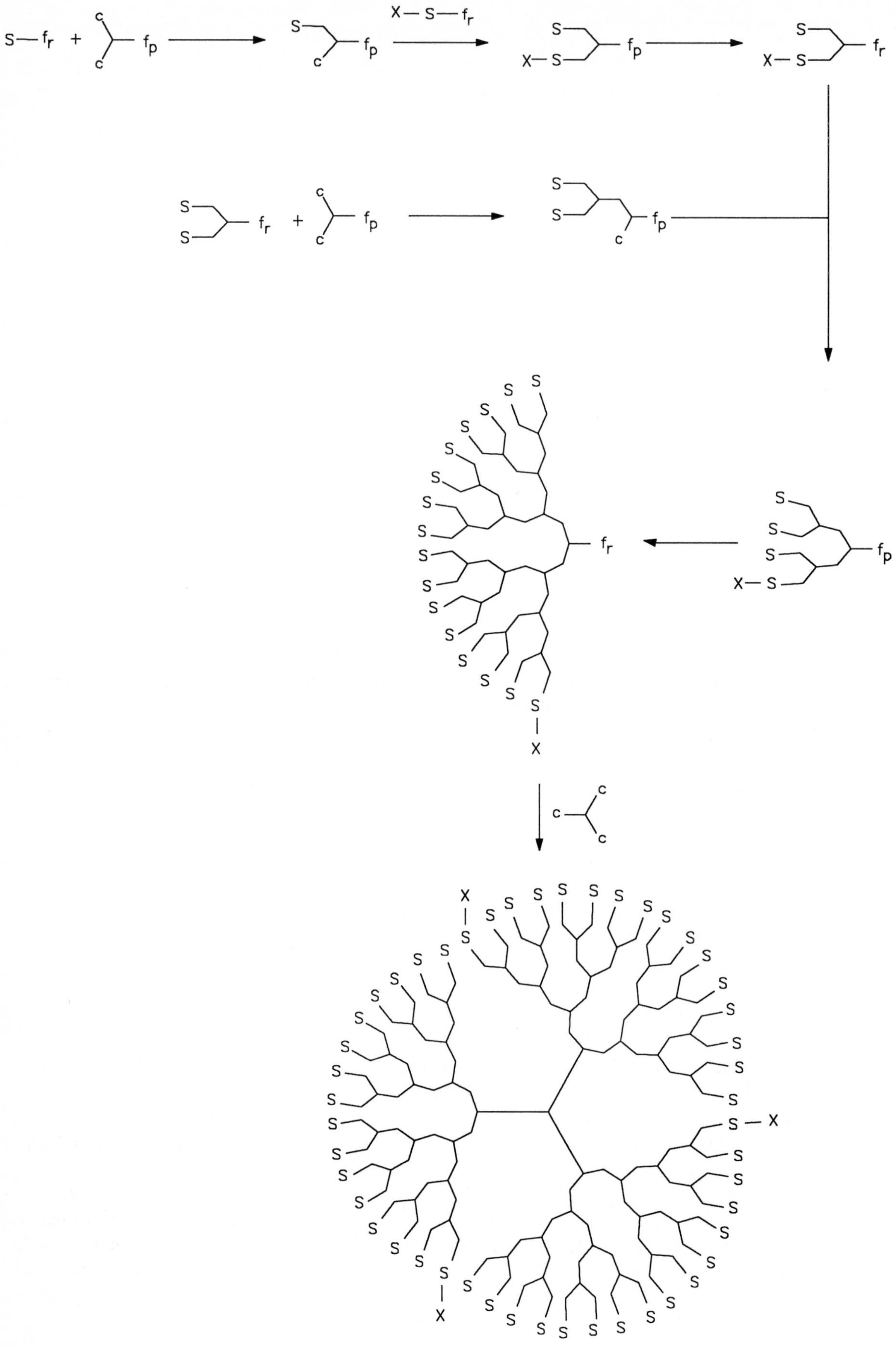

Scheme 5.16. General strategy for the construction of dendritic wedges possessing single, differentiated surface groups.

ing these dendritic amphiphiles, as micellar substrates, indicated a significant increase in the aqueous saturation concentration of polycyclic aromatics.

Fréchet et al.[60,61] reported the control of surface functionality with respect to the number and placement of terminal groups using the convergent method and ethereal connectivity. Dendrons were constructed with single cyano, bromo or carboxymethyl moieties at the surface. The preparative strategy for mono- di-, and tri-substituted cascades is illustrated in Scheme 5.16, in which c, f_p, f_r, S, and X represent coupling sites, protected functionality, reactive functionality, surface groups, and surface functionality, respectively. An example of this methodology, utilizing termini differentiated building blocks is given in Scheme 5.17. Thus, beginning with cyano substituted benzylbromide **68** and phenoxybenzyl alcohol **69**, the fourth tier cyanohalide cascade **70** was constructed in 5 steps.

Scheme 5.18 illustrates the preparation of mono- (**72**), di-(**73**), and tri-surface modified (**74**) dendrimers. Hence using the fourth tier, cyano-substituted cascade **71** and the unsubstituted analog (not depicted) for attachment to the trihydroxy core **26** the desired substituted dendrimers (**72–74**) were obtained.

Fréchet et al.[62a] further investigated the functionalization of the poly(benzyl ether) dendritic superstructure or framework (Scheme 5.19). For example, treatment of mono-alcohol **70** {[G-4]-OH; fourth tier alcohol} with a 1:3 mixture of *n*-butyllithium and potassium *tert*-pentoxide (THF, hexane, $-85\,°C$) gave the polypotassium salt **75**. Quenching with electrophiles that included D_2O, Me_3SiCl, $C_{18}H_{37}Br$ or CO_2 provided the functionalized dendrimer **76**. However, electrophilic addition was not site-specific, although multisite substituted dendrimers were produced.

Inoue et al.[63] reported the attachment of preconstructed polyethereal dendrons, prepared using the method of Fréchet,[27] to an eight-directional porphyrin core (**77**) to form the dendritic porphyrin **78** (Scheme 5.20). Attachment of the wedges afforded a series of photochemically active porphyrins buried within an increasingly crowded environment. Characterization of these macromolecules included the use of SEC, FAB MS and the photochemical investigation of the corresponding zinc complexes. For the larger dendrimers, access to the core by relatively small quenchers, such as vitamin K_3, was facile, whereas with larger quenchers, such as the first tier dendrimer, core access was effectively denied. Fluorescence quenching of these dendrimers using methyl viologen has also been investigated by Aida et al.[62b]

L'abbé et al.[64] combined the novel trithiolate core (**79**), prepared in three steps from triacylbenzene, with Fréchet's first and second generation ethereal benzyl bromides, to yield the rigid, trialkyne core dendrimers **80** and **81**, respectively (Scheme 5.21). These symmetrical products were structurally supported by NMR and MS techniques.

Schlüter et al.[65,66] have recently used rod-shaped polymers[67a] such as poly(*p*-phenylene)s and poly([1.1.1]propellane)s as cores for the attachment of convergently generated dendrons. These new branched macromolecules possess a rigid backbone wrapped with structural wedges, which become increasingly more dense toward the outer cylindrical surface. The Pd-catalyzed copolymerization of dibromobiphenyl derivatives with the Fréchet-type ethereal dendrons with a substituted aryl diboronic acid afforded the dendritic coated poly(*p*-phenylene) rod.

Polyether dendrons possessing -$(CH_2)_3$-connectors thereby eliminating reactive benzyl positions, have also been reported as "building blocks for functional dendrimers."[67b] Fréchet and Gisov[67c] have reported the preparation of hybrid „star-like" macromolecules possessing poly(aryl ether) dendrons attached to a four-directional, PEG molecule. These unique "dendritic stars" form conformationally differing unimolecular micelles depending on the polarity of the surrounding environment. Branched (polyaryl ether) dendrons have been attached to 1,10-phenthroline cores and complexed with Ruthenium.[67d]

5.4.3 1 → 2 *Aryl*-Branched, *Amide* Connectivity

Miller and Neenan[3] reported the preparation (Scheme 5.22) of a polyamido cascade **82**. The bis(acid halide) (**83**) of 5-nitroisophthalic acid was treated with aniline to gene-

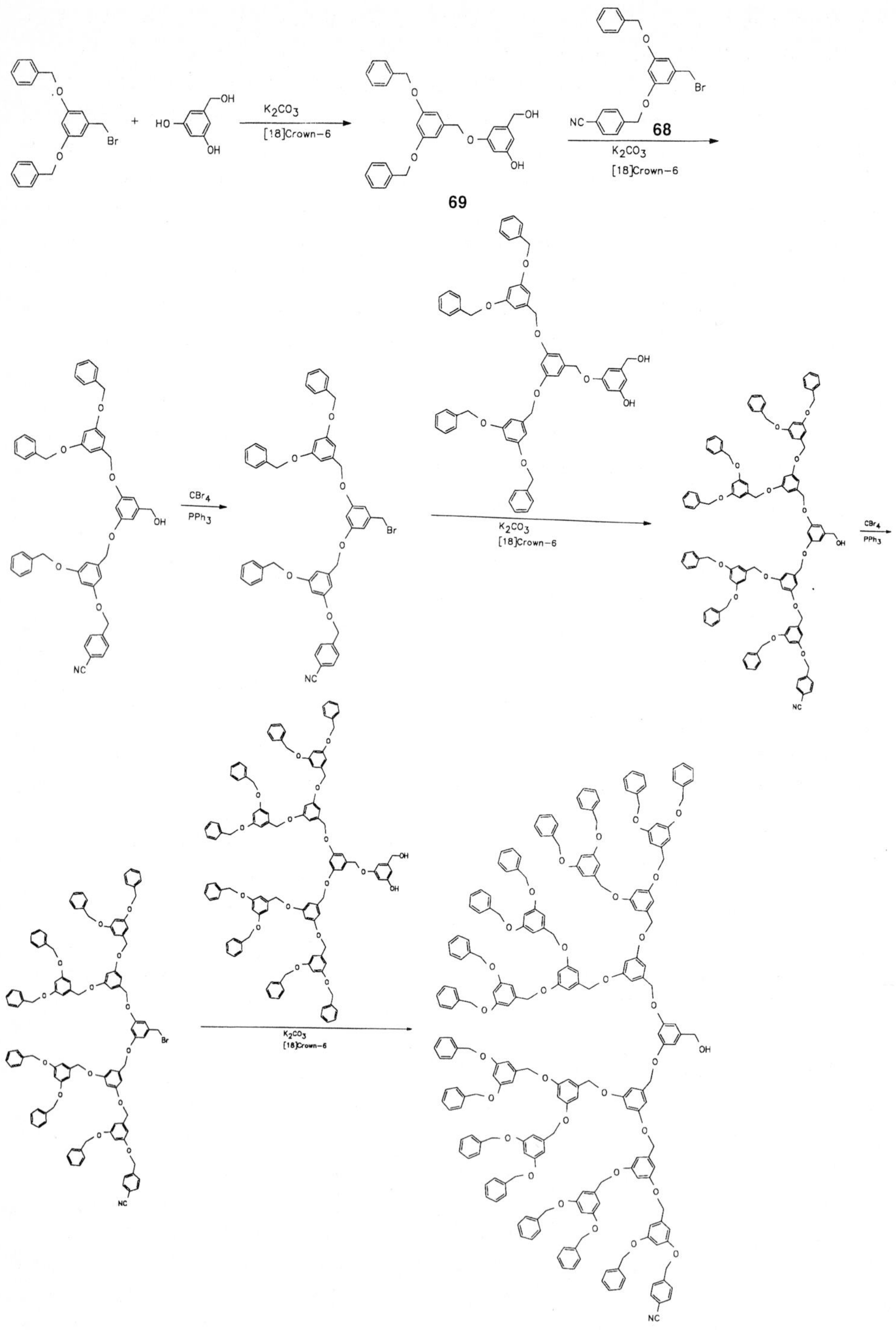

Scheme 5.17. Preparation of a poly(aryl ether) dendron possessing a single nitrile terminus.

rate bisamide **84**, which was catalytically reduced to give the disubstituted aniline **85**. Addition of three equivalents of amine **85** to 1,3,5-tris(chlorocarbonyl)benzene (**86**) then yielded the polyaromatic amidocascade (**82**).

Feast et al.[68–70a] reported the convergent construction of a series of aromatic polyamide dendrimers exactly analogous to the amide-based macromolecules described by Miller and Neenan.[3] The synthetic transformations employed for wedge construction were also similar except that reduction of the aromatic nitro group was effected by treatment with NaBH$_4$ and SnCl$_2 \cdot$ H$_2$O instead of catalytic reduction (see: Scheme 5.22).

Analysis of the ^{1}H NMR of the dendrimers suggested that, due to sterically restricted branch rotation, the interior secondary amide protons, which are equidistant from the core, were in different chemical environments. Solubility of the first, second, and third tier cascades was recorded at 24, 298, and 40 g/L, respectively, in THF at 25 °C.

Kraft[70b] has reported the preparation of small dendrimers possessing 1,3,4-oxadiazole spacers, which were constructed after dendrimer preparation via dehydration (POCl$_3$ or ClSO$_3$H) of hydrazide [-C(=O)NHNHC(=O)-] connectors. These assemblies were found to form π-stacks in solution as suggested by NMR and VPO. Other branched oxadiazoles habe been reported.[70c]

5.4.4   1 → 2 *Aryl*-Branched, *Ether* and *Amide* Connectivity

Newkome et al.[71] reported the convergent construction of an aromatic cascade based on a building block developed for the use in a divergent synthetic method (Scheme 5.23), utilizing the Miller and Neenan thought process.[3] Treatment of 5-nitroisophthaloyl dichloride (**83**) with two equivalents of the amine building block[72] **87** gave the hexaester dendron, which was catalytically reduced (PtO, H$_2$) to yield aromatic amine **88**. Monomer **88** was treated with triacyl core **86** affording the three-directional dendrimer **89**, which was hydrolyzed to the corresponding polyacid **90**. The related two-directional cascades (**91**) were also reported[71] employing terephthaloyl dichloride (**92**), as the core.

Shinkai et al.[73, 74] have reported the construction of 'crowned arborols' that incorporated diaza crown ethers as spacers, and three-directional, aromatic branching centers (Scheme 5.24). In this case, the convergent construction was found to be more effective than the divergent approach. Tetraoxadiazacrown monomer **93** was prepared from *N*-benzyloxycarbonyldiazacrown ether (**94**) and a monoacyl chloride **95**, affording the diprotected intermediate **96**, which was catalytically debenzylated (Pd/C, H$_2$) to give diester **93** or hydrolyzed (aq.base) to afford diacid **97**. Monomer **97** was then transformed to tetraester **98**, by means of the mixed anhydride method, with two equivalents of the

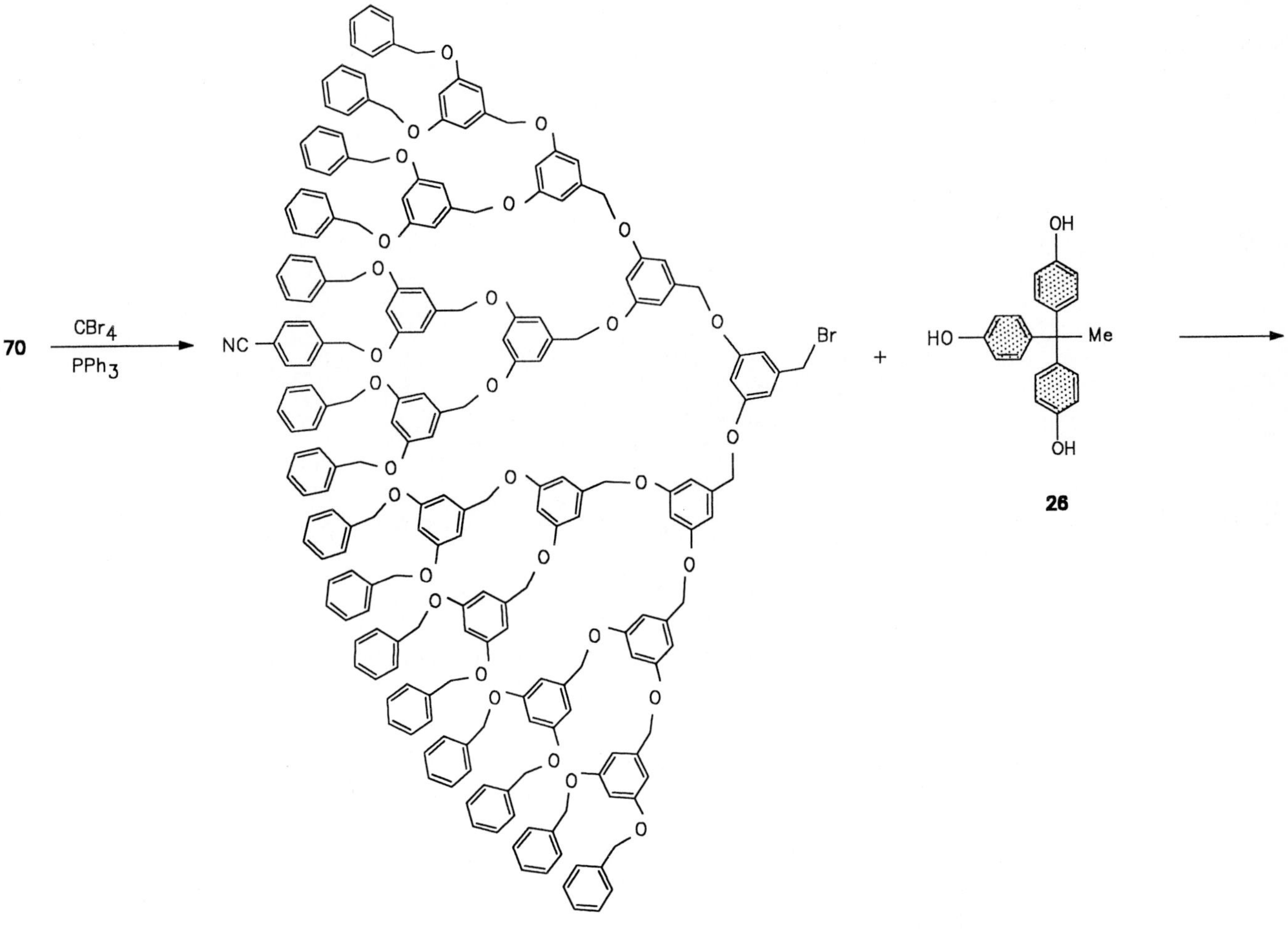

Scheme 5.18. (Continued on page 132.)

71

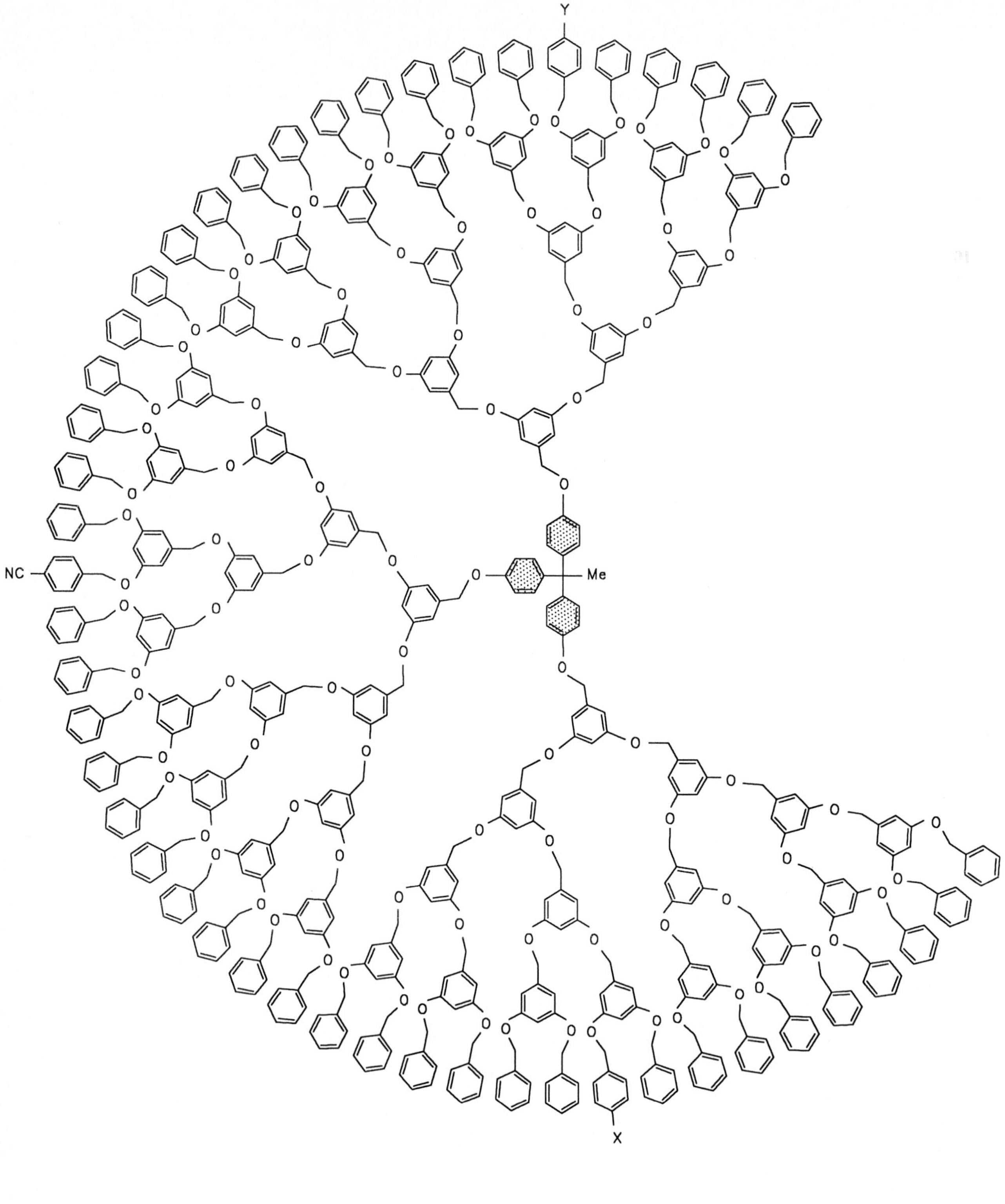

72   X = Y = H

73   X = H, Y = CN

74   X = Y = CN

Scheme 5.18. Statistical preparation of dendrimers possessing 1, 2, or 3 nitrile surface groups.

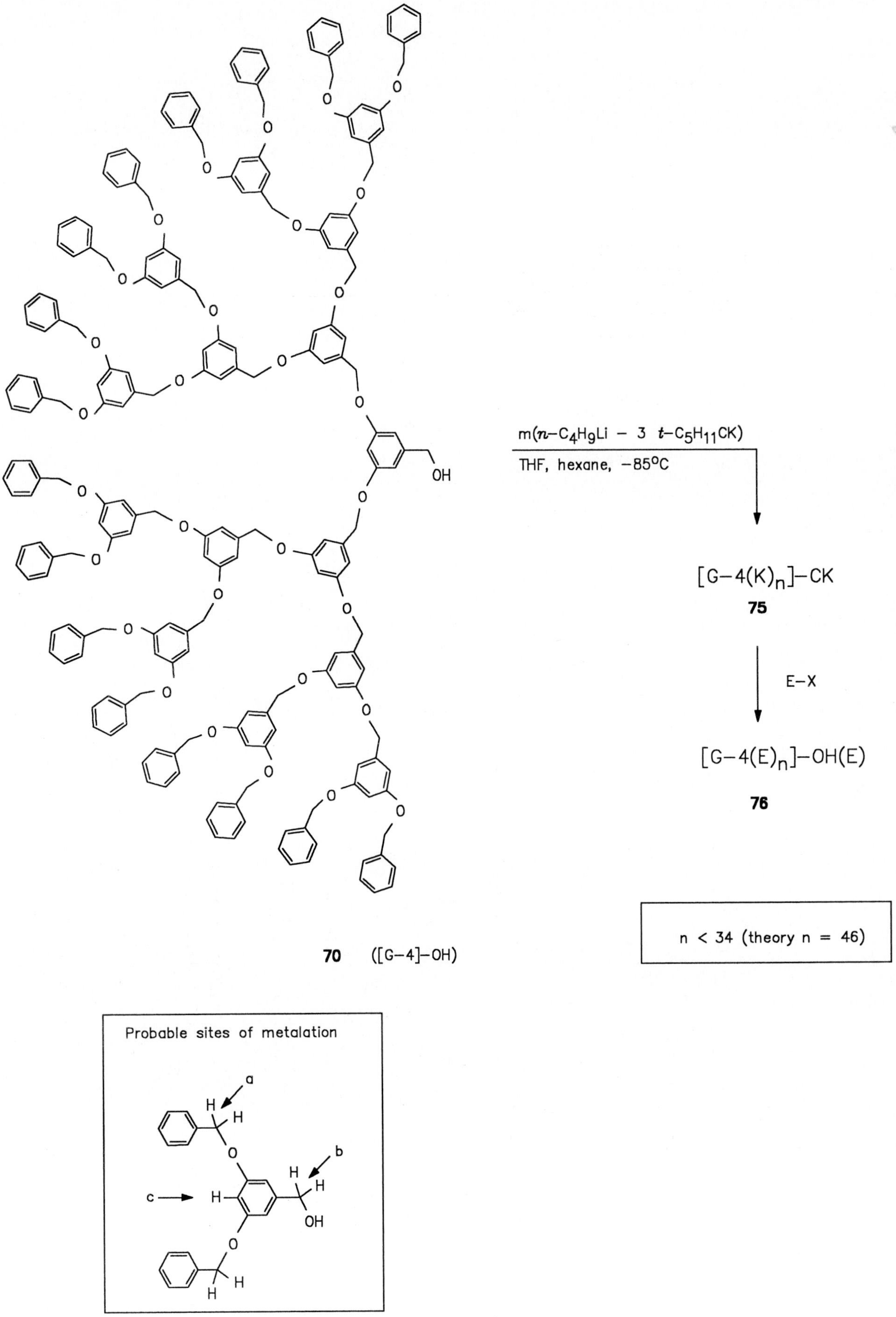

Scheme 5.19. Superstructure modification of poly(benzyl ether) dendrimers via metalation, followed by electrophile addition.

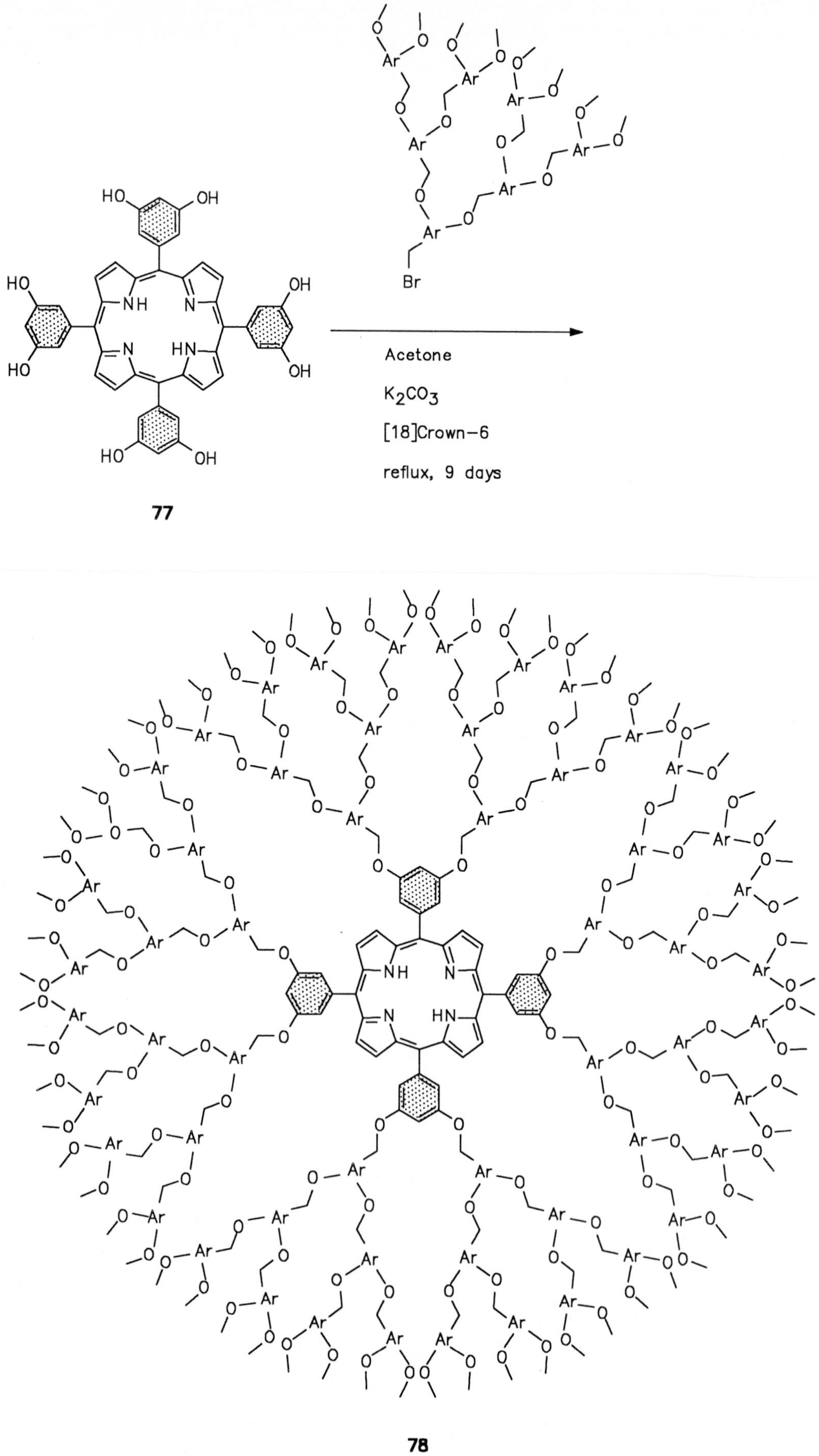

Scheme 5.20. Porphyrin modification via the attachment of Fréchet-type wedges.

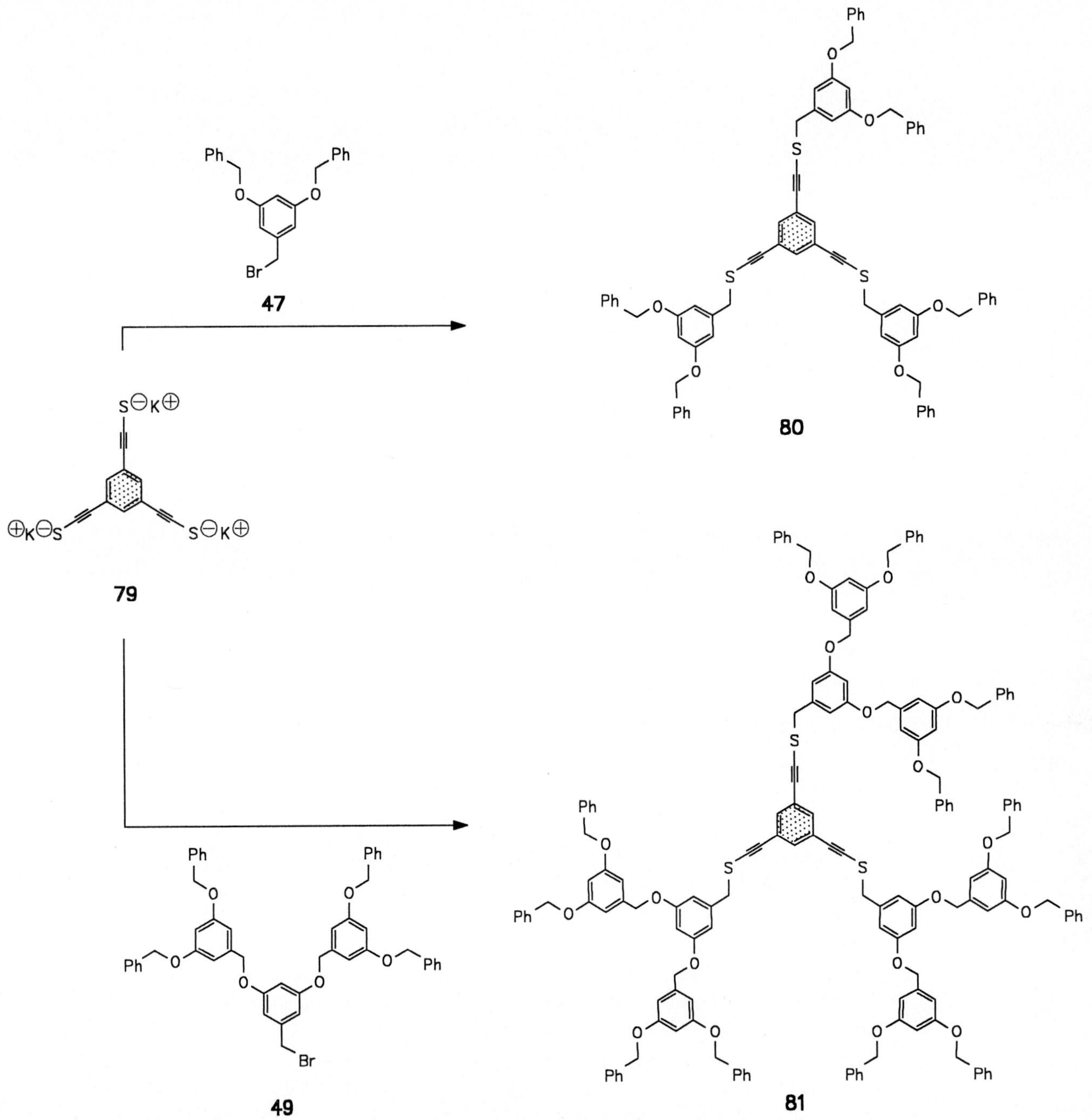

Scheme 5.21. Introduction of rigid, trialkyne cores via the convergent method.

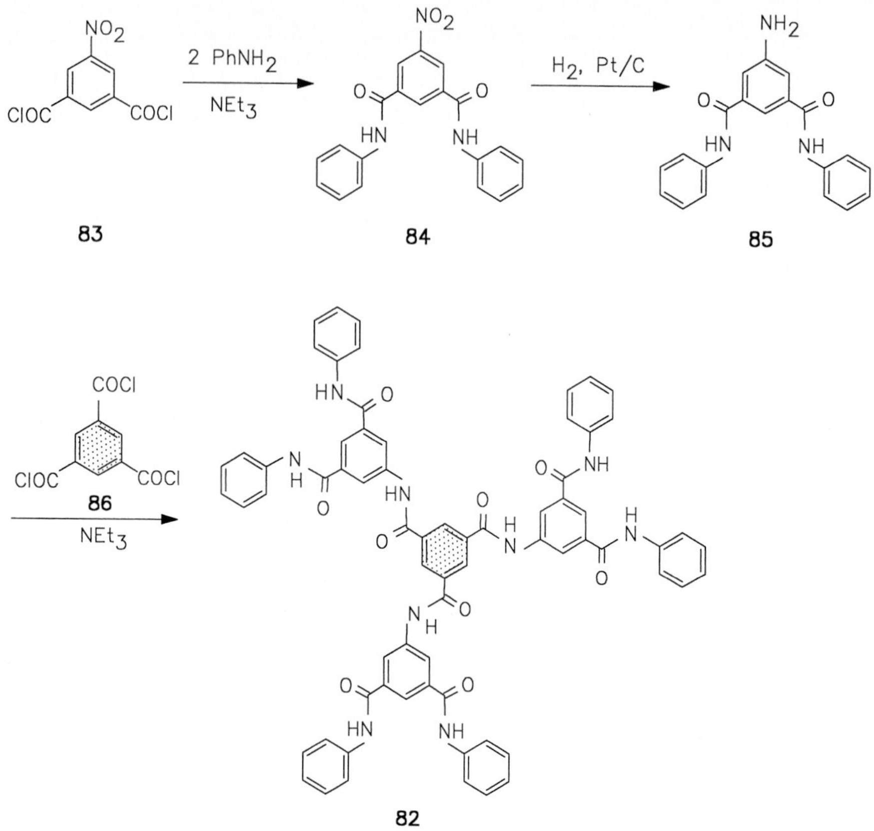

Scheme 5.22. Miller and Neenan's[3] arylamide-based sequence.

corresponding amine building block **93**. The resultant dendron **98** was attached to the tri-acyl core **86** affording the second generation polyethereal crown cascade **99**.

A related reduced series (Figure 5.5) was also reported.[74] Selectivity of these crowned arborols (**100–102**) towards alkali metal cation binding was examined and allosteric as well as conformational binding effects were studied. Also they showed that the 'sterically less crowded' arborols (e.g., **102**) are better for the dissolution of myoglobin in organic solvents, such as DMF. This example affords an interesting entrance to internal metal ion complexation at a specific loci; see Section 8.3.

Although strictly not a dendritic system, Agar et al.[75] have reported the preparation of copper(II) phthalocyaninate substituted with eight 12-membered tetraaza macrocycles as well as its nickel(II), copper(II), cobalt(II), and zinc(II) complexes. Thus, the use of the 1,4,7-tritosyl-1,4,7,10-tetraazacyclododecane offers a novel approach to the 1 → 3 branching pattern and a locus for metal ion encapsulation.

5.4.5 1 → 2 *Aryl*-Branched, *Ether* and *Urethane* Connectivity

Spindler and Fréchet[76] demonstrated the construction of cascade polymers possessing both the ethereal and urethane bonds as well as the growth of two generations in a single synthetic operation (Scheme 5.25). Their one-pot, multistep approach utilizes the preformed first generation alcohol (**103**), which is treated with the 3,5-diisocyanatobenzyl chloride (**104**) to build the second generation biscarbamate (**105**) possessing a benzyl chloride focal group. Without further purification, 3,5-dihydroxybenzyl alcohol (**48**) is added giving rise to the third generation alcohol (**106**). As expected, the judicious selection and use of specific monomers are critical to the successful application of this accelerated approach to dendritic materials.

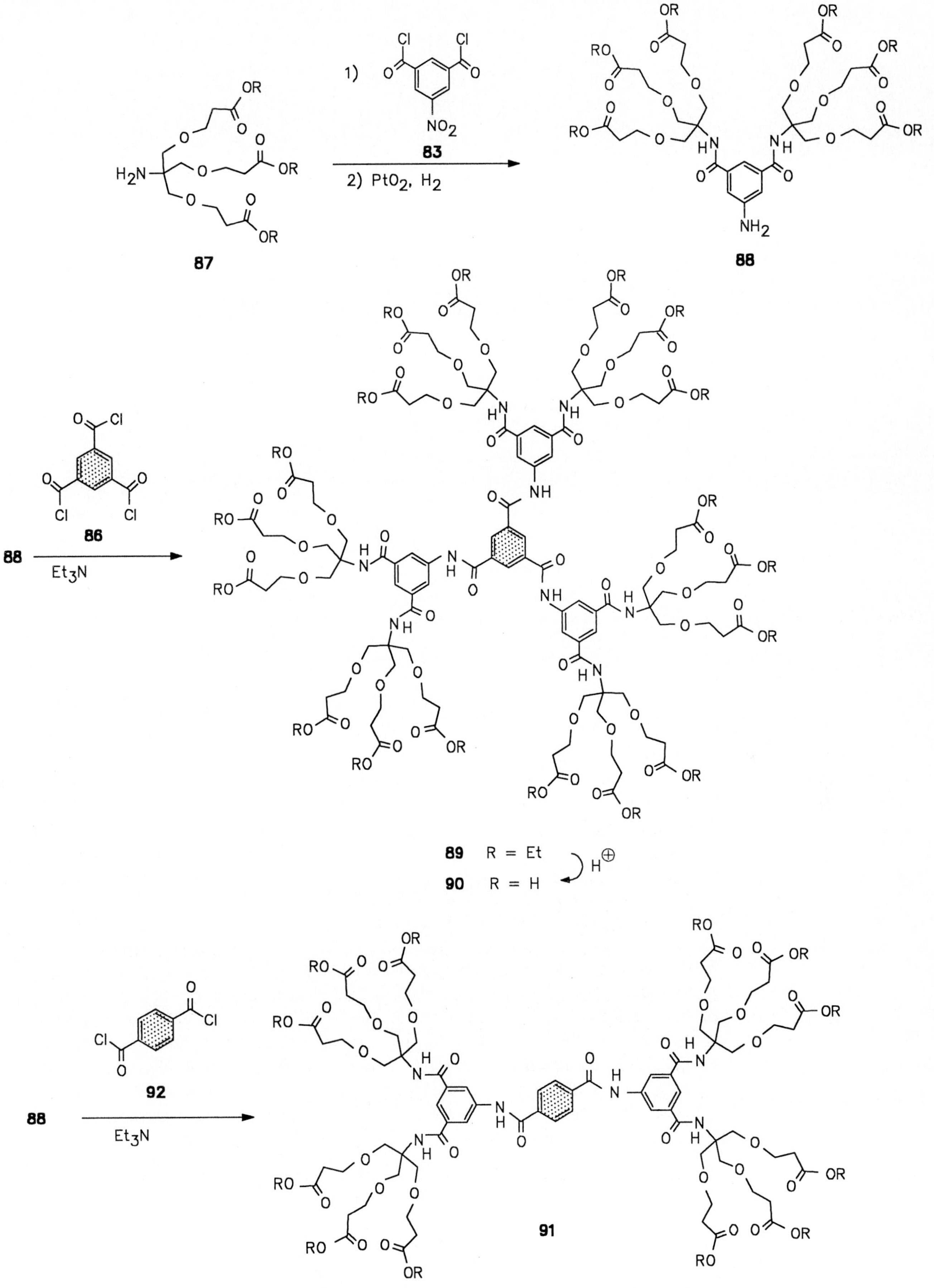

Scheme 5.23. Employment of a flexible aminotriester for the construction of dendrimers possessing arylamide-based cores.

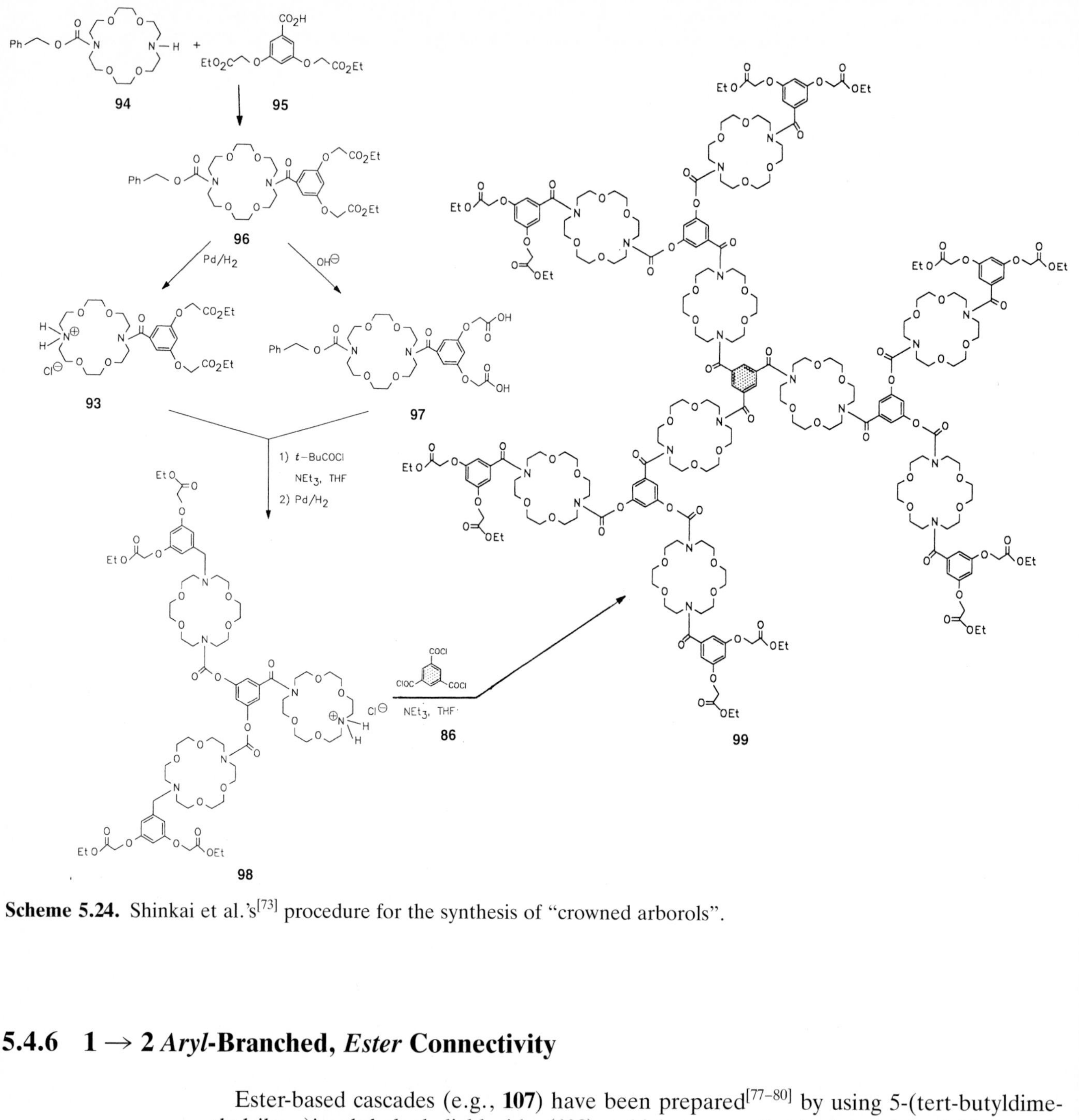

Scheme 5.24. Shinkai et al.'s[73] procedure for the synthesis of "crowned arborols".

5.4.6   1 → 2 *Aryl*-Branched, *Ester* Connectivity

Ester-based cascades (e.g., **107**) have been prepared[77–80] by using 5-(tert-butyldime-thylsiloxy)isophthaloyl dichloride (**108**), which was synthesized in high yield from 5-hydroxy-isophthalic acid (Scheme 5.26). The dendron wedges were prepared by treatment of siloxane **108** with phenol to give bis(aryl ester) **109**, which was hydrolyzed, or desilylated (HCl, acetone), to generate a new phenolic terminus. Treatment of this free phenolic moiety with monomer **108**, followed by hydrolysis, afforded the next tier (**110**). Repetition of the sequence followed by reaction of the free focal phenols with a triacyl chloride core, (e.g., **86**), afforded the fourth tier dendrimer **107** of the polyester aryl series. It was noted that the choice of base (*N,N*-dimethylaniline) used in the final esterification was critical, since with pyridine bases (pyridine or 4-(dimethylamino)pyridine) facile transesterification resulting in branch fragmentation occurred.

Characterization of the series included the standard ^{13}C and ^{1}H NMR spectroscopies and gel permeation chromatography. The melting point of the first tier ester (176–178 °C), was found not to differ greatly from that of the fourth tier ester (192–203 °C) even though the molecular weights differ by a factor of nearly 10. Haddleton et al.[81] have demonstrated the utility of MALDI MS on these aromatic polyether dendrimers.

Bryce et al.[82,83] have created new materials based on the incorporation of highly functionalized tetrathiafulvene (TTF) derivatives. Cascade peripheral modification was

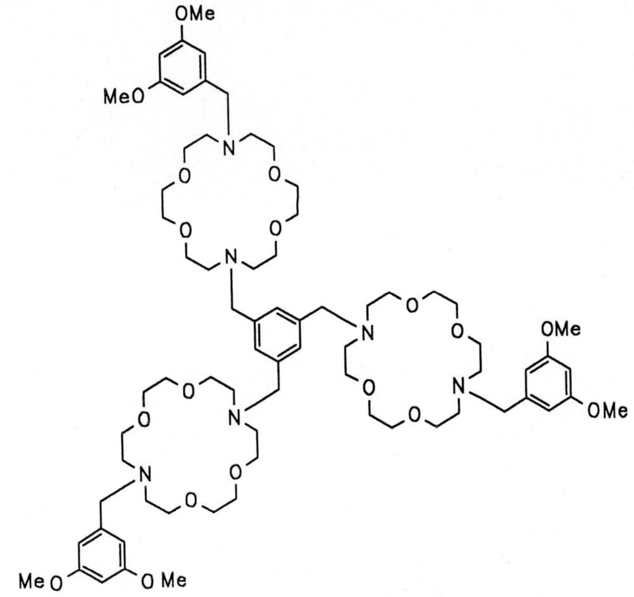

100

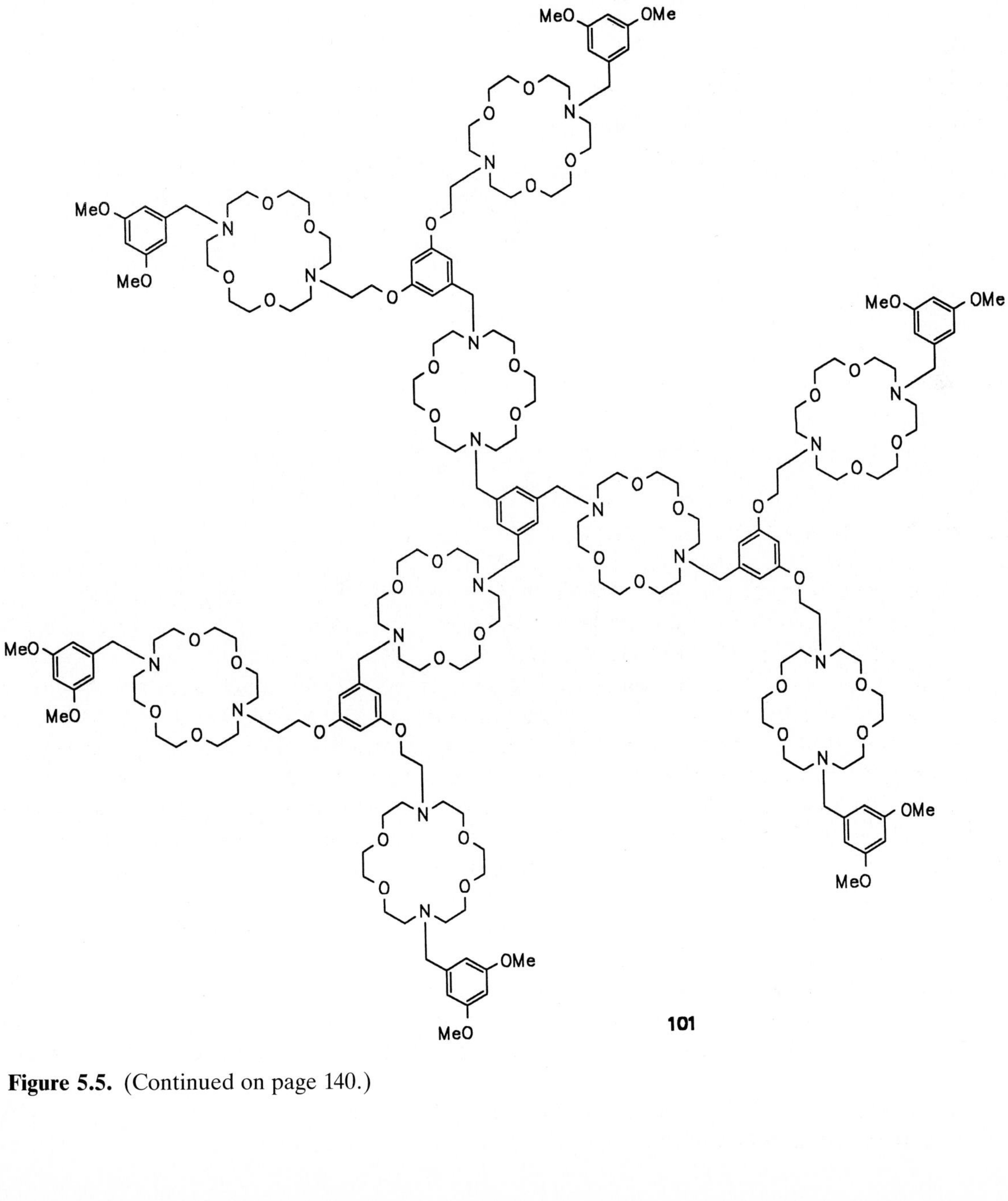

101

Figure 5.5. (Continued on page 140.)

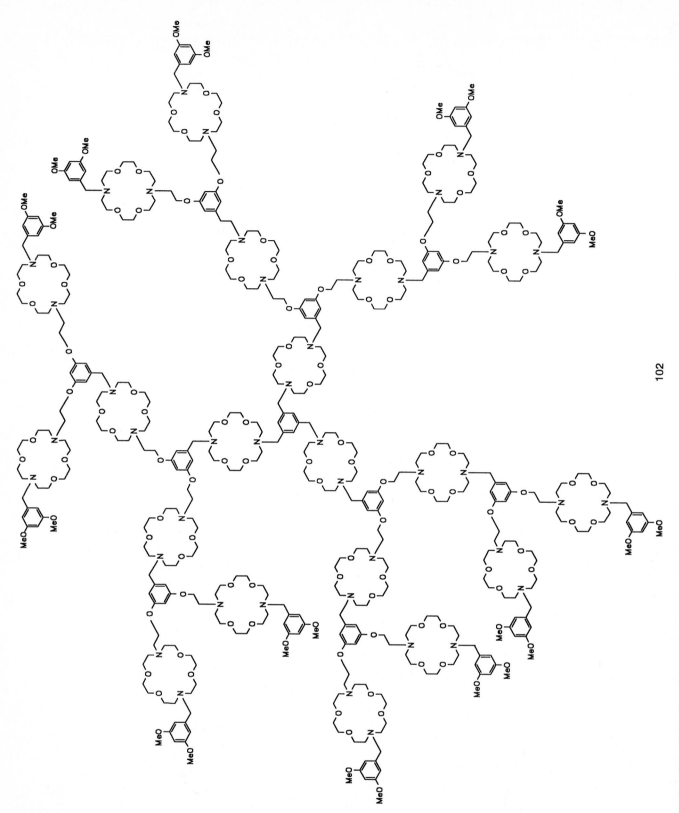

Figure 5.5. Reduction of the carbonyl moieties of the "crowned arborols" (Scheme 2.24) afforded these more lipophilic "crowned" dendrimers.

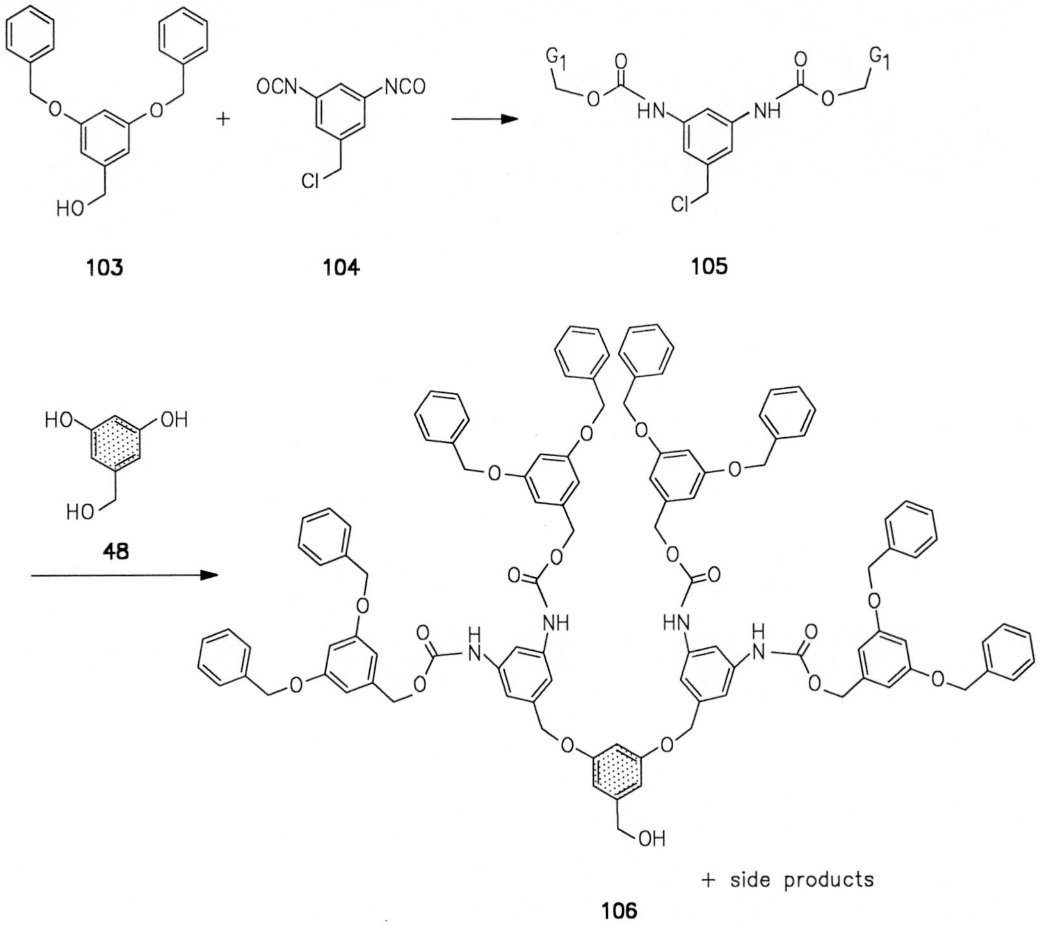

Scheme 5.25. Construction of two generations using a single-pot, multistep approach.

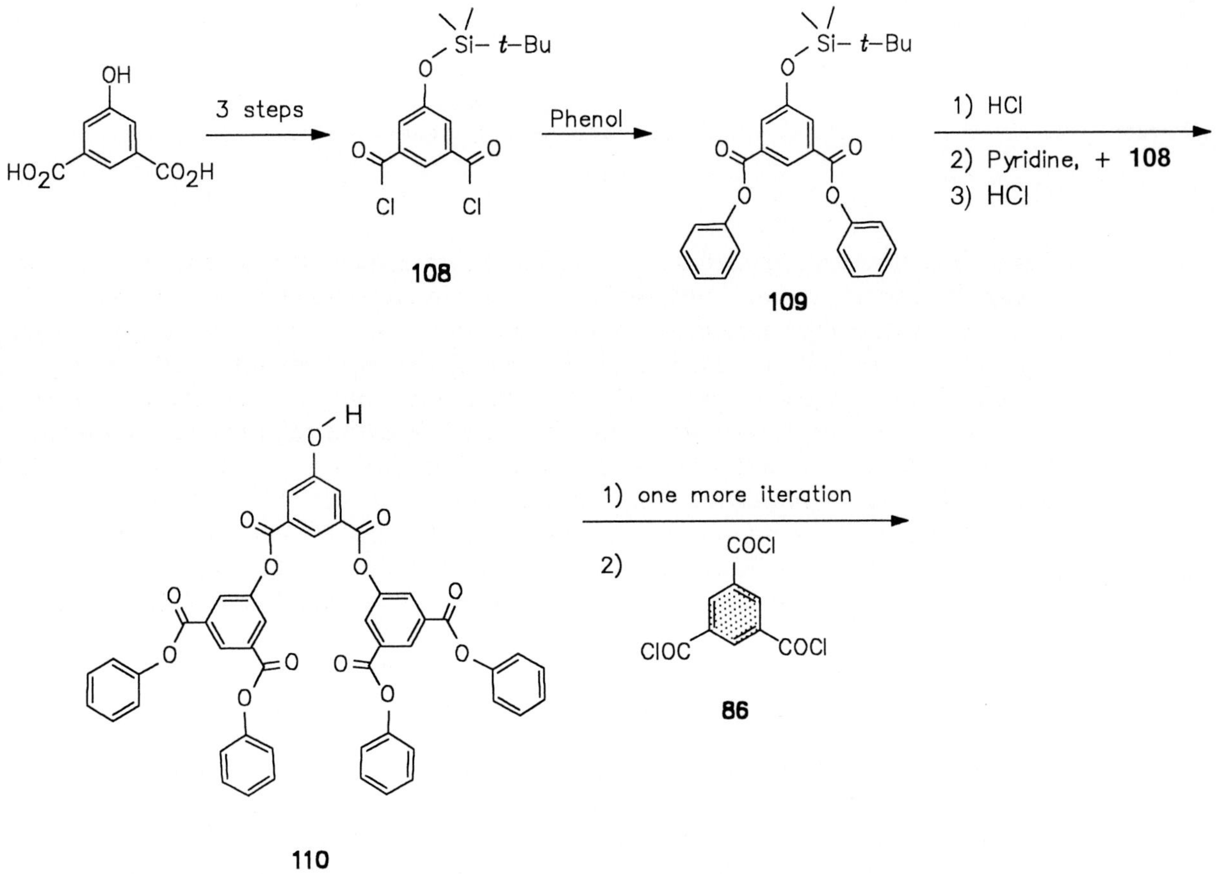

Scheme 5.26. (Continued on page 142.)

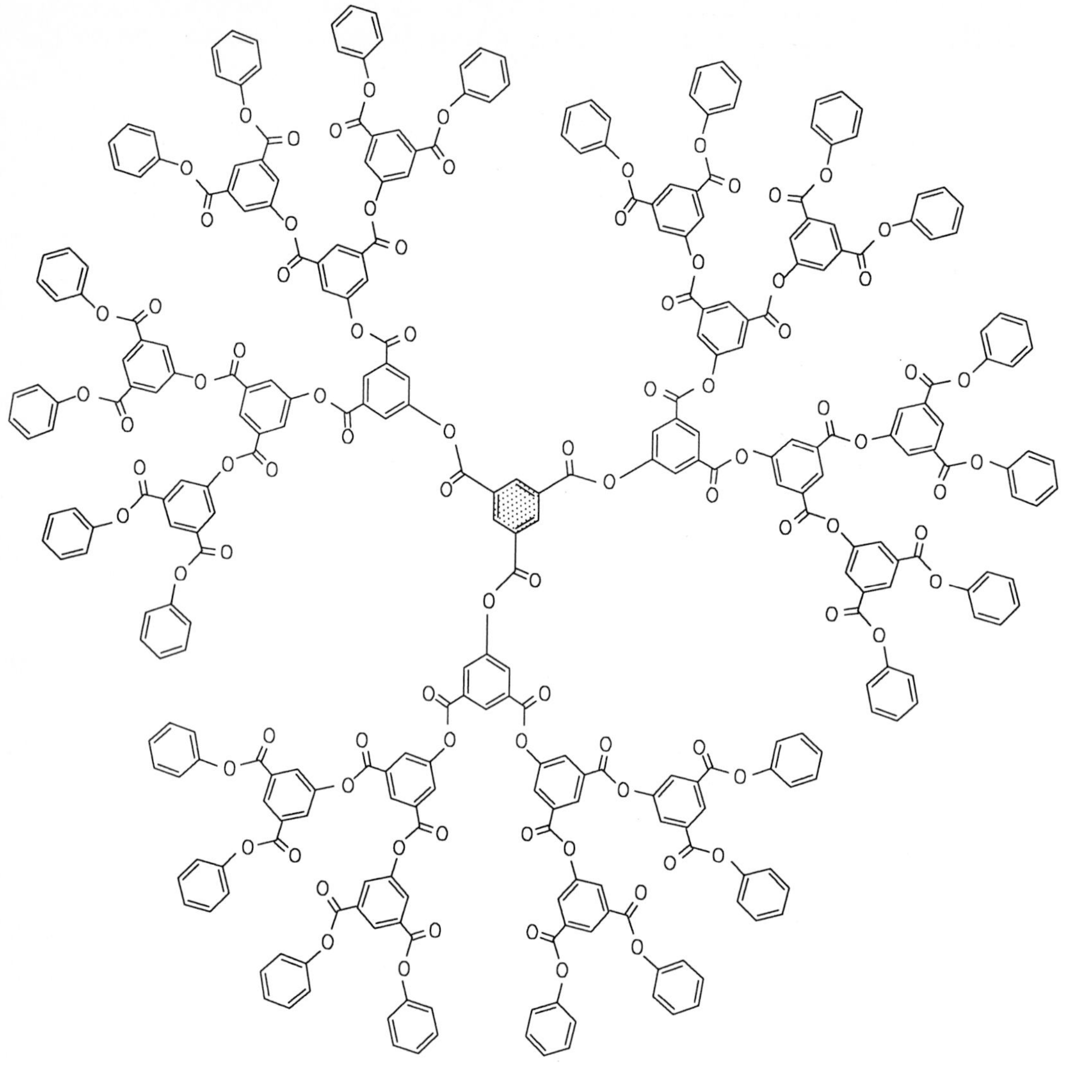

107

Scheme 5.26. Employment of silane protective groups has facilitated the preparation of polyester-based dendrimers.

envisioned to provide redox-active materials for areas such as molecular self-assembly and molecular electronics. Tetrathiafulvalene is of interest due to numerous factors that include: (1) sequential and reversible oxidation to the monocation and dication radical species, (2) the ability to effect oxidation potential change via substituent choice, (3) radical thermodynamic stability, and (4) stacking-propensity. The preparation (Scheme 5.27) of the TTF-capped dendrons initially utilized 5-(*tert*-butyldimethylsiloxy)isophtha-loyl chloride (**108**),[77] which upon treatment with the TTF-methanol[84] derivative (**111**) gave diester **112**. Deprotection of the phenol moiety (*n*-Bu₄N⁺F⁻, THF) of diester **112**, followed by reaction with 0.5 equivalents of bis(acid chloride) **108** and further desilyla-tion afforded the tetra-TTF dendron **113**.[85] Reaction of three equivalents of hexaester **113** with 1,3,5-tris(chlorocarbonyl)benzene (**86**) gave (32 % overall from **112**) the dodeca-TTF dendrimer **114**.

These TTF-dendrimers were reportedly stable when stored below 0 °C; however, within seven days at 25 °C in the absence of light and under an inert atmosphere (argon), notable decomposition was observed. The redox behavior of dodeca-TTF dendrimers **114**, as well as the corresponding lower generations, exhibited two well-defined, quasi-reversible couples at $E^{1/2} \approx +0.45$ and $E^{1/2} \approx +0.85$ V (vs. Ag/AgCl). Treatment of the TTF dendrimers with iodine in CH₂Cl₂ led to oxidation of the fulvalene groups, as indica-ted by the appearance of a broad absorption in the UV spectrum ($\lambda_{max} = 590$ nm). This signal is consistent with formation of the TTF cation radical. From their studies, these

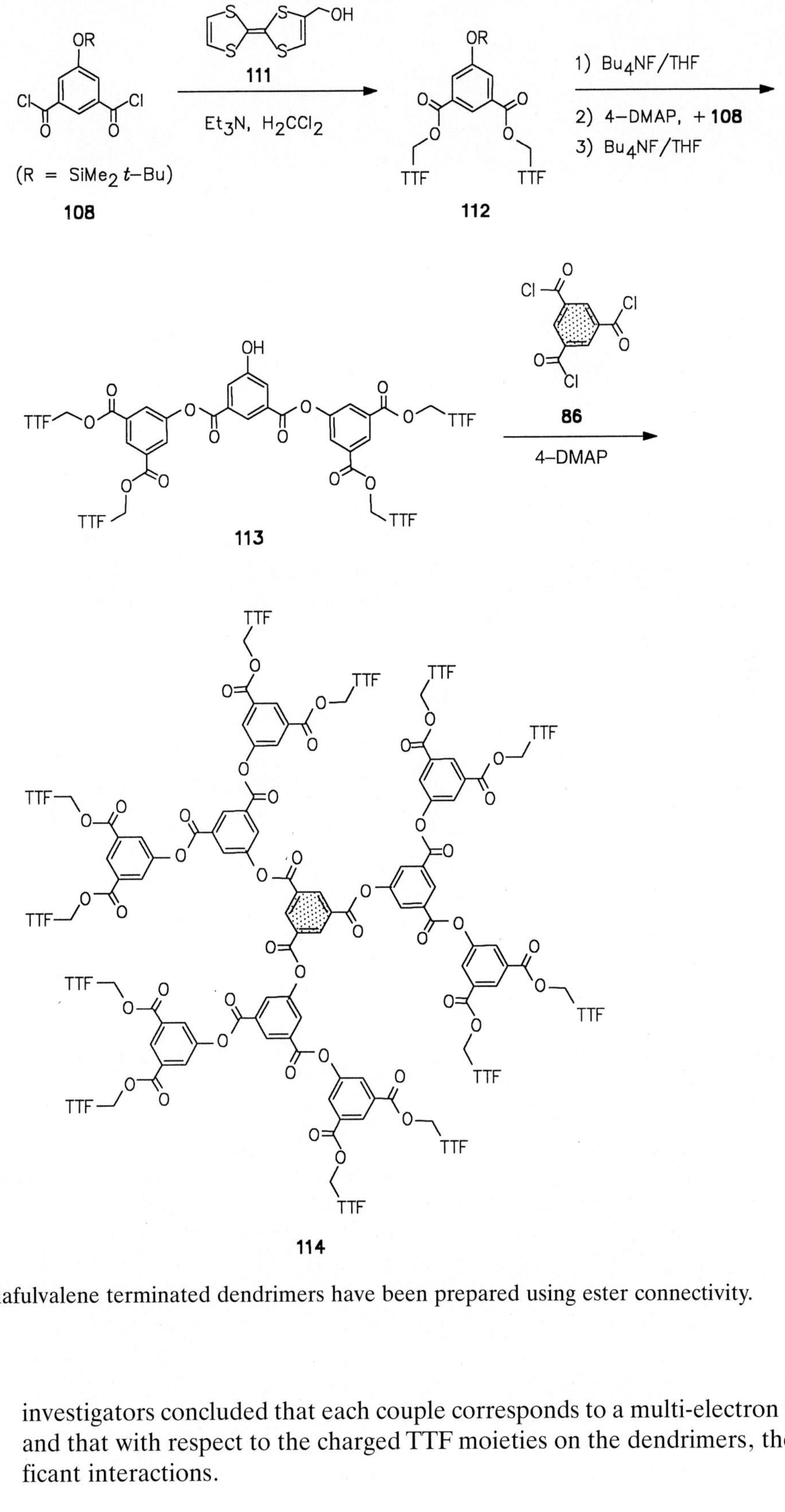

Scheme 5.27. Tetrathiafulvalene terminated dendrimers have been prepared using ester connectivity.

investigators concluded that each couple corresponds to a multi-electron transfer process and that with respect to the charged TTF moieties on the dendrimers, there are no significant interactions.

5.4.7   1 → 2 *Aryl*-Branched, *Ether* and *Ester* Connectivity

Fréchet et al.[86] reported a 'branched-monomer approach' to the convergent synthesis of dendritic macromolecules; this approach permits an accelerated growth by the replacement of the simplest repeat unit with a larger repeat unit of the next generation. In essence, the traditional AB$_2$ monomer is replaced with an AB$_4$ unit. Scheme 5.28 depicts the transformation of the AB$_4$ unit (**115**) to the trimethylsilyl protected tetraester **116**

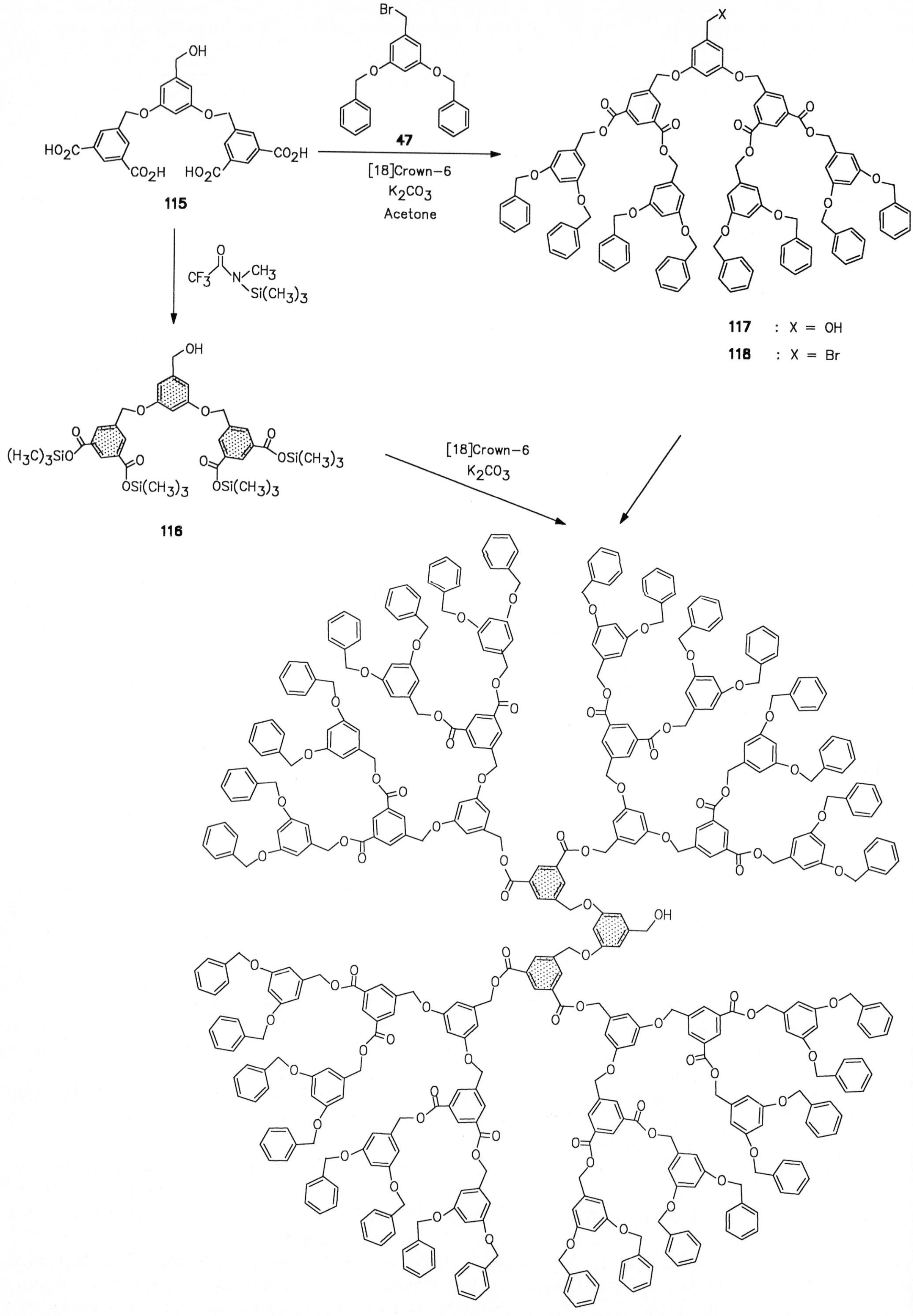

119

Scheme 5.28. The "branched-monomer approach" leads to dendrimers possessing differentiated generation connectivity.

(CF₃CONMe(SiMe₃) and to the benzyloxy terminated building block **117** by treatment ([18]crown-6, K₂CO₃, acetone) with bromide **47**. Activation by CBr₄ and P(C₆H₅)₃ of the alcoholic moiety in **117** to the corresponding bromide **118** and treatment with 0.25 equivalents of tetrakis(trimethylsiloxane) **116** generated the desired dendrimer **119**. The incorporation of the internal hydrolyzable ester linkages affords entry to internal hydrolytic sites.

The use of a terminal 3,5-bis(benzyloxy)benzoic acid in the convergent process with internal ester connectivity permitted the catalytic deprotection (H₂/Pd−C) of the benzyl groups. The resultant spherical macromolecule possessed the reactive phenolic functionality, which facilitated aqueous solubility and functionalization.[87]

Hawker and Fréchet[88] created novel mix and match combinations utilizing dendritic polyesters and polyethers. The resultant globular block architectures were generated by the controlled location of various ester and ether groups affording either radial or concentric patterns. Thus, a dendritic "segment"-block copolymer is prepared by the attachment of radially alternating dendritic segments from the core; whereas, a dendritic "layer"−block is created by concentric alteration of ester and ether layers from the core. Since the termini and core are similar in each case, the glass transition temperatures for the various block copolymers were controlled by the relative proportion of ether and ester building blocks rather than their precise geometries.

5.4.8   1 → 2 *Aryl*-Branched, *Ether* and *Ketone* Connectivity

Morikawa et al.[89] have prepared a series of poly(ether ketone) dendrimers by a convergent approach utilizing 3,5-bis(4-fluorobenzoyl)anisole as the building block. Initial conversion of the fluoro groups to phenyl ether moieties is followed by deprotection (AlCl₃) of the methyl ether moiety; nucleophilic aromatic substitution using the fluoro building block completes the general scheme. Four generations are easily constructed by repetition of this procedure; the cascade polymers each possess a narrow molecular mass distribution.

5.4.9   1 → 2 *Aryl*-Branched, *Ethyne* Connectivity

Moore and Xu[90,91] reported the convergent preparation of phenylacetylene dendritic wedges (Scheme 5.29) by a repetitive strategy involving a Pd-catalyzed coupling of a terminal alkyne to an aryl halide possessing a trimethylsilyl-capped alkyne. The concept is illustrated by considering the following transformations: two equivalents of phenylacetylene (**120**) are attached to a silyl-protected 3,5-dibromophenylacetylene[92,93] (**121**), via cross-coupling using Pd(dba)₂. After deprotection with methanolic carbonate, the process was either repeated to give the next higher generation (path a) or the terminal alkyne was removed from the growth sequence and subjected to core attachment (path b). Dendritic wedge synthesis proceeded as expected through the second generation; however, two factors, dendron stability and steric inhibition to reaction at the focal point, were operative. Modest gains in the yield of the third tier dendron were realized by using *p*-substituted peripheral phenyl units such as 4-MeOC₆H₅ or 4-(*tert*-Bu)C₆H₅. Even higher yields of the third generation wedge **122** were effected by the use of an elongated monomer **123** (Figure 5.6) indicating that increased spacer length should allow facile construction of large, rigid dendrimers.

Moore and Xu[94] later reported improvements in phenylacetylene dendron and dendrimer preparation. Essentially, a dendritic construction set (**124–135**; Figure 5.7) was prepared by using the previously described[90,94] aryl halide−alkyne coupling combined with trimethylsilyl alkyne masking as well as an aryldiethyltriazene to aryl iodide transformation.[95] The aryldiethyltriazene moiety served as a protected aryl iodide that could be introduced by triazene treatment with MeI at 110 °C in a sealed tube. Specifically, the use of peripheral 3,5-di-*tert*-butylphenyl units, rigorous exclusion of molecular oxygen from the Pd-mediated, alkyne−aryl halide cross-coupling reaction, and particularly, the employment of optimized conditions (35–40 °C, 2 d) allowed the preparation (37 %) of

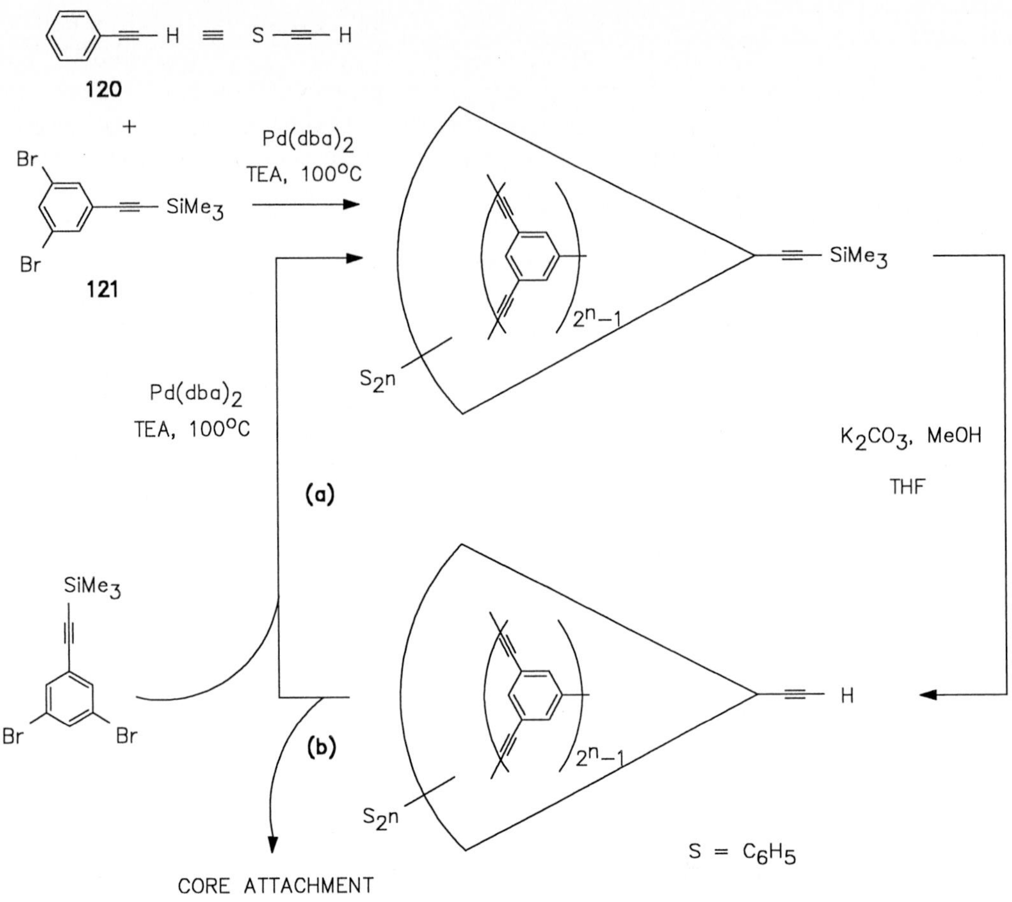

Scheme 5.29. Moore et al.'s[90] iterative procedure for the preparation of poly(aryl alkyne) dendrimers.

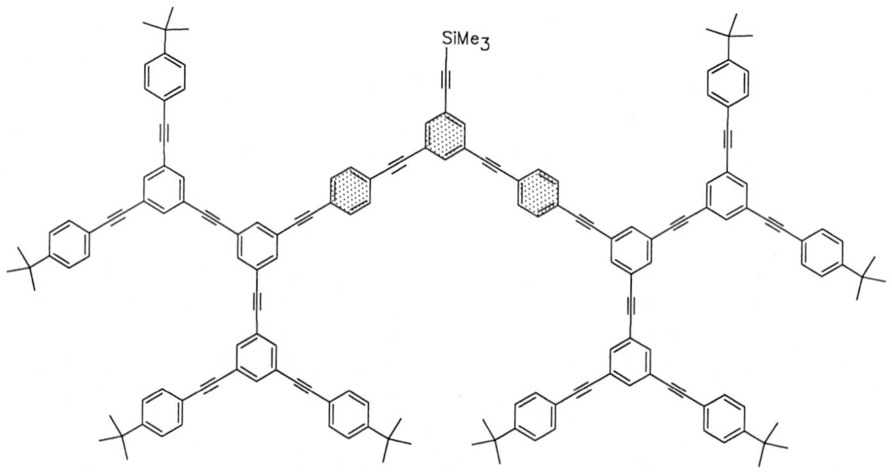

122

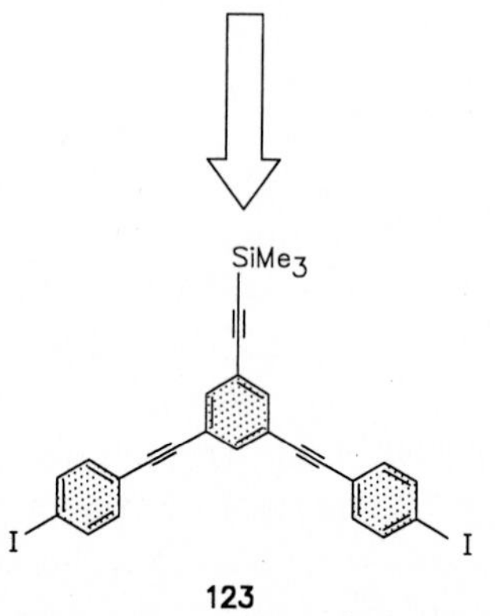

123

Figure 5.6. Use of extended building blocks facilitated the construction of large, rigid dendrons.

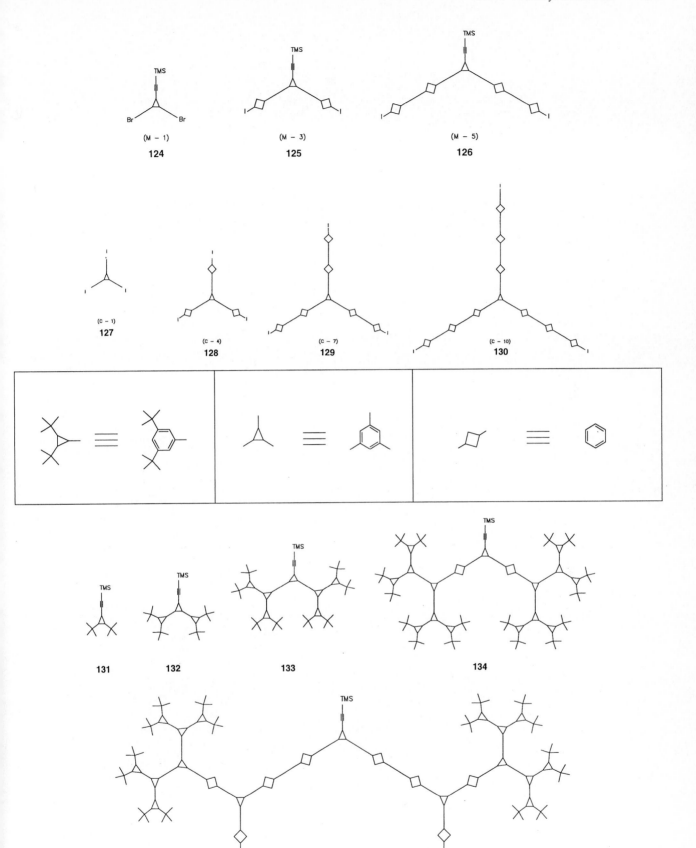

Figure 5.7. A dendritic construction set was created based on Moore et al.'s[94] phenylacetylene chemistry. (Aryl connectors consist of — ≡ — bonds.)

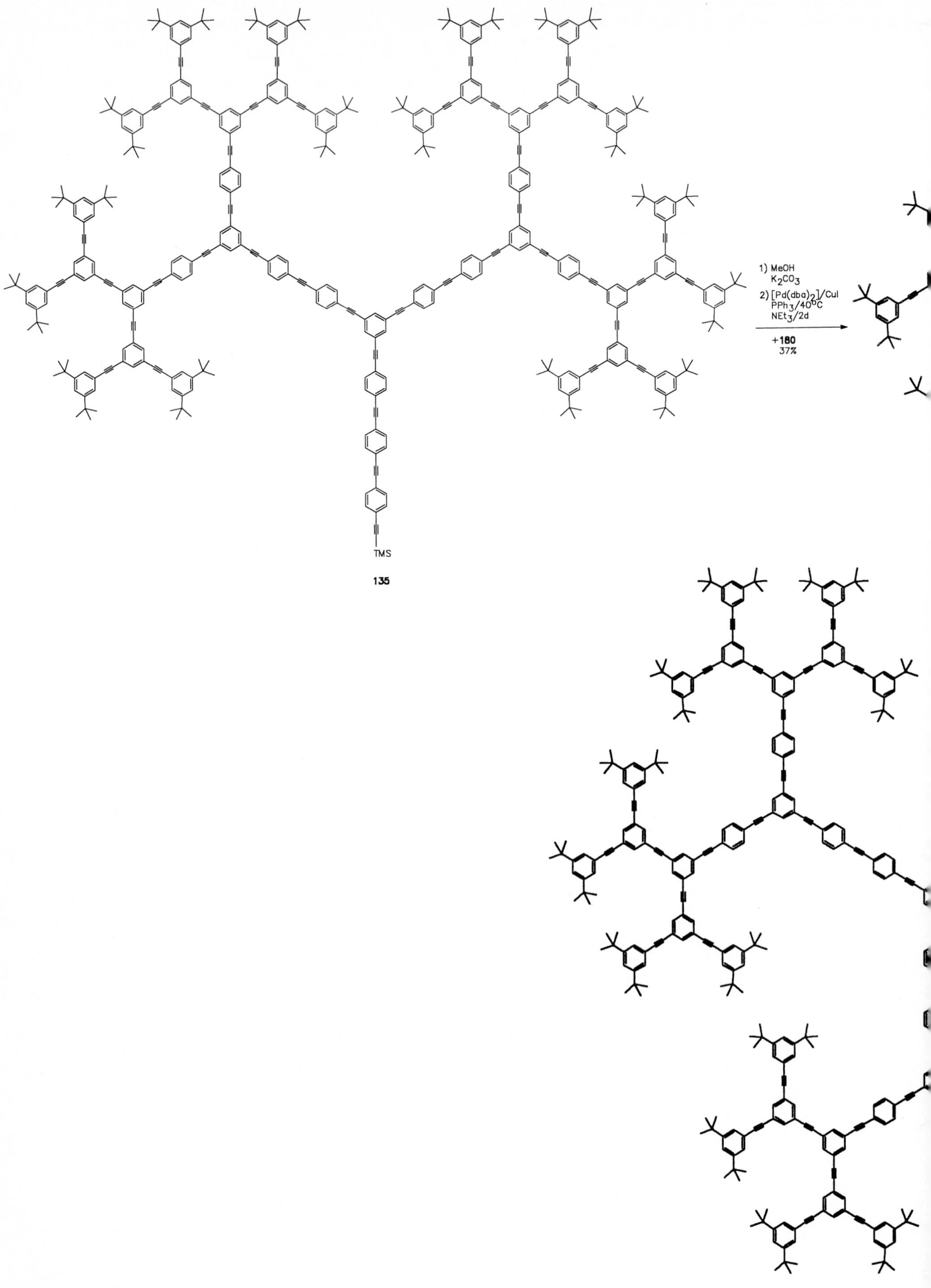
TMS
135
1) MeOH
K2CO3
2) [Pd(dba)2]/CuI
PPh3/40°C
NEt3/2d
+180
37%

Scheme 5.30. Synthesis of **137** by the *SYNDROME* mode of construction.

137

the '94-mer' dendrimer **136** (Figure 5.8). With the increased alkyl surface (i.e., aryl 3,5-di-*tert*-butyl groups), all of the dendrimers were readily soluble in pentane at 25 °C. Characterization included a discussion of the ^{1}H NMR COSY-45 spectrum of the hydrocarbon dendrimer **136**. Mass spectrometry has been shown[96] to be an excellent tool to characterize their purity and structural composition.

Moore et al.[97,98] subsequently employed the concept of dendritic spacer elongation for the construction of a series of phenylacetylene cascades (e.g., **137**; Scheme 5.30) with the longest being 12.5 nm in diameter. Hence, reaction of an extended aryl triiodide core (i.e., **130**), prepared by repetitive addition of 1-(4-ethynylphenyl)-3,3-diethyltriazene to 1,3,5-triiodobenzene, with three equivalents of dendron **135**, afforded the fourth generation '127-mer' **137** (Scheme 5.30). The method of avoidance of steric inhibition to dendritic construction via the use of increasingly longer spacer moieties gave rise to the acronym *SYNDROME* (*SYN*thesis of *D*endrimers by *R*epetition *O*f *M*onomer *E*nlargement). These dendrimers, as noted by the authors[97,99] are 'shape-persistent' and 'dimension-persistent', which can play an important role in development of molecular frameworks, that require rigorous control of functional group juxtapositions.

Moore et al.[100] expanded the use of the complimentary protection−deprotection sequence, facilitated by the diethyltriazine and trimethylsilane groups, to construct phenylacetylene oligomers. Up to 16 arylalkyne units were successfully connected (overall yield 67 %; 2251 amu). The method, 'demonstrating a high level of chain length control',

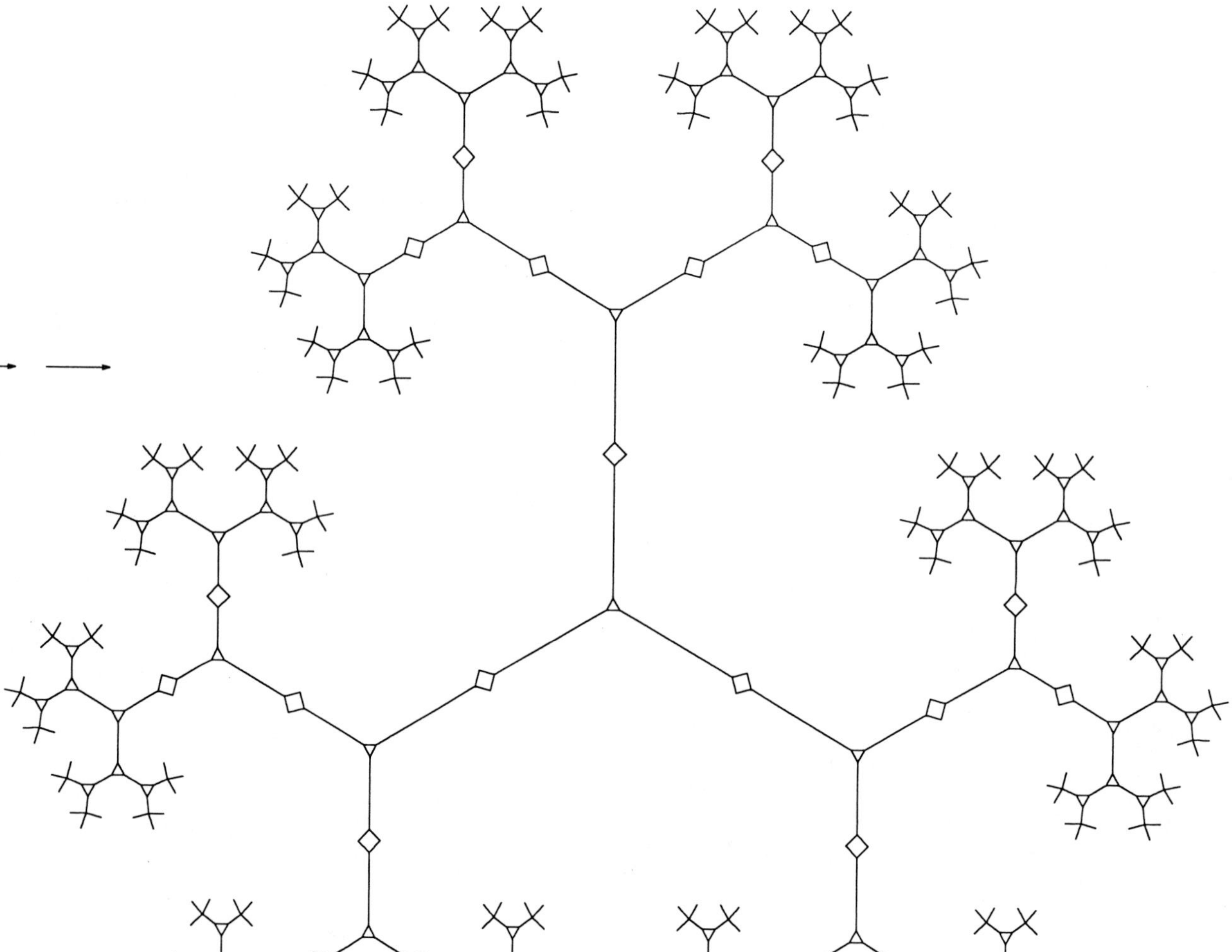

136

Figure 5.8. An idealized compact tridendron **136**.

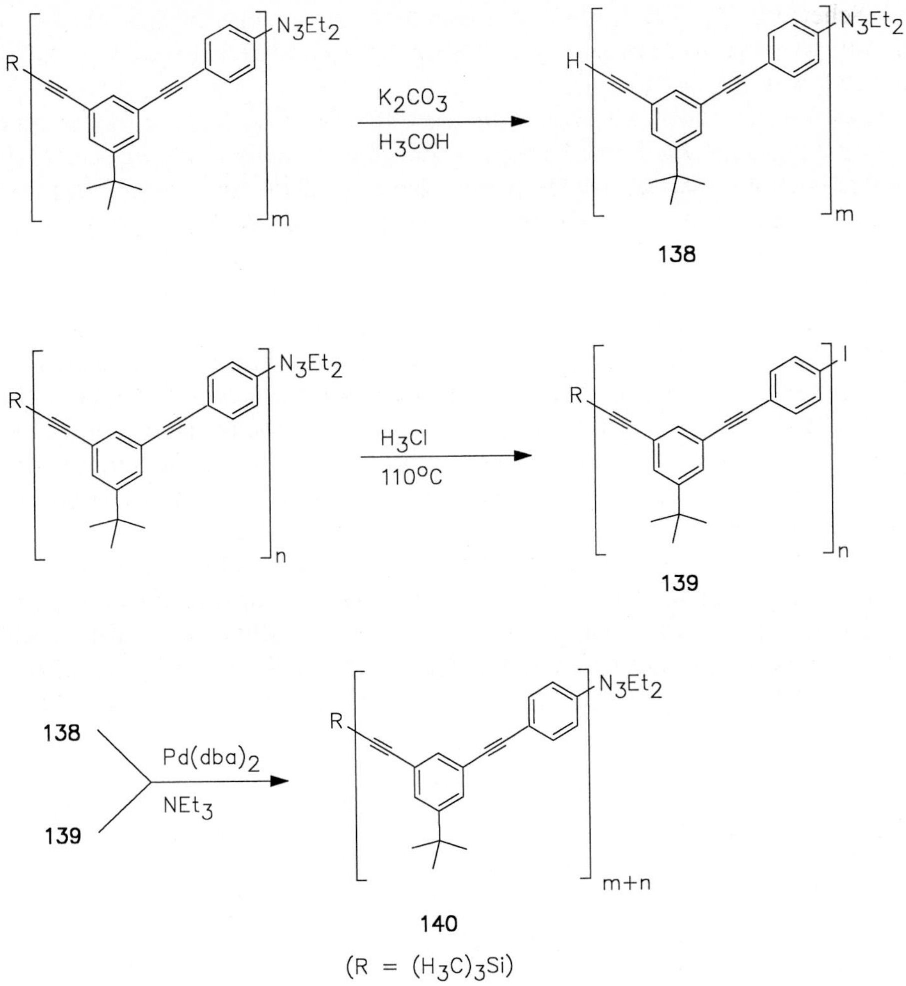

Scheme 5.31. An iterative procedure for the construction of phenylacetylene oligomers.

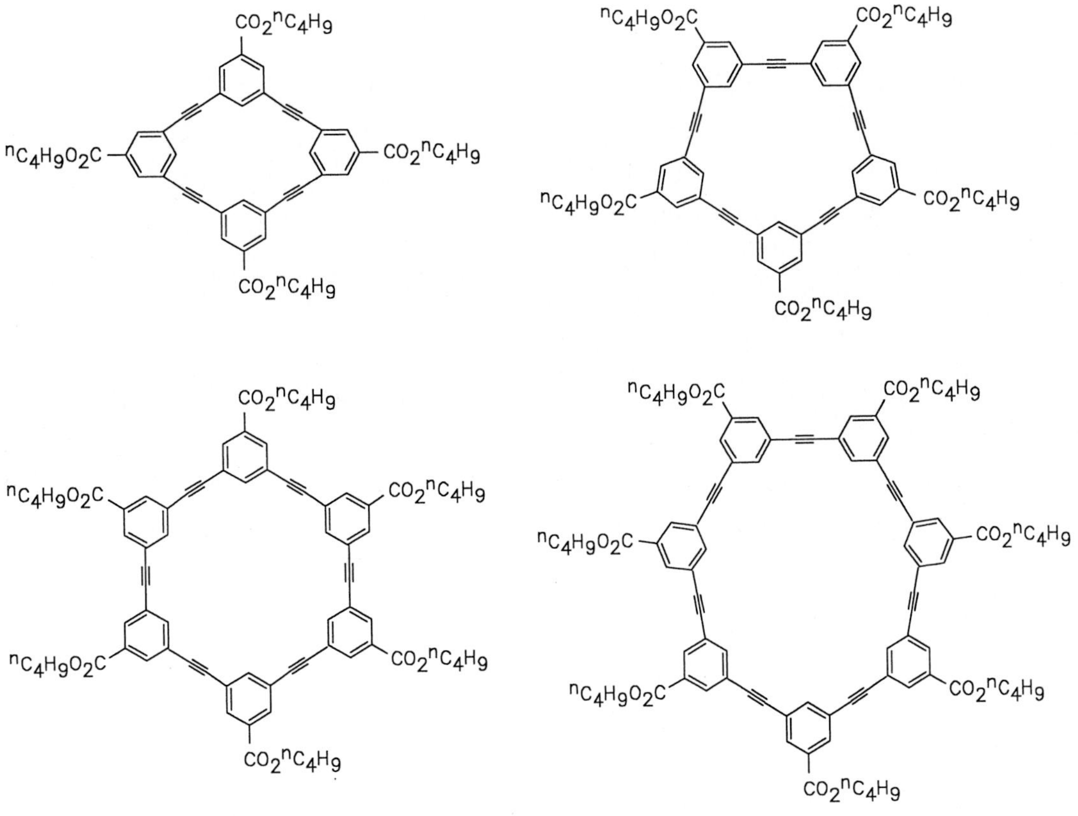

Figure 5.9. The versatile technology by Moore et al.[101] has been used for the creation of *phenylacetylene macrocycles* PAMs).

is illustrated in Scheme 5.31, whereby units of equal length (i.e., **138** and **139**; *m = n =* 8) were coupled via the standard Pd(bda)$_2$ terminal alkyne−aryliodide reaction to afford the 16-phenylacetylene oligomer **140**.

Moore and coworkers[101a] further reported the use of the iterative scheme for the synthesis of 'geometrically-controlled and site-specifically-functionalized *p*henyl*a*cetylene *m*acrocycles' (*PAM*s). Examples are depicted in Figure 5.9. The solid-state characteristics of the hydrocarbon frameworks 'offer potential for producing a set of modular building blocks to rationally assemble molecular crystals and ligand crystals'. A freely hinged macrotricycle possessing a molecular cavity with dimensions of $36 \times 12 \times 12$ Å (5184 Å^3) has also been constructed.[101b]

Kawaguchi and Moore[102] have described the construction of 'ball & chain' dendritic polymers for comparative purposes with 'monodendron-based' copolymers (Figure 5.10; **141** & **142**, respectively). Architecturally, the 'ball & chain' design features a spherical or globular dendrimer, possessing a single functionally differentiated terminal group connected to a linear polymeric chain. This is in contrast to the monodendron topology, which arises from connection of the linear polymer to the focal point of a dendritic wedge. The convergent strategy employed for the preparation of 'site-specified' phenylacetylene dendrimers included construction of (1) dendritic wedges with 3,5-di-*tert*-butyl phenyl terminated groups, (2) site-specified wedges containing one carboxaldehyde moiety at the periphery, and (3) spherical dendrimers with one functionally differentiated terminal group. The strategy is similar to Fréchet's procedure for the control of surface functionality.[61]

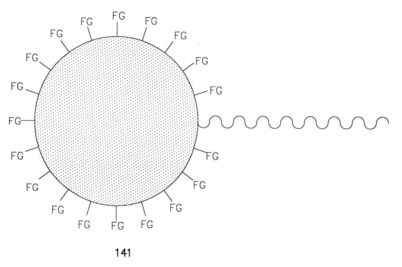

141

"Ball and chain" copolymer architecture

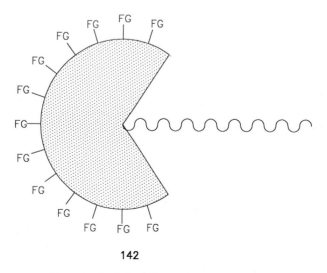

142

Monodendron−based copolymer

Figure 5.10. "Ball and chain"-type dendrimers have been constructed for comparison to monodendronbased macromolecules.

Bis- and tetrakis-terminated alkyne cores (**143** & **144**, respectively) were prepared via standard iterative transformations[95] (Scheme 5.32). These hydrido-terminated cores were then used for the preparation of the *tert*-butyl dendritic wedges (e.g., **145**; Scheme 5.33) containing up to 63 aryl 1 → 2 branching centers.

Scheme 5.34 shows the transformations used for the synthesis of dendritic wedges (**147**) that are *di*functionalized with aryl iodide and trimethylsilyl protected arylalkyne moieties at the focal region. Reaction of 3,5-dibromobenzotriazene (**148**) with one equivalent of isopropoxy-protected acetylene **149**, followed by treatment with base, gave the 3-bromo-5-ethynylaryltriazene (**150**). Higher selectivity of Pd-catalyzed alkynylation of aryl iodides (e.g., **146**) afforded the difunctionalized wedge (**151**).

Monocarboxaldehyde wedges (**156**) were prepared analogously (Scheme 5.35) by coupling one equivalent each of 4-ethynylbenzenecarboxaldehyde (**152**) and 1,3-di-*tert*-butyl-5-ethynylbenzene (**153**) to an aryl dibromide (**148**) to afford the monoaldehyde diazene building block (**154**). Conversion of the diazene moiety to an aryl iodide and treatment with trimethylsilylacetylene, followed by silane removal, allowed the preparation of ynealdehyde **155**. Reaction of difunctional building block **154** with the corresponding aryl iodide derived from **145** afforded the monofunctionalized phenylacetylene dendrimer **156**.

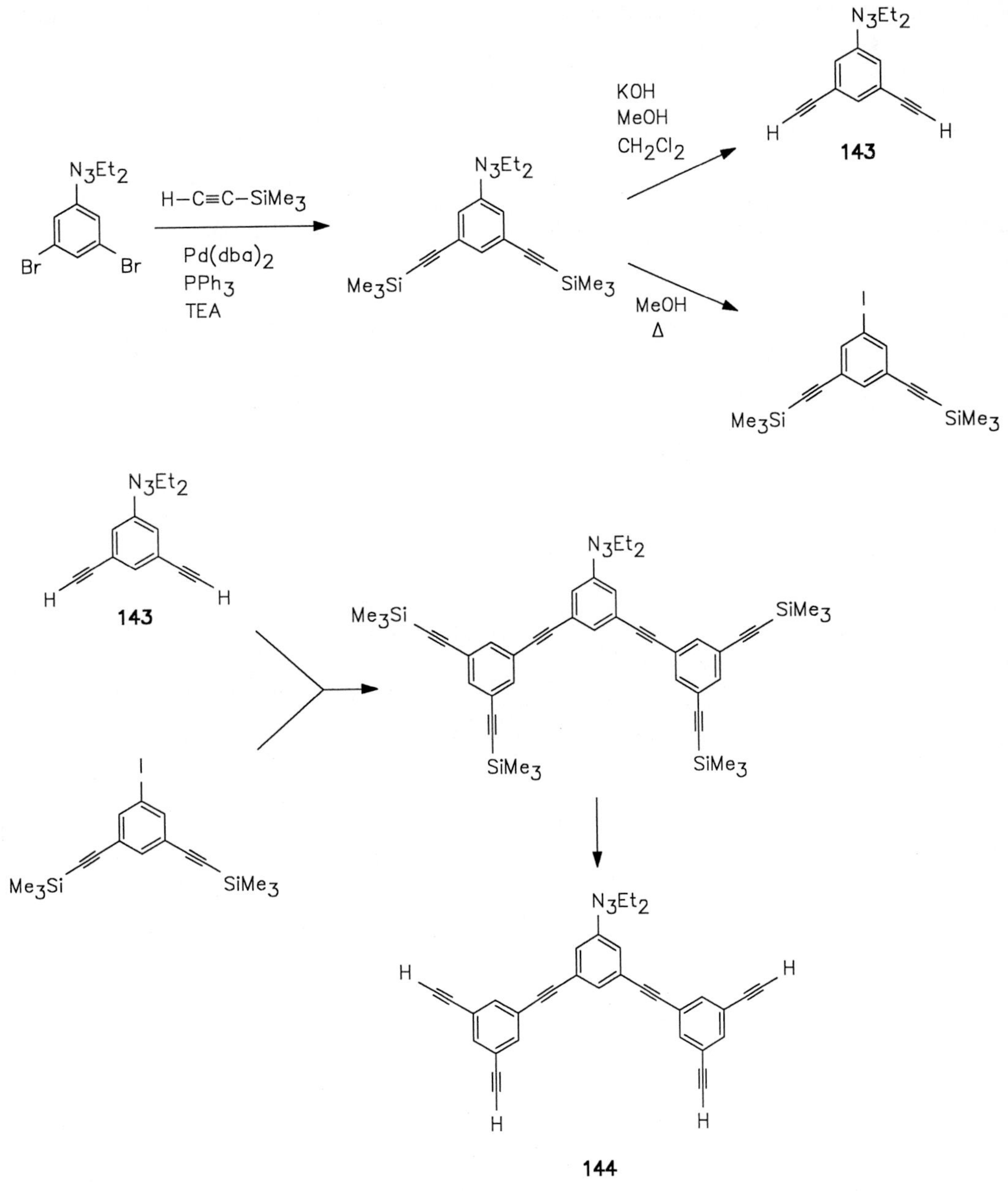

Scheme 5.32. Preparation of polyalkyne terminated building blocks.

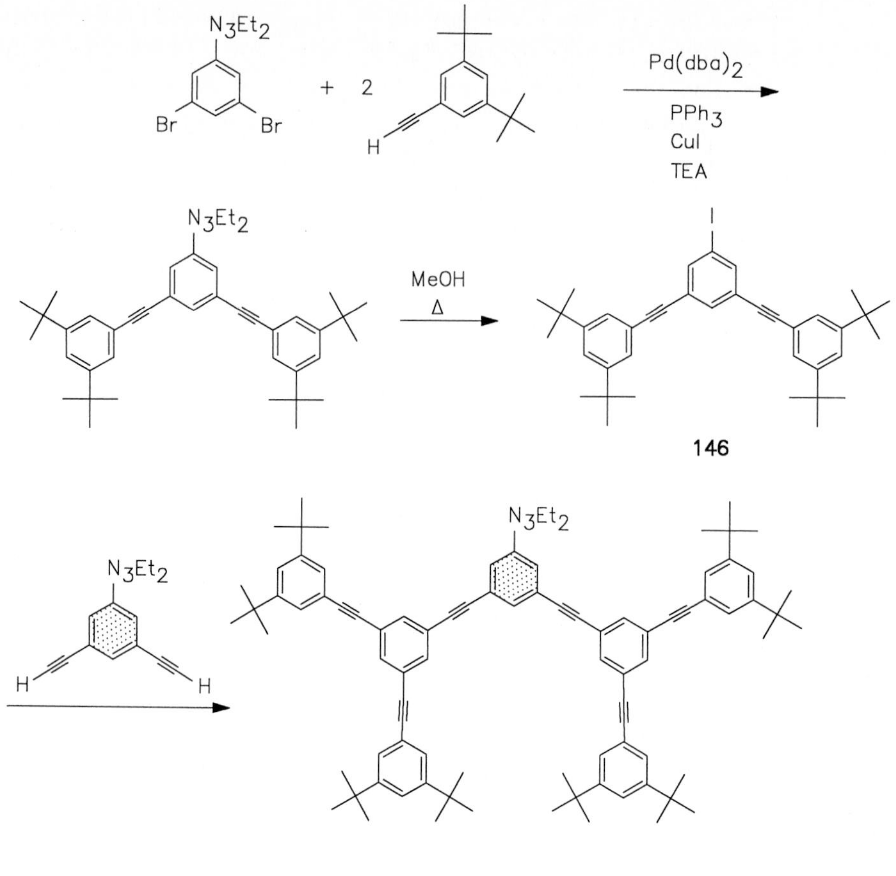

Scheme 5.33. Construction of dendrons possessing diethyltriazene focal groups.

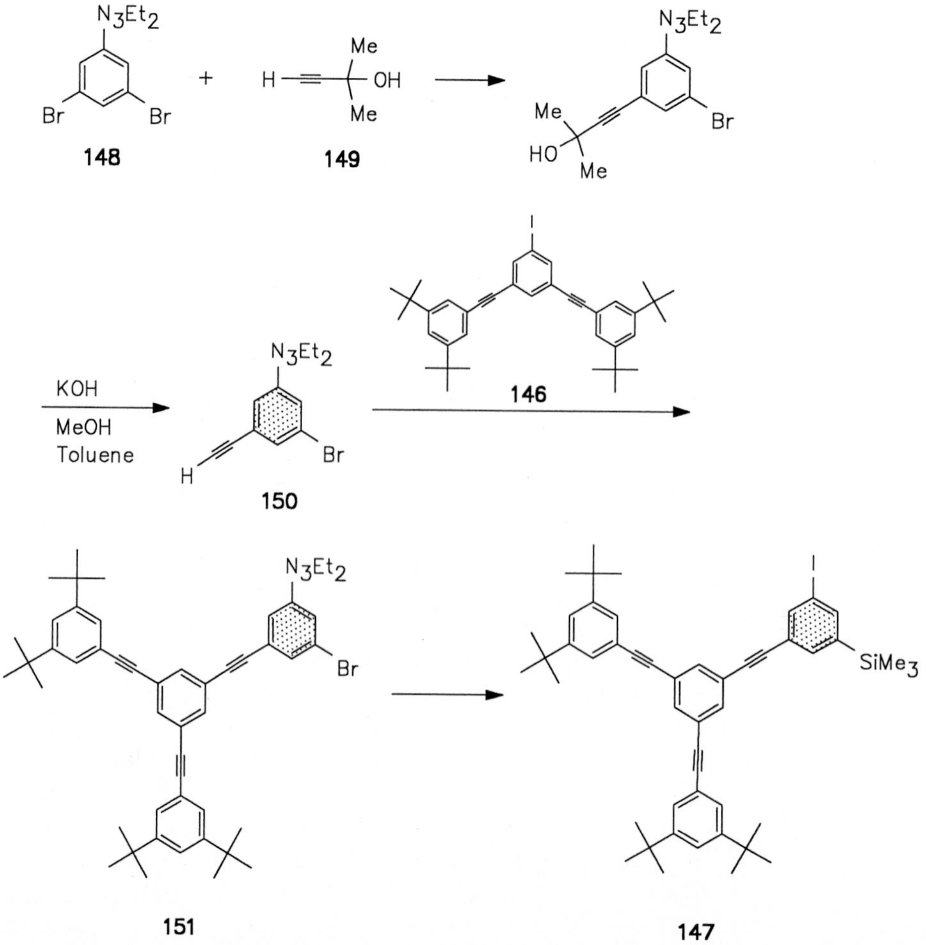

Scheme 5.34. Preparation of dendrons possessing dual functionality at the focal region.

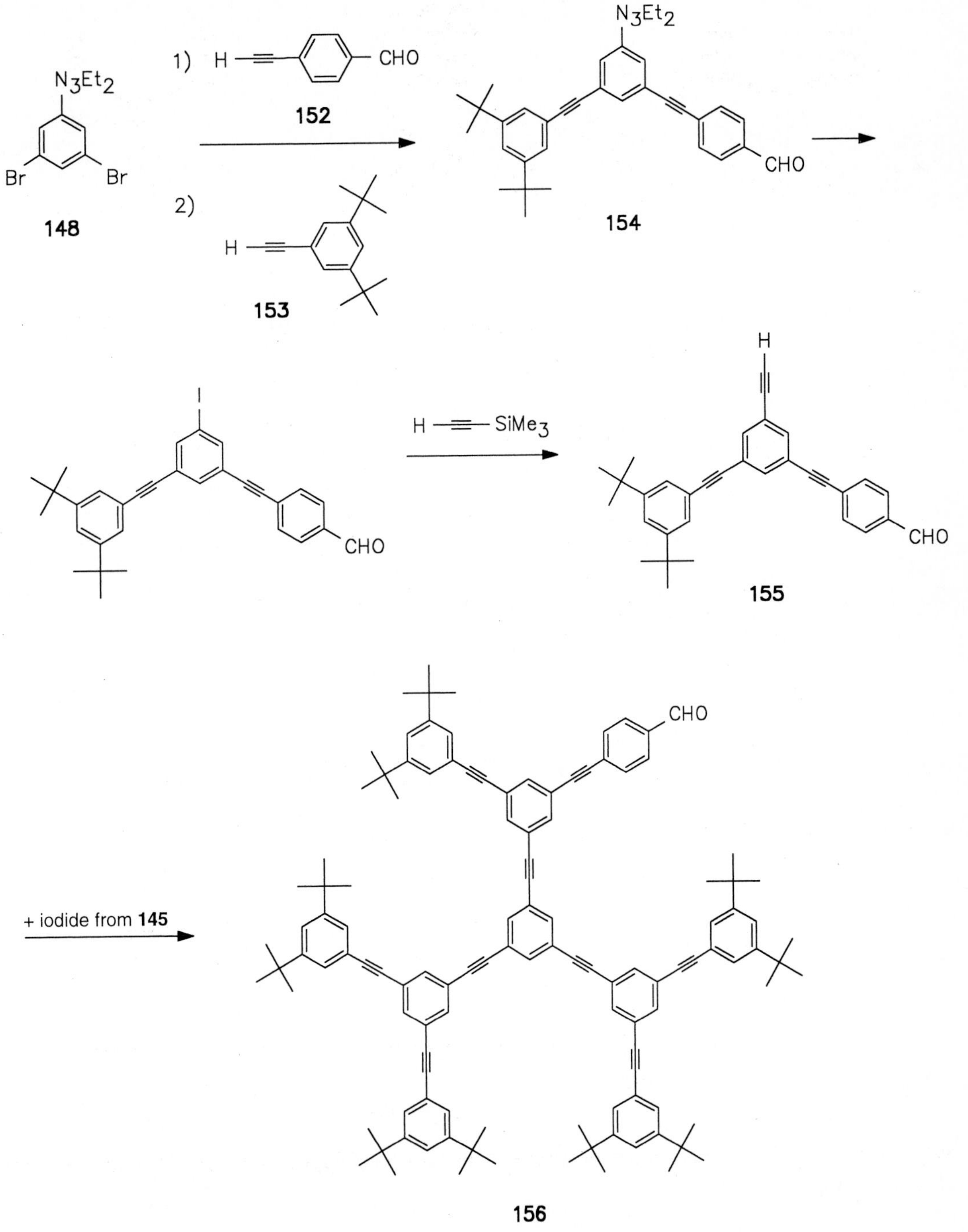

Scheme 5.35. Sequence for the construction of "site-specified" phenylacetylene dendrimers.

Using this technology, the preparation of the fifth tier, monoaldehyde dendrimer was reported.[102] Subsequently, living poly(methyl methacrylate) (PMMA, prepared by group transfer polymerization) was treated with the site-specific cascades. Polydispersities determined for the copolymer were similar to those recorded for the living polymer when smaller dendrimers were used, whereas the use of larger dendrimers for copolymer formation leads to a dependence of polydispersity on the dendrimer.

Xu and Moore[103] proposed the use of their dendritic arylacetylenes, as directional molecular antennas. The directional transduction of energy through a convergent pathway is conveniently incorporated in the macromolecular framework of **157**. The three-directional branching centers are nodes in the molecular electronic wave functions and the phenylacetylene connectors are localized regions of extended π-conjugation. A luminescent probe, perylene, was appended at the dendron's core in order to test the hypothesis of molecular harvesting of light energy. Energy is collected at the outer surface and funneled through the convergent cascade until it reaches the single fluorescent emitter. Figure 5.11 depicts the collection and transmission route through the dendron. Prelimi-

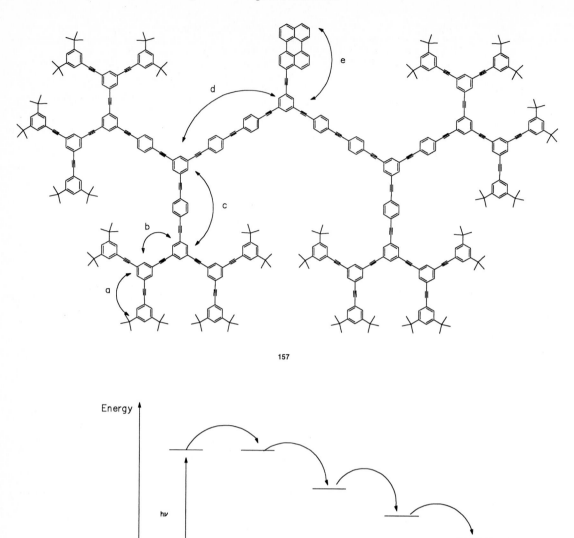

157

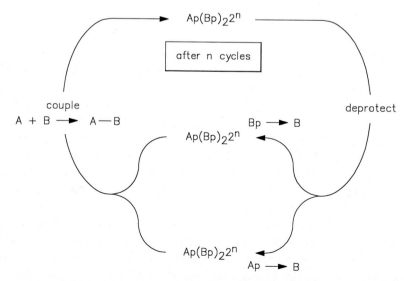

Figure 5.11. Rigid phenylacetylene dendrimers have been employed as "molecular antennas" for the directional transduction of energy.

Figure 5.12. Moore et al.'s[105] scheme for "double exponential dendrimer growth".

nary photophysical properties of dendrimer **157** indicate that perylene emission after excitation at a wavelength (312 nm) corresponding to the absorption maximum of the peripheral aryl groups yields an enhanced intensity when compared with that of 3-ethynylperylene alone.

Moore et al.[104–106a] reported a 'double exponential dendrimer growth', which is an accelerated convergent process for the synthesis of monodendrons by means of a bidirectional procedure, where the degree of polymerization follows Equation (5.1).

$$dp = s^{2n} - 1 \qquad\qquad (5.1)$$

This is a dramatic increase in the average polymer growth considering that for a $1 \rightarrow 2$ branching pattern and using a convergent procedure, the degree of polymerization follows Equation (5.2).

$$dp = 2^{(n+1)} - 1 \qquad\qquad (5.2)$$

This double exponential growth scheme is represented by the reaction−deprotection cyclic diagram shown in Figure 5.12. The significance of the strategy is reinforced by a comparison of dendritic growth by examination of the degree of polymerization using this procedure versus simple convergent growth. Theoretically, the degree of polymerization using the double exponential method could potentially reach 250 after three generations (or after only nine synthetic steps), while by a simple convergent route this degree of polymerization would take 21 synthetic steps or after 7 generations. The advantages are analogous to the "double stage convergent method"[26] but also the hypercore grows concomitantly with peripheral monodendrons; the disadvantages are the need for a pair of orthogonal protecting groups and the rapid onset of steric crowding.

Martin et al.[106b] have employed low-dose techniques to obtain selected area electron diffraction and high-resolution electron microscopy of these phenylacetylene dendrimers.

5.5   1 → 2 *N*-Branched

5.5.1   1 → 2 *N*-Branched, *Amide* Connectivity

Uhrich and Fréchet[107] reported a convergent approach to polyamido cascades (Scheme 5.36). The iterative synthesis was accomplished by coupling two equivalents of a secondary amine **158** to the *N*-BOC-protected aminodicarboxylic acid (**159**). Subsequent deprotection (TFA) of bisamide **160** afforded a new secondary amine **161**, which in turn was acylated with 0.5 equivalents of the imini-protected diacid **159**. The procedure was repeated three times affording the third generation polyamide **162**. Higher generations were not prepared due to decreasing reactivity at the focal amine, which was attributed to steric inhibition as well as difficulties with purification.

Tomalia et al.[108] described the preparation of PAMAM cascades[109] by the construction of dendritic wedges (see Section 4.3.2) possessing a reactive focal group for a convergent-type construction. The resultant branched macromolecules are prepared by an iterative treatment of an aminoalcohol with methyl acrylate, followed by ethylenediamine, which afforded the multigram synthesis of $1 \rightarrow 2$ amino-branched dendrimers. Dendritic termini were transformed to diverse groups (-CO_2H, -CN, -$CONH_2$, -OH, or -SH) via standard methods from either a CO_2Me or NH_2 moiety. Implementation of these dendrons, as building blocks, in dendrimer syntheses with three-directional cores, e.g. **86** (Scheme 5.27), allows the preparation of polytermini differentiated dendrimers. Peripheral differentiation results from statistical reactivity and hence product distribution relies on collision frequency.

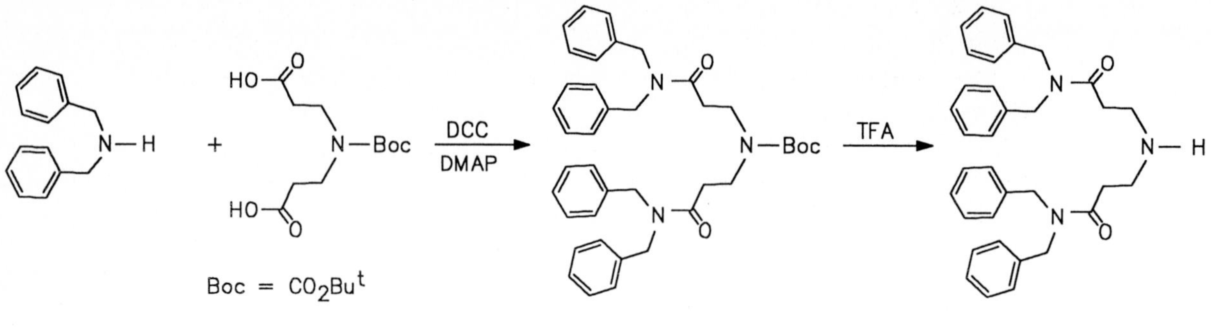

158 **159** **160** **161**

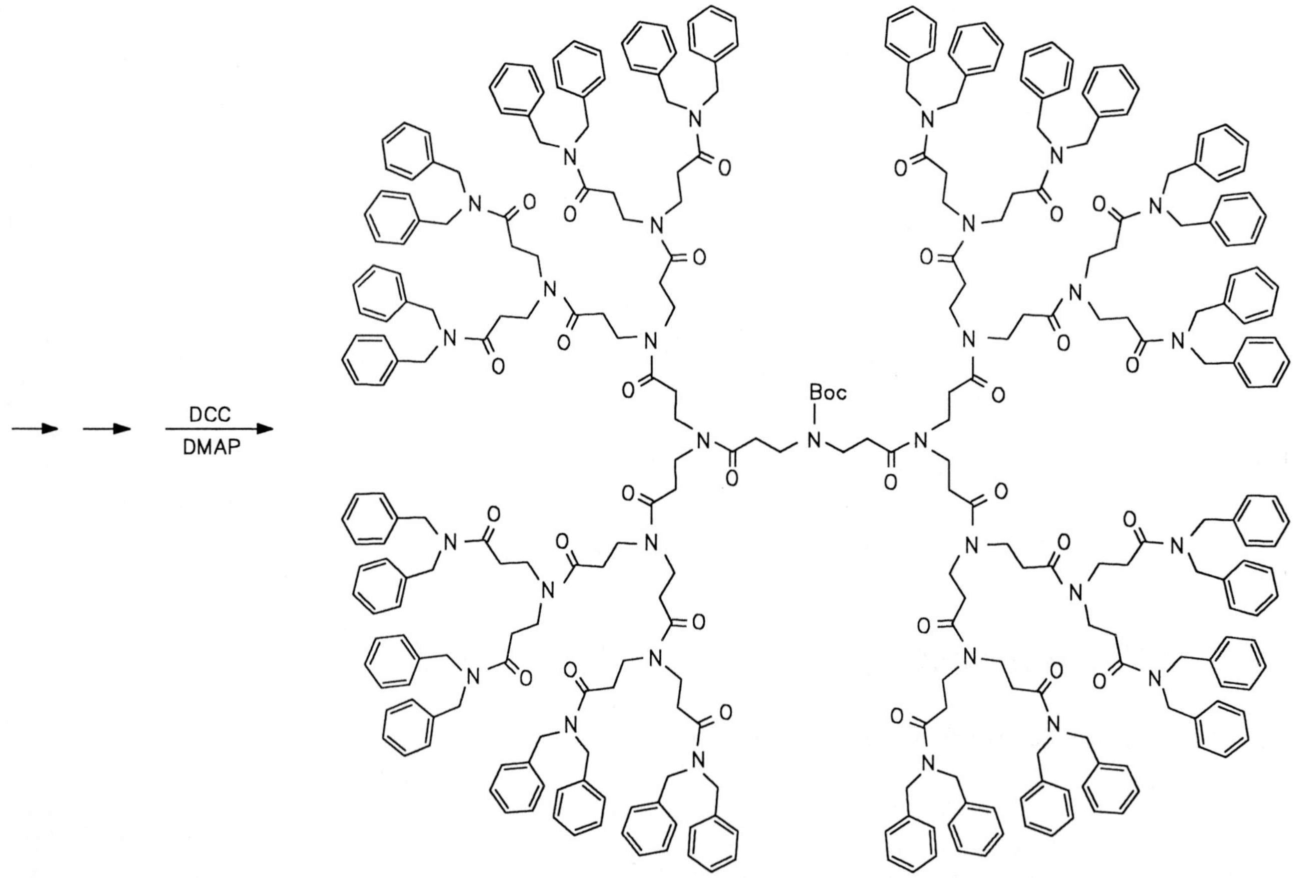

162

Scheme 5.36. Construction of *N*-benzyl terminated amide-based dendrimers.

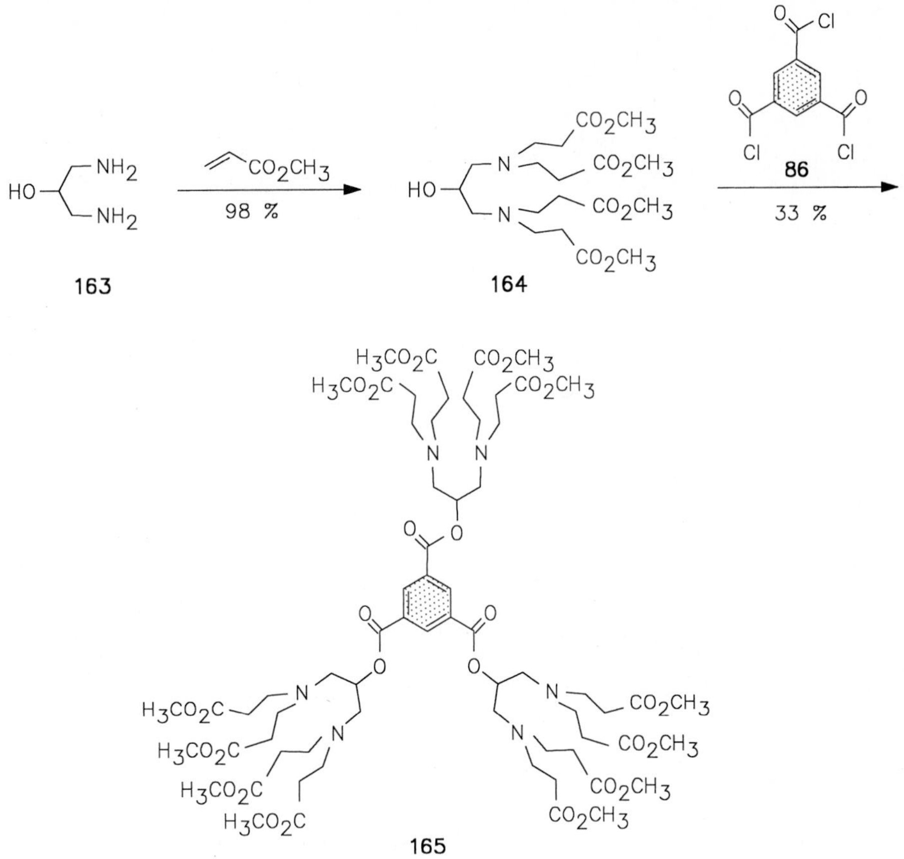

Scheme 5.37. Preparation of small, aminoester-based dendrimers.

5.6 1 → 2 *C*- & *N*-Branched, *Ester* Connectivity

Mitchell et al.[110] reported the use of 1,3-diaminopropan-2-ol (**163**) as a building block that incorporates both 1 → 2 *C*- as well as 1 → 2 *N*-branching. Michael-type addition of alkyl acrylates (Scheme 5.37) to diamine **164** provided the unique alcohol, which was then treated with the acyl chloride **86** to give dodecaester **165**. Using this technology and the appropriate α,β-unsaturated carbonyl, i.e., simple alkyl or phenyl acrylates, a series of ester terminated, medium sized dendrimers was prepared. Size exclusion chromatography did not give reasonable relative molecular masses for the higher molecular weight members in this series; however, FAB MS correlated precisely with the calculated masses of the dendrimers.

5.7 1 → 2 *Si*-Branched, *Silyloxy* Connectivity

Imai et al.[111, 112] synthesized an initial series of branched polysiloxane polymers based on tris[(phenyldimethylsiloxy)dimethylsiloxy]methylsilane, as the core, and bis[(phenyldimethylsiloxy)methylsiloxy]dimethylsilanol, as the building block. The construction of a series of polysiloxane dendrimers based on allylbis[4-(hydroxydimethylsilyl)phenyl]methylsilane (**166**), as the key building block, was then reported by Morikawa and co-workers.[113] The general procedure is depicted in Scheme 5.38. Conversion of an allyl moiety to an (*N,N*-diethylamino)silane (**167**) was accomplished by Pt-mediated hydrosilylation, followed by reaction with HNEt$_2$. Treatment of monomer **166** with two equivalents of silylamine **167** afforded the second generation tetranitrile **168**. Analogously, the polysiloxane "wedges" possessing 8 and 16 terminal groups, corresponding to third and fourth generation dendrons, respectively, were prepared. The attachment of the third generation wedge to the tris(hydroxysilyl)silane, as a core, afforded a three-directional cascade (not shown). The larger polymers showed a higher glass transition temperature (T_g) than the smaller homologues in a range from −61 °C to 36 °C.

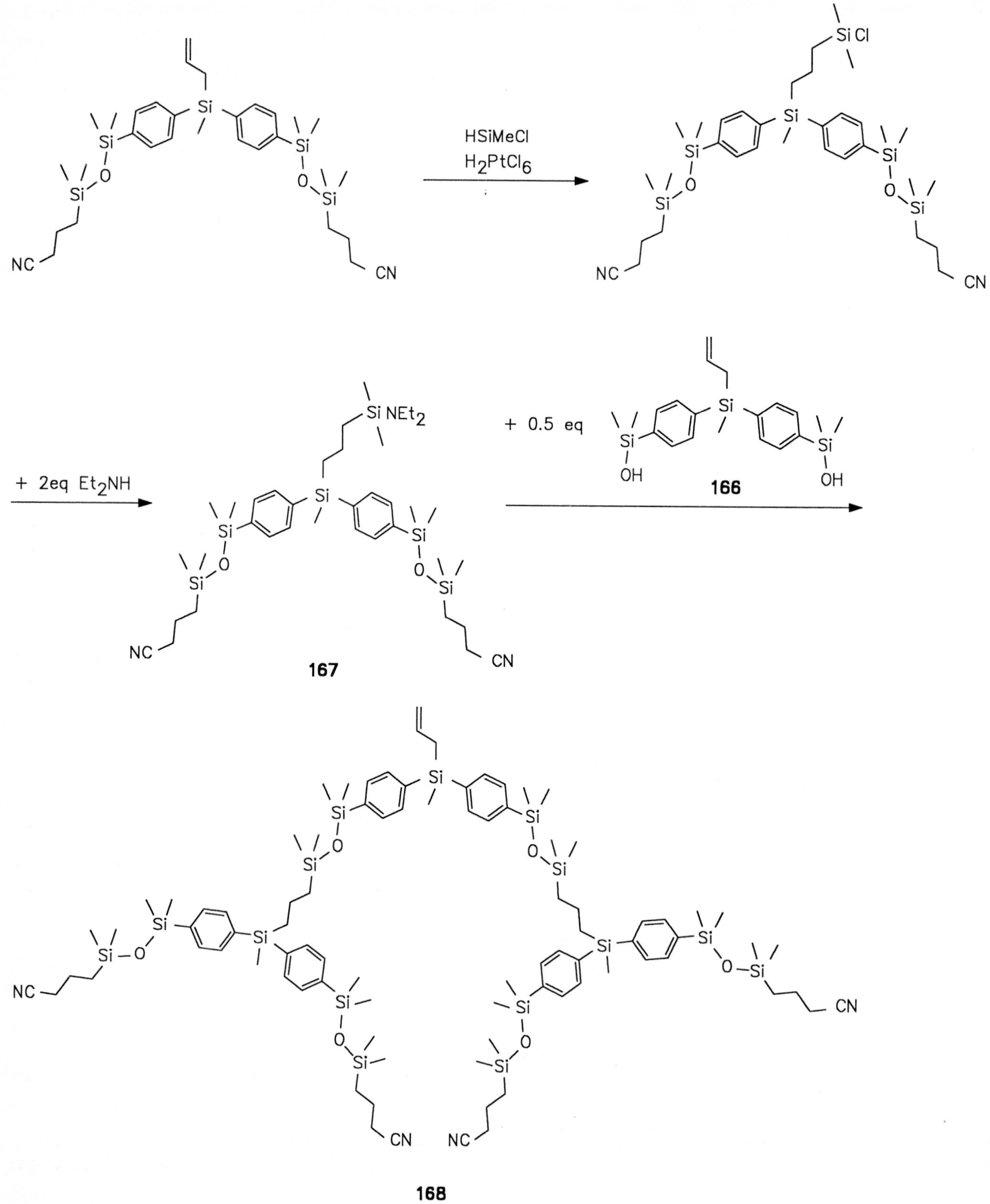

Scheme 5.38. The convergent preparation of polysiloxane dendrimers also employing silicon as a branching center.

5.8 References

[1] C. Hawker, J. M. J. Fréchet, *J. Chem. Soc., Chem. Commun.* **1990,** 1010.

[2] J. M. J. Fréchet, C. J. Hawker, A. E. Phillippides, US Pat. 5,041,516 (Aug. 20, **1991**).

[3] T. M. Miller, T. X. Neenan, *Chem. Mater.* **1990,** *2,* 346.

[4] J. M. J. Fréchet, *Science* **1994,** *263,* 1710.

[5] J. M. J. Fréchet, C. J. Hawker, K. L. Wooley, *J. Macromol. Sci., Pure Appl. Chem.* **1994,** *A31,* 1627.

[6] A. Rajca, *J. Org. Chem.* **1991,** *56,* 2557.

[7] A. Rajca, *J. Am. Chem. Soc.* **1990,** *112,* 5889.

[8] A. Rajca, *J. Am. Chem. Soc.* **1990,** *112,* 5890.

[9] A. Rajca in *Advances in Dendritic Macromolecules,* Vol. 1, (Ed.: G. R. Newkome), JAI Press, Greenwich, Connecticut, **1994,** p.133.

[10] A. Rajca, *Chem. Rev.* **1994,** *94,* 871.

[11] A. Rajca, *Adv. Mater.* **1994,** *6,* 605.

[12] A. Rajca, S. Utamapanya, *J. Am. Chem. Soc.* **1993,** *115,* 2396.

[13] K. Nakatani, J. Y. Carriat, Y. Journaoux, O. Kahn, F. Lloret, J. P. Renard, Y. Pei, J. Sletten, M. Verdaguer, *J. Am. Chem. Soc.* **1989,** *111,* 5739.

[14] A. Caneschi, D. Gatteschi, R. Sessoli, P. Rey, *Acc. Chem. Res.* **1989,** *22,* 392.

[15] H. Iwamura, *Adv. Phys. Org. Chem.* **1990,** *26,* 179.

[16] D. A. Dougherty, *Acc. Chem. Res.* **1991,** *24,* 88.

[17] A. Rajca, S. Utamapanya, S. Thayumanavan, *J. Am. Chem. Soc.* **1992,** *114,* 1884.

[18] S. Utamapanya, A. Rajca, *J. Am. Chem. Soc.* **1991,** *113,* 9242.

[19] a) A. Rajca, S. Utamapanya, J. Xu, *J. Am. Chem. Soc.* **1991,** *113,* 9235. b) K. Inoue, T. Hayamizu, H. Iwamura, D. Hashizume, Y. Ohashi, *J. Am. Chem. Soc.* **1996,** *118,* 1803.

[20] A. Rajca, S. Utamapanya, *J. Am. Chem. Soc.* **1993,** *115,* 10688.

[21] A. Rajca, S. Janicki, *J. Org. Chem.* **1994,** *59,* 7099.

[22] A. Rajca, S. Utamapanya, *Mol. Cryst. Liq. Cryst.* **1993,** *223,* 305.

[23] A. Rajca, S. Rajca, R. Padmakumar, *Angew. Chem.* **1994,** *106,* 2193; *Angew. Chem., Int. Ed. Engl.* **1994,** *33,* 2091.

[24] A. Rajca, S. Rajca, S. R. Desai, *J. Am. Chem. Soc.* **1995,** *117,* 806.

[25] K. Matsuda, N. Nakamura, K. Takahasi, K. Inoue, N. Koga, H. Iwamura, *J. Am. Chem. Soc.* **1995,** *117,* 5550.

[26] K. L. Wooley, C. J. Hawker, J. M. J. Fréchet, *J. Am. Chem. Soc.* **1991,** *113,* 4252.

[27] C. J. Hawker, J. M. J. Fréchet, *J. Am. Chem. Soc.* **1990,** *112,* 7638.

[28] K. L. Wooley, C. J. Hawker, J. M. Pochan, J. M. J. Fréchet, *Macromolecules* **1993,** *26,* 1514.

[29] H. Nemoto, J. G. Wilson, H. Nakamura, Y. Yamamoto, *J. Org. Chem.* **1992,** *57,* 435.

[30] a) H. Nemoto, J. Cai, Y. Yamamoto, *J. Chem. Soc., Chem. Commun.* **1994,** 577. b) V. Percec, P. Chu, G. Ungar, J. Zhou, *J. Am. Chem. Soc.* **1995,** *117,* 11441.

[31] M. J. Damha, S. Zabarylo, *Tetrahedron Lett.* **1989,** *30,* 6295.

[32] R. H. E. Hudson, M. J. Damha, *J. Am. Chem. Soc.* **1993,** *115.* **2119.**

[33] T. M. Miller, T. X. Neenan, R. Zayas, H. E. Bair, *J. Am. Chem. Soc.* **1992,** *114,* 1018.

[34] T. M. Miller, T. X. Neenan, H. E. Bair, *Polym. Preprints* **1991,** *32*(3), 627.

[35] N. Miyaura, T. Yanagai, A. Suzuki, *Synth. Commun.* **1981,** *11,* 513.

[36] T. M. Miller, T. X. Neenan, R. Zayas, H. E. Bair, *Polym. Preprints* **1991,** *32*(3), 599.

[37] E. Reimann, *Chem. Ber.* **1969,** *102,* 2887.

[38] T. H. Mourey, S. R. Turner, M. Rubinstein, J. M. J. Fréchet, C. J. Hawker, K. L. Wooley, *Macromolecules* **1992,** *25,* 2401.

[39] R. L. Lescanec, M. Muthukumar, *Macromolecules* **1990,** *23,* 2280.

[40] P.-G. de Gennes, H. Hervet, *J. Phys. Lett.* **1983,** *44,* L 351.

[41] P. M. Saville, J. W. White, C. J. Hawker, K. L. Wooley, J. M. J. Fréchet, *J. Phys. Chem.* **1993,** *97,* 293.

[42] C. J. Hawker, P. J. Farrington, M. E. Mackay, K. L. Wooley, J. M. J. Fréchet, *J. Am. Chem. Soc.* **1995,** *117,* 4409.

[43] I. Gitsov, K. L. Wooley, J. M. J. Fréchet, *Angew. Chem.* **1992,** *104,* 1282; *Angew. Chem., Int. Ed. Engl.* **1992,** *31,* 1200.

[44] I. Gitsov, K. L. Wooley, C. J. Hawker, P. T. Ivanova, J. M. J. Fréchet, *Macromolecules* **1993,** *26,* 5621.

[45] I. Gitsov, J. M. J. Fréchet, *Macromolecules* **1993,** *26,* 6536.

[46] I. Gitsov, J. M. J. Fréchet, *Macromolecules* **1994,** *27,* 7309.

[47] I. Schipor, J. M. J. Fréchet, *Polym. Preprints* **1994,** *35*(2), 480.

[48] I. Gitsov, P. T. Ivanora, J. M. J. Fréchet, *Macromol. Rapid Commun.* **1994,** *15,* 387.

[49] I. Gitsov, K. L. Wooley, C. J. Hawker, J. M. J. Fréchet, *Polym. Preprints* **1991**, *32*(3), 631.
J. M. J. Fréchet, I. Gitsov, *Macromol. Symp.* **1995**, *98*, 441.

[50] C. J. Hawker, J. M. J. Fréchet, *Polymer* **1992**, *33*, 1507.

[51] K. L. Wooley, C. J. Hawker, J. M. J. Fréchet, F. Wudl, G. Srdanov, S. Shi, C. Li, M. Kao,
J. Am. Chem. Soc. **1993**, *115*, 9836.

[52] F. Diederich, L. Isaacs, D. Philp, *Chem. Soc. Rev.* **1994**, *23*, 243.

[53] C. J. Hawker, K. L. Wooley, J. M. J. Fréchet, *J. Chem. Soc., Chem. Commun.* **1994**, 925.

[54] C. J. Hawker, K. L. Wooley, J. M. J. Fréchet, *J. Am. Chem. Soc.* **1993**, *115*, 4375.

[55] K. L. Wooley, C. J. Hawker, J. M. J. Fréchet, *J. Am. Chem. Soc.* **1993**, *115*, 11496.

[56] C. J. Hawker, K. L. Wooley, J. M. J. Fréchet, *Polym. Preprints* **1991**, *32*(3), 623.

[57] E. M. Sanford, J. M. J. Fréchet, K. L. Wooley, C. J. Hawker, *Polym. Preprints* **1993**, *34*, 654.

[58] C. J. Hawker, K. L. Wooley, J. M. J. Fréchet, *J. Chem. Soc., Perkin Trans. 1* **1993**, 1287.

[59] C. J. Hawker, K. L. Wooley, J. M. J. Fréchet, *Polym. Preprints* **1993**, *34*, 54.

[60] C. J. Hawker, J. M. J. Fréchet, *Macromolecules* **1990**, *23*, 4726.

[61] K. L. Wooley, C. J. Hawker, J. M. J. Fréchet, *J. Chem. Soc., Perkin Trans. 1* **1991**, 1059.

[62] a) L. Lochmann, K. L. Wooley, P. T. Ivanova, J. M. J. Fréchet, *J. Am. Chem. Soc.* **1993**, *115*,
7043. b) R. Sadamoto, N. Tomioka, T. Aida, *J. Am. Chem. Soc.* **1996**, *118*, 3978.

[63] R.-H. Jin, T. Aida, S. Inoue, *J. Chem. Soc., Chem. Commun.* **1993**, 1260.

[64] G. L'abbé, B. Haelterman, W. Dehaen, *J. Chem. Soc., Perkin Trans. 1* **1994**, 2203. G. L'abbé,
W. Dehaen, B. Haelterman, D. Vangenengden, *Acros Org. Acta* **1995** *1*, 61.

[65] A.-D. Schülter, *Polym. Preprints* **1995**, *36*(1), 745.

[66] W. Claussen, N. Schulte, A.-D. Schülter, *Macromol. Rapid Commun.* **1995**, *16*, 89.

[67] a) R. Freudenberger, W. Claussen, A.-D. Schülter, H. Wallmeier, *Polymer* **1994**, *35*, 4496. b)
H.-F. Chow, I. Y.-K. Chan, C. C. Mak, M.-K. Hg, *Tetrahedron* **1996**, *52*, 4277. c) I. Gitsov,
J. M. J. Fréchet, *J. Am. Chem. Soc.* **1996**, *118*, 3785. d) S. Serroni, S. Campagna, A. Juris, M.
Venturi, V. Balzani, G. Denti, *Gazz. Chim. Ital.* **1994**, *124*, 123.

[68] P. M. Bayliff, W. J. Feast, D. Parker, *Polym. Bull.* **1992**, *29*, 265.

[69] S. C. E. Backson, P. M. Bayliff, W. J. Feast, A. M. Kenwright, D. Parker, R. W. Richards,
Macromol. Symp. **1994**, *77*, 1.

[70] a) Also see: S. C. E. Backson, P. M. Bayliff, W. J. Feast, A. M. Kenwright, D. Parker, R. W.
Richards, *Polym. Preprints* **1993**, *34*(1), 50. b) A. Kraft, *J. Chem. Soc., Chem. Commun.*
1996, 77. c) J. Bettenhausen, P. Strohriegl, *Polym. Preprints* **1996**, *37*(1), 504.

[71] G. R. Newkome, X. Lin, J. K. Young, *Synlett* **1992**, 53.

[72] G. R. Newkome, X. Lin, *Macromolecules* **1991**, *24*, 1443.

[73] T. Nagasaki, M. Ukon, S. Arimori, S. Shinkai, *J. Chem. Soc., Chem. Commun.* **1992**, 608.

[74] T. Nagasaki, O. Kimura, M. Ukon, S. Arimori, I. Hamachi, S. Shinkai, *J. Chem. Soc., Perkin
Trans 1* **1994**, 75.

[75] E. Agar, B. Bati, E. Erdem, M. Özdemir, *J. Chem. Res. (S),* **1995**, 16.

[76] R. Spindler, J. M. J. Fréchet, *J. Chem. Soc., Perkin Trans. 1* **1993**, 913.

[77] E. W. Kwock, T. X. Neenan, T. M. Miller, *Chem. Mater.* **1991**, *3*, 775.

[78] T. M. Miller, E. W. Kwock, T. X. Neenan, *Macromolecules* **1992**, *25*, 3143.

[79] E. W. Kwock, T. X. Neenan, T. M. Miller, *Polym. Preprints* **1991**, *32*(3), 635.

[80] E. W. Kwock, T. X. Neenan, T. M. Miller, *Chem. Mater.* **1991**, *3*, 775.

[81] H. S. Sahota, P. M. Lloyd, S. G. Yeates, P. J. Derrick, P. C. Taylor, D. M. Haddleton,
J. Chem. Soc., Chem. Commun. **1994**, 2445.

[82] M. R. Bryce, A. S. Batsanov, W. Devonport, J. N. Heaton, J. A. K. Howard, G. J. Moore, J.
P. Skabara, S. Wegener, in *Molecular Engineering for Advanced Materials,* (Ed.: J. Becher),
Kluwer, Dordrecht, **1994**.

[83] G. J. Marshallsay, T. K. Hansen, A. J. Moore, M. R. Bryce, J. Becher, *Synthesis* **1994**, 926.

[84] J. Garin, J. Orduna, S. Uriel, A. J. Moore, M. R. Bryce, S. Wegener, D. D. Yufit, J. A. K.
Howard, *Synthesis* **1994**, 489.

[85] M. R. Bryce, W. Devonport, A. J. Moorek, *Angew. Chem.* **1994**, *106*, 1862; *Angew. Chem.,
Int. Ed. Engl.* **1994**, *33*, 1761.

[86] K. L. Wooley, C. J. Hawker, J. M. J. Fréchet, *Angew. Chem.* **1994**, *106*, 123; *Angew. Chem.
Int. Ed. Engl.* **1994**, *33*, 82.

[87] C. J. Hawker, J. M. J. Fréchet, *J. Chem. Soc. Perkin Trans. 1* **1992**, 2459.

[88] C. J. Hawker, J. M. J. Fréchet, *J. Am. Chem. Soc.* **1992**, *114*, 8405.

[89] A. Morikawa, M.-a. Kakimoto, Y. Imai, *Macromolecules* **1993**, *26*, 6324.

[90] J. S. Moore, Z. Xu, *Macromolecules* **1991**, *24*, 5893.

[91] J. S. Moore, Z. Xu, *Polym. Preprints* **1991**, *32*(3), 629.

[92] D. L. Trumbo, C. S. Marvel, *J. Polym. Sci.* **1986**, *26*, 2311.

[93] N. A. Bumagin, A. B. Ponomarev, I. P. Beletskaya, *Izv. Adad. Nauk SSSR, Ser. Khim.* **1984**,
1561.

[94] Z. Xu, J. S. Moore, *Angew. Chem.* **1993,** *105,* 261; *Angew. Chem., Int. Ed. Engl.* **1993,** *32,* 246.

[95] J. S. Moore, E. J. Weinstein, Z. Wu, *Tetrahedron Lett.* **1991,** *32,* 2465.

[96] K. L. Walker, M. S. Kahr, C. L. Wilkins, Z. Xu, J. S. Moore, *Am. Soc. Mass Spectrom.* **1994,** *5,* 730.

[97] Z. Xu, J. S Moore, *Angew. Chem.* **1993,** *105,* 1394; *Angew. Chem., Int. Ed. Engl.* **1993,** *32,* 1354.

[98] Z. Xu, Z.-Y. Shi, W. Tan, R. Kopelman, J. S. Moore, *Polym. Prints* **1993,** *34*(1), 130.

[99] Z. Xu, J. S. Moore, *Polym. Preprints* **1993,** *34*(1), 128.

[100] J. Zhang, J. S. Moore, Z. Xu, R. A. Aguirre, *J. Am. Chem. Soc.* **1992,** *114,* 2273.

[101] a) J. Zhang, D. J. Pesak, J. L. Ludwick, J. S. Moore, *J. Am. Chem. Soc.* **1994,** *116,* 4227. b) Z. Wu, J. S. Moore, *Angew. Chem., Int. Ed. Engl.* **1996,** *35,* 297. c) C. J. Buchko, P. M. Wilson, Z. Xu, J. Zhang, J. S. Moore, D. C. Martin, *Polymer* **1995,** *36,* 1817.

[102] T. Kawaguchi, J. S. Moore, *Polym. Preprints* **1994,** *35*(2), 872.

[103] Z. Xu, J. S. Moore, *Acta Polym.* **1994,** *45, 83.*

[104] T. Kawaguchi, J. S. Moore, *Polym. Preprints* **1994,** *35*(2), 669.

[105] T. Kawaguchi, K. L. Walker, C. L. Wilkins, J. S. Moore, *J. Am. Chem. Soc.* **1995,** *117,* 2159.

[106] a) Z. Xu, M. Kahr, K. L. Walker, C. L. Wilkins, J. S. Moore, *J. Am. Chem. Soc.* **1994,** *116,* 4537. b) C. J. Buchko, P. M. Wilson, Z. Xu, J. Zhang, J. S. Moore, D. C. Martin, *Polymer* **1995,** *36,* 1817.

[107] K. E. Uhrich, J. M. J. Fréchet, *J. Chem. Soc., Perkin Trans. 1* **1992,** 1623.

[108] D. A. Tomalia, D. R. Swanson, J. W. Klimash, H. M. Brothers III, *Polym. Preprints* **1993,** *34*(1), 52.

[109] D. A. Tomalia, H. Baker, J. Dewald, M. Hall, G. Kallos, S. Martin, J. Roeck, J. Ryder, P. Smith, *Polym. J.* **1985,** *17,* 117.

[110] L. J. Twyman, A. E. Beezer, J. C. Mitchell, *J. Chem. Soc., Perkin Trans. 1* **1994,** 407.

[111] A. Morikawa, M.-a. Kakimoto, Y. Imai, *Macromolecules* **1991,** *24,* 3469.

6 Synthetic Methodologies: One-Step (Hyperbranched) Procedures

6.1 General Concepts

Whereas the well-characterized, perfect (or nearly so) structures of dendritic macromolecules, constructed in discrete stepwise procedures have been described in the preceding chapters, this Chapter reports on the related, less than perfect, hyperbranched polymers, which are synthesized by means of a direct, one-step polycondensation of A_xB monomers, where $x \geq 2$. Flory's prediction and subsequent demonstration[1,2] that A_xB monomers generate highly branched polymers heralded advances in the creation of idealized dendritic systems; thus the desire for simpler, and in most cases more economical, (one-step) procedures to the hyperbranched relatives became more attractive.

Such one-step polycondensations afford products possessing a high degree of branching, but are not as idealized as the stepwise constructed dendrimers. The supramolecular assemblies and micellar properties of these hyperbranched polymers offer synthetic and physical insights as well as noteworthy comparative relationships to the flawless dendritic analogues. The degree of branching of these hyperbranched polymers is generally in the range 55–70 % and is independent of their molecular weights.

Synthetic high molecular weight polymers with spherical symmetry have also been created[3a] by a graft-on-graft procedure (chloromethylation, followed by anionic grafting), which results in tree-like structures, analogous to these cascade hyperbranched materials. The term "arborescent graft polymers"[3b] has been used to describe them. In general, grafting side chains of comparable molecular weight on a linear core forms a "comb-branched" structure leading to materials with increasing globular or spherical shape as generations increase. Such polymers are obtained with molecular weights ideally increasing geometrically as denoted;

$$M = M_b + M_bf + M_bf^2 + ... = \sum_{x=0}^{G+1} M_bf^\infty,$$

where M_b is the molecular mass per branch, f is the branching functionality, which remains constant for each generation G. For the graft-on-graft procedure, high molecular weights ($> 10^6$) were realized after three graftings with $M_w/M_n \approx 1.1 - 1.3$ at each generation. Tomalia, Hedstrand, and Ferrito[3c] have described the synthesis of poly-(ethyleneimine)-based "comb-burst dendrimers"; Monte Carlo simulations[3d] of these structures have been reported.

6.2 1 → 2 *Aryl*-Branched

6.2.1 1 → 2 *Aryl*-Branched and Connectivity

Kim and Webster[4–6] reported the facile one-step conversion[7] of (3,5-dibromophenyl)boronic acid[8] (1) in the presence of a catalytic amount of tetrakis(triphenylphosphine)palladium(0) under reflux conditions in aqueous carbonate to give the hyperbranched polyphenylene 3 (Scheme 6.1).[9,10] An alternative route utilized the mono-Grignard[4] (2), prepared from 1,3,5-tribromobenzene with activated magnesium[11] and bis(triphenylphosphine)nickel chloride. This procedure is advantageous for the large scale runs. Based on ^{13}C NMR spectroscopy, the degree of branching was estimated to be ca. 70 %. The molecular mass of 3 was qualitatively found to be a function of the organic solvent employed in this one-step process. Polymerization with Ni(II) gave polymers of $\overline{N}_n$ in the range 2000–4000 often with greater polydispersity than the boronic acid route. Molecu-

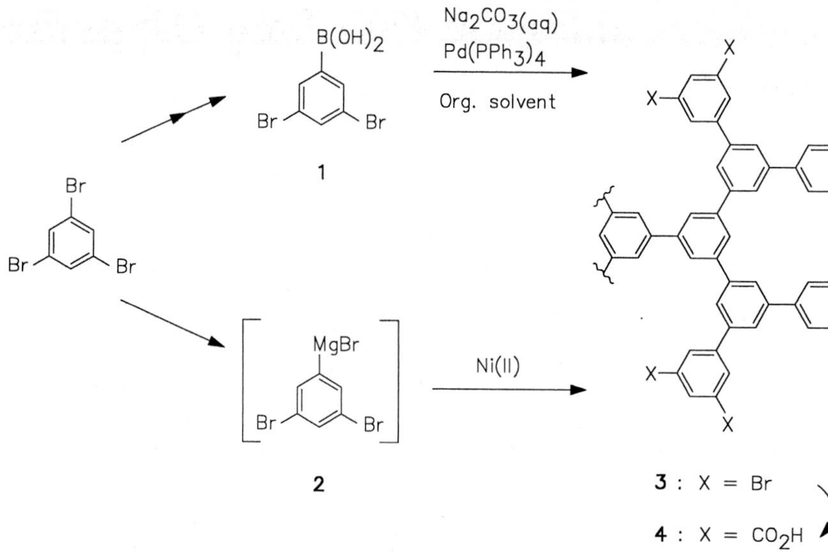

Scheme 6.1. Transition metal mediated preparation of polyphenylene macromolecules.

lar weight limitations may result from steric hindrance at the organometallic center or intramolecular cyclization(s). The effect of different terminal groups on the glass transition of these polyphenylenes and triphenylbenzenes has been reported.[12]

Diverse derivatives were prepared from the polylithio—polyphenylene, which, due to stability, was generated in less than quantitative yields by a metal—halogen exchange process. Electrophiles that were allowed to react with the polylithiated polyphenylene include CO_2, CH_3OCH_2Br, $(CH_3)_2CO$, DMF, $C_6H_5C(O)CH_3$, CH_3OH, $(CH_3)_2SO_4$, and $(CH_3)_3SiCl$.

Reaction of the polybromide **3** with the anion of 2-methyl-3-butyn-2-ol was also examined. In this case, "polymer reactivity seems to be enhanced when a small amount of the bromide groups had reacted with the reagent." The authors speculated [6] that "accelerated reactivity of a partially converted polymer could be one characteristic of highly branched materials." This supposition is supported by Wooley and coworkers[13] as they noted that the convergent coupling of dendritic wedges to a hexavalent core proceeded to completion with no evidence of partially substituted cores based on GPC experiments, thus, indicating an accelerated reactivity of partially transformed branched macromolecules. The phenomenon was attributed to "localized polarity or microenvironmental effects" favoring building block connectivity. These observations lend support to the assertion of complete surface transformation for many *divergently* constructed dendrimers.

When diphenyl ether or 1-methylnaphthalene was used as solvent, the polyphenylenes were determined to possess higher M_n values with xylene. No molecular weight increase was obtained by the addition of added monomer at or toward the end of the polymerization. The highest molecular weight was obtained using nitrobenzene as the solvent.

In order to investigate the unimolecular micellar behavior of the water-soluble, lithium salt of polycarboxylic acid **4**, the ^{1}H NMR spectrum of **4** in the presence of an NaOAc/H_2O solution of *p*-toluidine was recorded. Absorptions associated with the guest molecule(s) shifted upfield and were dramatically broadened.

Webster et al.[14a] expanded on the preparation of the polyphenylenes to develop the one-pot synthesis of hyper-crosslinked poly(triphenylcarbinol). Thus, reaction of 4,4′-dilithiobiphenyl with $(CH_3)_2CO_3$ ($-80\,°C \rightarrow 25\,°C$, THF) afforded trityl alcohol-based polymer. The absence of carbonyl or methoxycarbonyl NMR resonances led to the speculation that the polymer grows via a branched convergent process.

Repetitive Diels-Alder reactions utilizing bis-*ortho*-quinodimethane units have been employed[14b] for the construction of polymeric and oligomeric [60]-fullerenes with molecular weights up to 80,000 amu. The potential of using this method for dendrimer construction was suggested.

6.2.2   1 → 2 *Aryl*-Branched, *Ester* Connectivity

Fréchet et al.[15] reported the high yield, one-step, reproducible preparation of the hyperbranched aromatic polyester (**5**) with controllable molecular weights via the self-condensation of 3,5-bis(trimethylsiloxy)benzoyl chloride (**6**), which was synthesized from 3,5-dihydroxybenzoic acid by silylation (Me_3SiCl, Et_3N), followed by treatment with thionyl chloride and a catalytic amount of tetramethylammonium chloride.

Polymerization[15,16] of acid chloride **6** (Scheme 6.2) was effected thermally; a polystyrene equivalent weight average molecular mass of 184,000 amu for the polyester was obtained (91 %) using a reaction temperature of 200 °C with a catalytic amount of DMF or trimethylamine hydrochloride. The thermal stability was reported to be analogous to that of similar linear materials; however, the solubility was found to be enhanced. The degree of branching was determined to be between 55 and 60 % and contained a large number of free phenolic groups, both internal and external; their functionalization was possible. The molecular weight increases with higher reaction temperature, longer reaction times, and increased levels of catalyst.

Interesting comparisons have been made[17] between dendritic and the hyperbranched structures; the thermal properties (glass transition temperature and thermogravimetric analysis) were independent of architecture and their solubilities were comparable, but greater than that shown for linear counterparts.

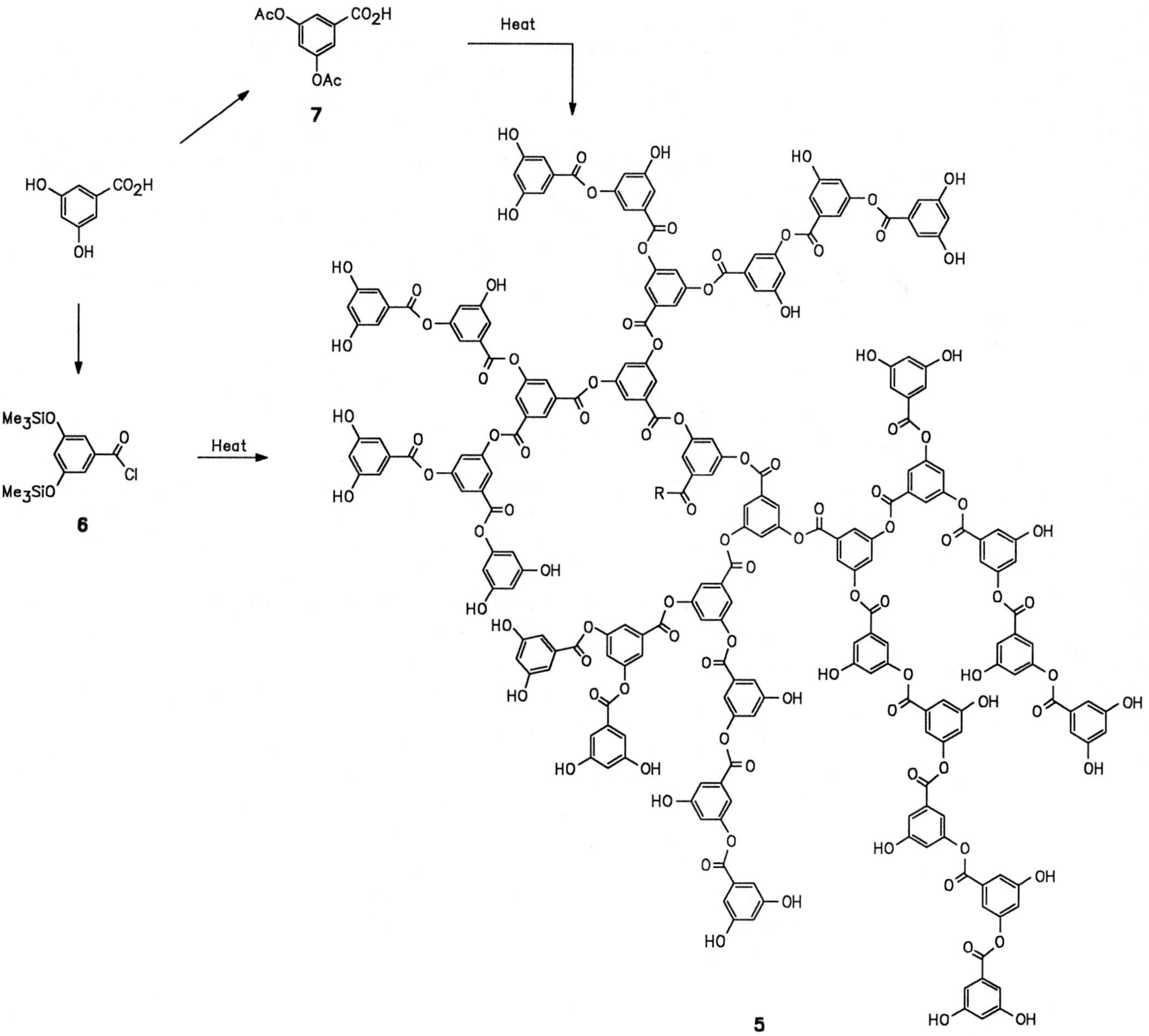

Scheme 6.2. Condensation of AB_2-type monomers afforded high molecular weight (ca. 180,000 to 1,000,000 amu) polyesters.

Turner, Voit, and Mourey[18,19] later reported analogous polyester hyperbranched macromolecules (**5**) via the thermal polymerization of A$_2$B monomer **7** (Scheme 6.2). The diacetate was prepared by the treatment of 3,5-dihydroxybenzoic acid with acetic anhydride.

Polymer structures derived from this building block were supported by similar spectral characterization as that described for polymers obtained from the acyl chloride **6**. Condensation of bisacetate **7** below 170 °C was found to be "slow"; whereas at 250 °C, the rate was substantially increased. Products (possessing $\overline{M}_w > 1,000,000$ amu) were much less sensitive to starting material (i.e., **7**) purity than those synthesized employing TMS-monomer **6**.

When monomer **7** was heated in vacuo, it was the quality and duration of the vacuum that had the biggest affect on the resultant weight. $\overline{M}_w$ ranged from 5,000 to 800,000 amu depending on the temperature, time, and vacuum. Catalysts such as magnesium or *p*-toluenesulfonic acid did not change the course of the reaction.

A related series of hyperbranched polymers possessing high molecular weight (20,000–50,000 amu) was also created[20,21] by the melt condensation of 5-acetoxy- (**8**) or 5-(2-hydroxyethoxy)-(**9**) isophthalic acids (Scheme 6.3). Polymerization of diacid **8** was effected in two stages: (1) melting at 250 °C combined with removal of acetic acid with the aid of an inert gas, and (2) application of a vacuum at the onset of solid state formation. Refluxing the resultant acid-terminated, ester-linked polymer **10** in THF/H$_2$O decomposed the labile anhydride cross-links, which were generated under the reaction conditions.

The degree of branching has been defined[22] as the sum of dendritic units plus terminal units divided by the sum of dendritic units plus terminal and linear units, and in this case was determined to be ca. 50 %. Also, the acetoxy monomer was copolymerized with various AB-type monomers, e.g., 3-(4-acetoxyphenyl)propionic acid.

Isophthalic acid **9**, prepared by ethoxylating the phenolic group of 5-hydroxyisophthalic acid with ethylene oxide, was polymerized at 190 °C using Bu$_2$SnAc$_2$ as the catalyst. The resulting carboxylic acid terminated hyperbranched polymer (not shown) was readily soluble in typical organic solvents. Due to a lower condensation temperature than that employed for the polymerization of diacid **8**, evidence of anhydride bond formation was not observed.

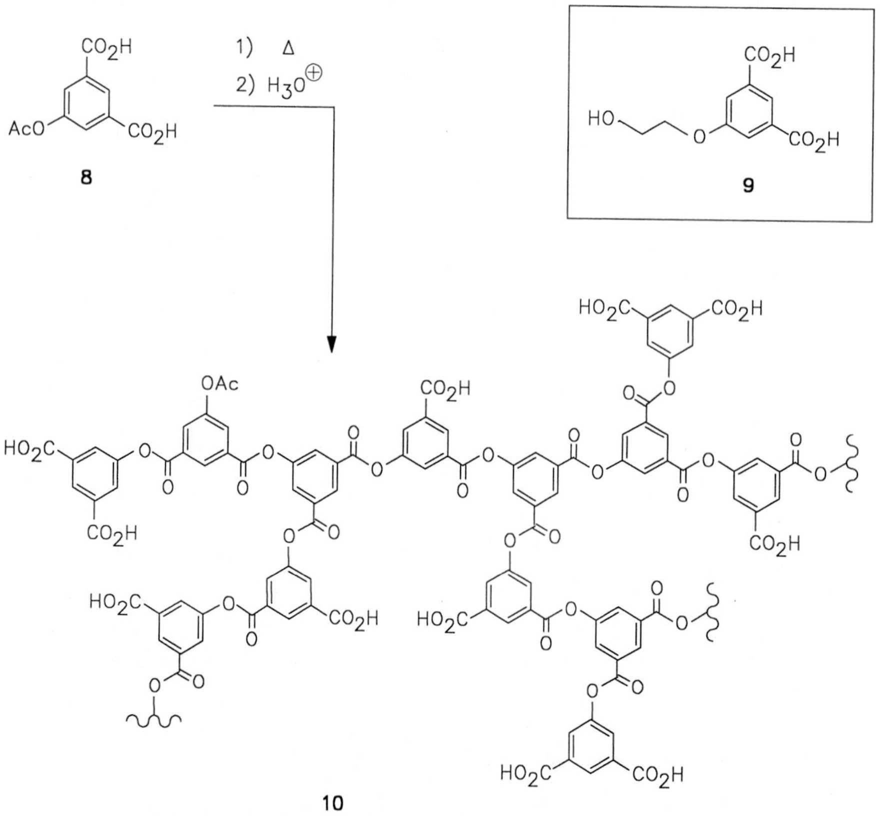

Scheme 6.3. Melt condensation of isophthalic acid derivatives results in polyester formation.

Massa, Voit, and coworkers[23] conducted a survey of the phase behavior of blends of these polyester hyperbranched polymers with linear polymers. Blend miscibility of a hydroxyl terminated polyester was comparable to that of poly(vinylphenol) indicating strong H-bonding interactions, whereas miscibility of an acetoxy terminated analog decreased relative to the hydroxy derivative.

In this series, a relatively clean condensation process was demonstrated[24–26a] by reactions of the silylated carboxylic acids with acetylated phenol moieties. This procedure avoids the presence of acidic protons as well as reduces the effects of acid-catalyzed side reactions.[26b]

6.2.3 1 → 2 *Aryl*-Branched, *Ether* Connectivity

A one-step synthesis (Scheme 6.4) of the hydroxyl terminated, dendritic polyether **11** from 5-(bromomethyl)-1,3-dihydroxybenzene (**12**), that was in turn easily prepared from 1,3-dihydroxy-5-(hydroxymethyl)benzene (**13**) upon treatment with triphenylphosphine and carbon tetrabromide, has been reported.[27] Bromomethyl monomer **12** was polymerized upon addition to an acetone suspension of potassium carbonate and [18]crown-6 (a ubiquitous method of connectivity for a convergent, iterative sequence).[28] The rate of monomer addition did not significantly affect the polymer characteristics and weight average molecular weights exceeded 10^5 amu. The ^{1}H NMR spectroscopic analysis indicated that *C*- as well as *O*-alkylation occurred during polymerization to various degrees in the range 11–32 % when parameters such as reaction time, concentration, and solvent were varied; the least amount of *O*-alkylation occurred using prolonged reaction times (92 h) and acetone as a solvent. Polymer characterization also included molecular weight determination using SEC and low-angle laser light scattering (LOLLS).

Miller, Neenan, et al.[29, 30] reported a general single-pot method (Scheme 6.5) for the preparation of poly(arylether) hyperbranched macromolecules (**14 a–d**), that are functionally analogous to linear poly(arylether) engineering plastics.[29–31] The hyperbranched polymers were generated from phenolic A₂B-type monomers (**15 a–d**), that were converted to the corresponding sodium phenoxides (**16 a–d**; NaH, THF) and subjected to polymerization.

The procedure utilized phenoxide monomers possessing two carbonyl, sulfonyl or tetrafluorophenyl activated aryl fluoride moieties that are displaced during building block connectivity. Reaction conditions included short duration (0.5–2 h) and moderate temperature (100–180 °C).

Polymerizations involving monomers **15 a,b** were insensitive to the reaction temperature, whereas self-reaction of monomers **15 c,d** at 140 °C afforded insoluble gels, presumably due to acetylene cross-linking; however at 100 °C, fully soluble polymers were realized. A related procedure (bromide displacement via phenoxide ion) using 2,4-dibromophenol or 2,4,6-tribromophenol as the starting material has also been reported.[5]

6.2.4 1 → 2 *Aryl*-Branched, *Ether* and *Ketone* Connectivity

Hawker et al.[32,33] described the structural characterization of hyperbranched macromolecules, obtained from the polymerization of monomers related to the AB₂ systems such as 3,5-difluoro-4′-hydroxybenzophenone and 3,5,-dihydroxy-4′-fluorobenzophenone. The new related families of poly(ether ketone)s possess the same internal linkages as well as terminal groups but differ in the degree of branching. Interestingly, as the degree of branching increases, the solubility of the polymer increases, although the viscosity decreases; there is no change in thermal characteristics.

6.2.5 1 → 2 *Aryl*-Branched, *Amide* Connectivity

Kim[5,34] reported two related types of hyperbranched aromatic polyamides; each employed amide bond formation for building block connectivity and was based on three-

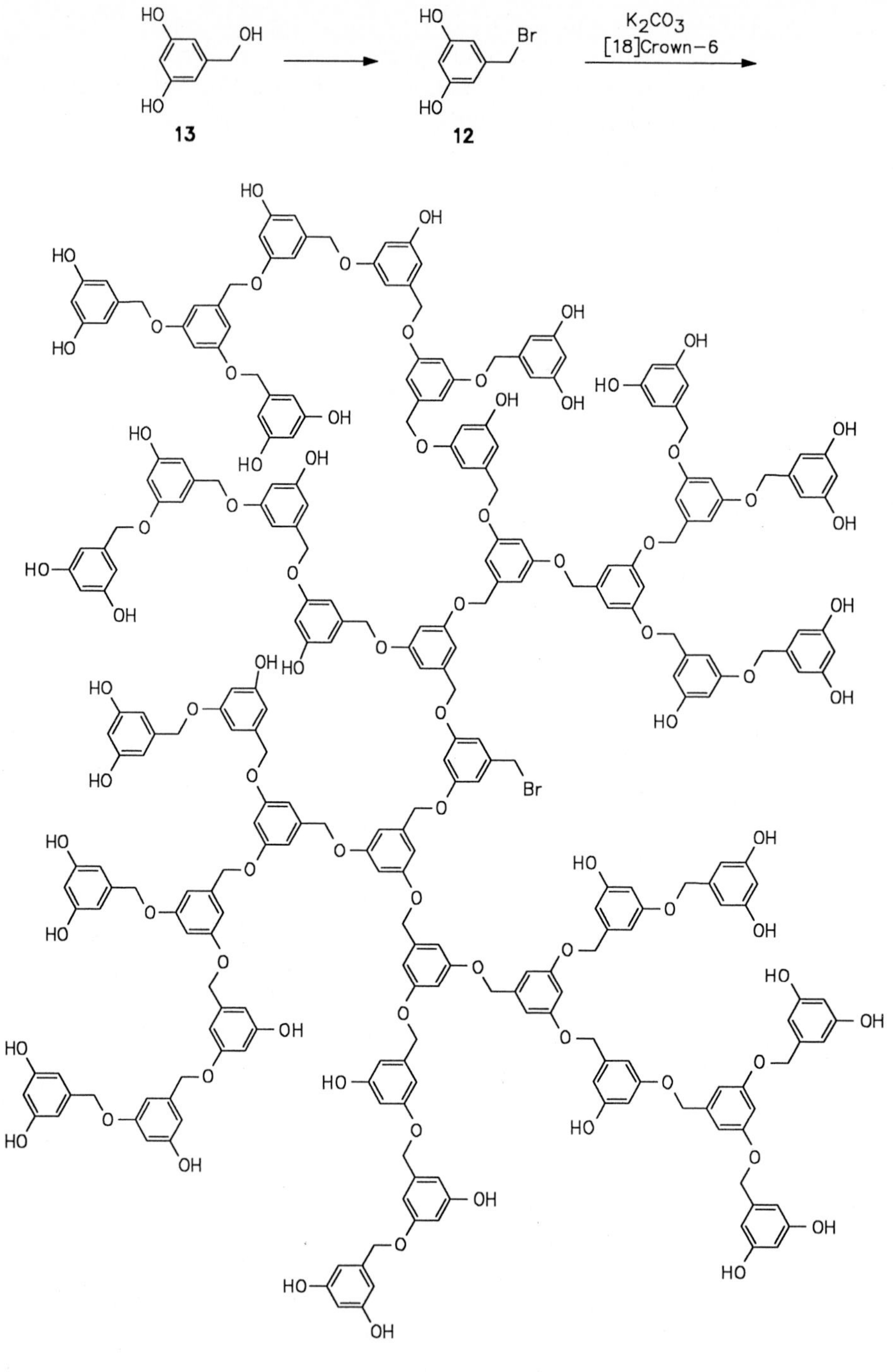

Scheme 6.4. Preparation of a hyperbranched analog of Fréchet's convergently constructed, benzyl ether-based dendrimers.

directional, aryl branching centers (Scheme 6.6). Polymerization occurred upon the neutralization of either amino acyl chloride hydrochlorides **17** or **18** to afford the amine or carboxylic acid terminated dendrimers **21** or **22**, respectively. Similar polymerizations, utilizing the sulfinyl-masked, amino acid chlorides **19** or **20** were effected via hydrolysis of the amine protecting group. Lyotropic liquid crystalline behavior was manifested in birefringence exhibited by the carboxylic acid terminated polymer (**22**) in either *N*-methylpyrrolidinone (NMP) or *N,N*-dimethylformamide (DMF) possessing > 40 wt %. The corresponding poly(methyl ester) also displayed birefringence. Gel permeation chromatography indicated molecular weights in the range 24,000–46,000 amu when the eluent was NMP/LiBr/H$_3$PO$_4$/THF; however, elution with pure DMF indicated aggregates of 700,000–1,000,000 amu.

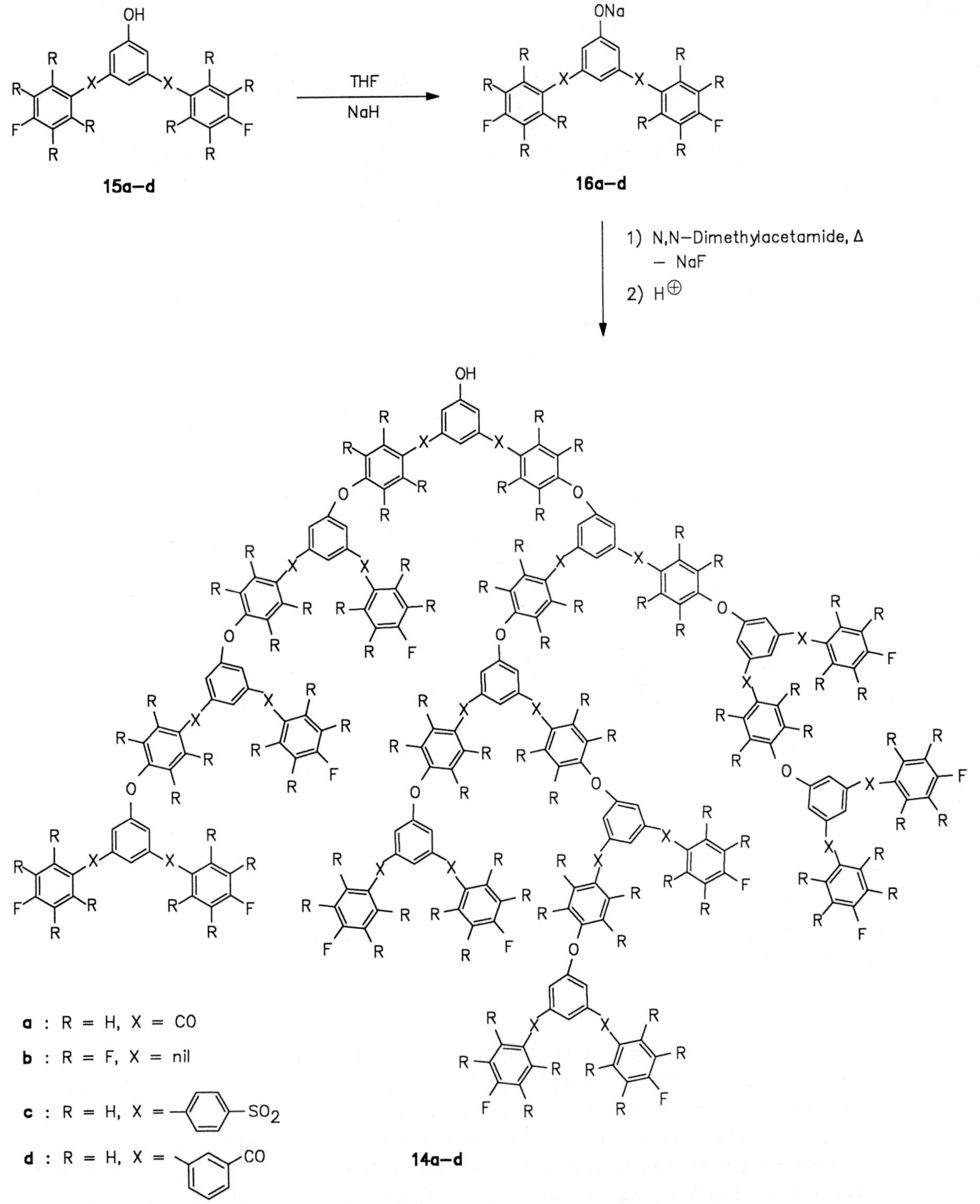
15a–d
THF
NaH
16a–d
1) N,N–Dimethylacetamide, Δ
– NaF
2) H⊕
a : R = H, X = CO
b : R = F, X = nil
c : R = H, X = SO2
d : R = H, X = CO
14a–d

Scheme 6.5. Single-pot preparation of branched analogs of poly(aryl ether) engineering plastics.

Reichert and Mathias[35] prepared related branched aramids, to those of Kim,[5,34] from 3,5-dibromoaniline (**23**) under Pd-catalyzed carbonylation conditions (Scheme 6.7). These brominated hyperbranched materials (**24**) were insoluble in solvents such as DMF, DMAc, and NMP, in contrast to the polyamine and polycarboxylic acid terminated polymers that Kim synthesized, which were soluble. This supports the observation that surface functionality plays a major role in determining the physical properties of hyperbranched and dendritic macromolecules.[4,36] A high degree of cross-linking could also significantly effect solubility. When a four-directional core was incorporated into the polymerization via tetrakis(4-iodophenyl)adamantane,[37] the resultant hyperbranched polybromide (e.g., **25**) possessed enhanced solubility in the above solvents, possibly as a result of the disruption of crystallinity and increased porosity.

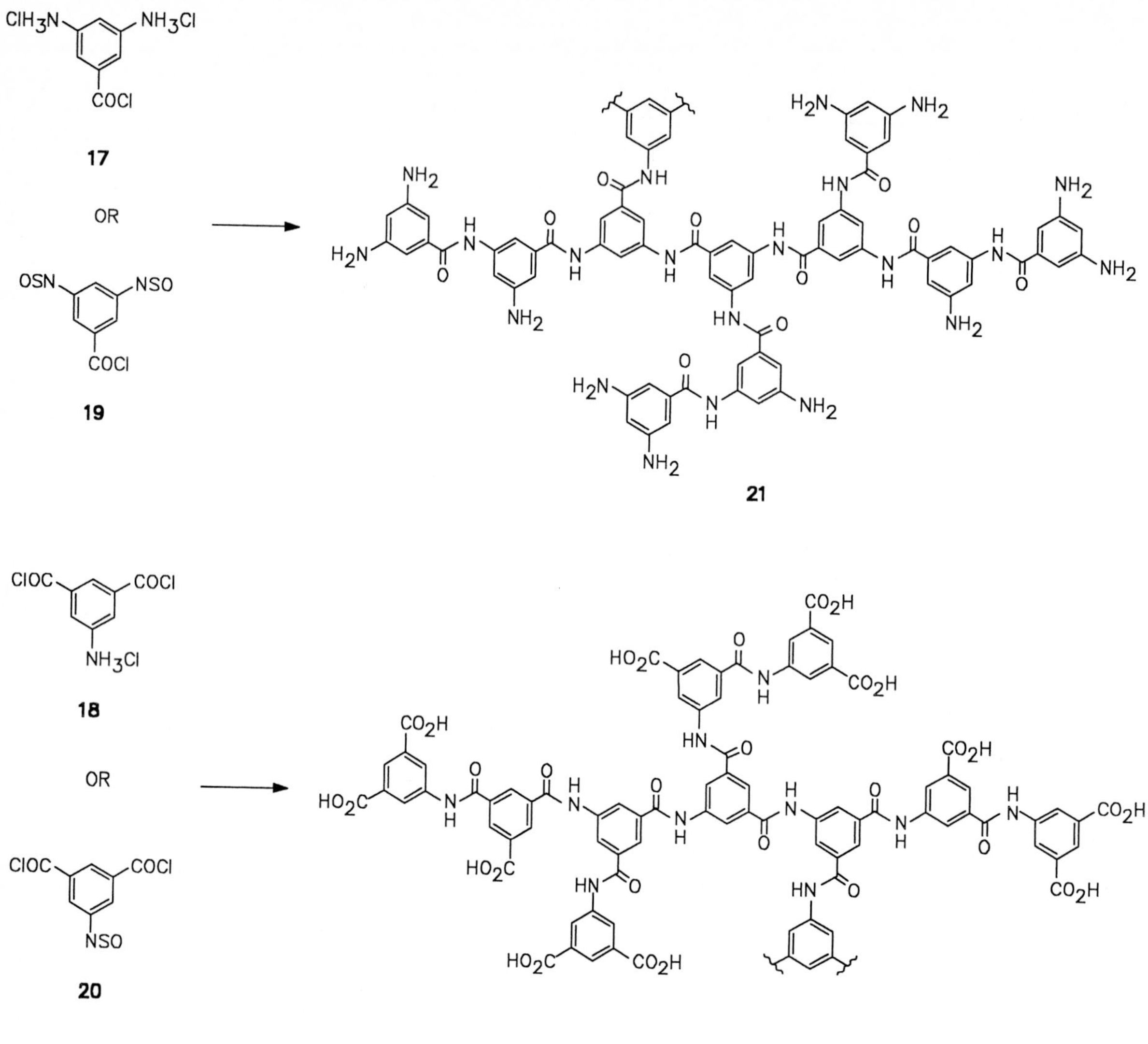

Scheme 6.6. Procedures for the preparation of poly(aryl amide) polymers. The carboxylic acid terminated polymers exhibited lyotropic liquid crystalline behavior.

1,3,5,7-Tetrakis(4-iodophenyl)adamantane has been transformed to the corresponding tetraacetylene derivatives,[37] which were subsequently polymerized via a stepwise process to afford highly cross-linked, hyperbranched polymers.

The solid-phase preparation of hyperbranched dendritic polyamides, based on 3,5-di-aminobenzoic acid, was attempted but deemed to "have severe limitations", such as the critical condensation reaction could not be forced to completion. The accessibility of chain ends within the solid support was suggested[38] to be the major obstacle to high yield conversion.

6.2.6   1 → 2 *Aryl*-Branched, *Carbamate* Connectivity

Spindler and Fréchet[39] prepared hyperbranched polyurethanes by step-growth polymerization (Scheme 6.8) of protected, or "blocked", isocyanate AB$_2$ monomers. The method is dependent on the thermal dissociation of a carbamate unit into the corresponding isocyanate and alcohol moieties.[40,41] Decomposition temperatures range from ca. 250 °C for alkyl carbamates to ca. 120 °C for aryl carbamates.[40]

Thus, 3,5-bis[(benzoxycarbonyl)imino]benzylalcohol (26; R = H) was heated in refluxing THF to give the bisisocyanate benzylalcohol and two equivalents of phenol (Scheme 6.8a). Self-addition of the isocyanate moieties with the alkyl alcohol group subsequently gave the hyperbranched carbamate 27 (Scheme 6.8b).

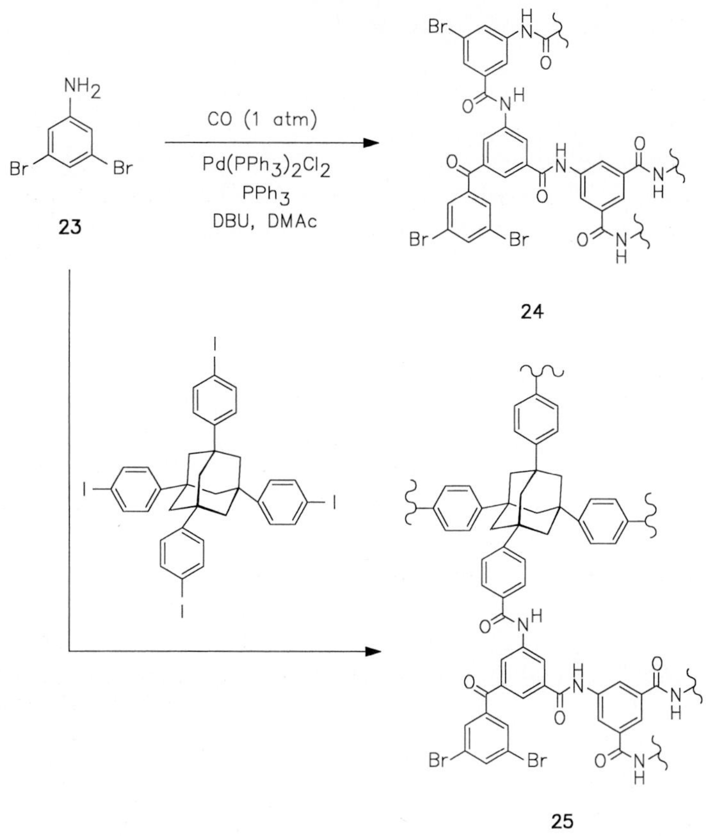

Scheme 6.7. Concomitant carbonylation and hyperbranched polyaramide formation. Rigid, tetrahedral, adamantane-based cores provided enhanced solubility characteristics.

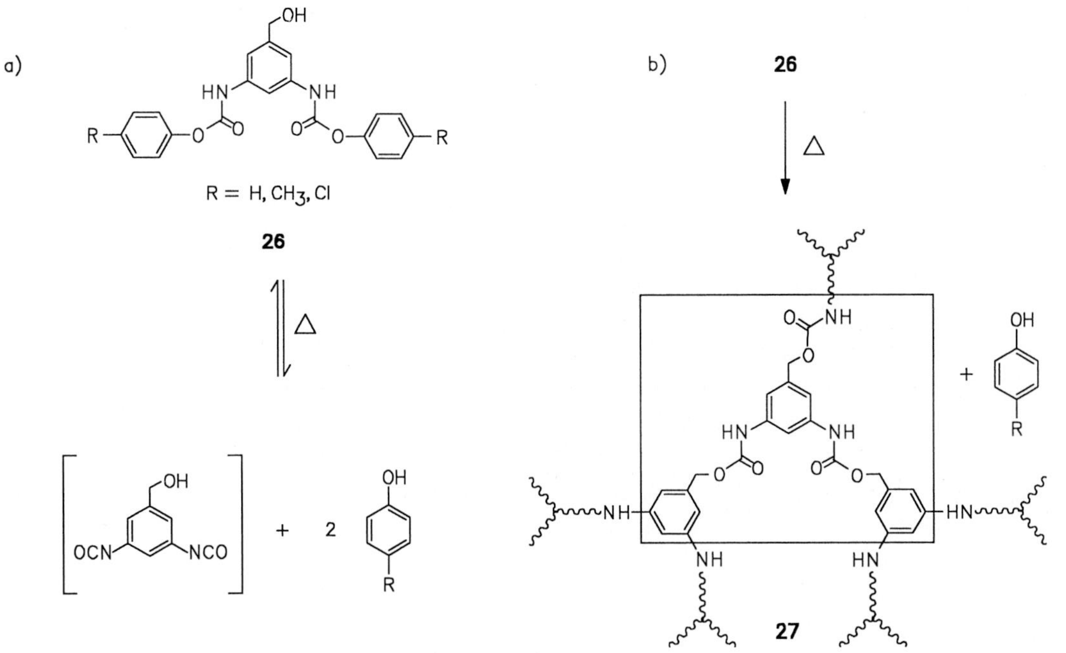

Scheme 6.8. Synthesis of polyurethanes based on the thermal decomposition of a "blocked" *bis*isocyanate monomer.

Addition of an alcohol (e.g., p-H$_3$CC$_6$H$_4$CH$_2$OH, p-NO$_2$C$_6$H$_4$CH$_2$OH, H$_3$COCH$_2$-CH$_2$OCH$_2$CH$_2$OH, H$_3$C(CH$_2$)$_5$OH, or CH$_3$(CH$_2$)$_9$OH) at the early stages of polymerization reduced the amount of cross-linking side reactions that are characteristic of isocyanate transformations by trapping reactive surface groups. This method of "end-capping" of the hyperbranched polymers yielded polyurethanes that were soluble in common organic solvents such as THF, DMSO, and DMF with conversions in the range 28–83 %.

6.2.7 1 → 2 *Aryl*-Branched, *Urethane* Connectivity

Kumar and Ramakrishnan[42] demonstrated that the thermal decomposition (107 °C) of 3,5-dihydroxybenzoyl azide gave rise to the labile 3,5-dihydroxyphenylisocyanate, which afforded (at 110 °C in dry DMSO with a catalytic amount of dibutyltin dilaurate) the polyurethane in 95 % yield. The polydispersity is greater than two, which is indicative of the poor mobility of the monomers; the polymers were soluble in aqueous base, confirming the presence of free phenolic termini.

6.2.8 1 → 2 *Aryl*-Branched, *Ether* and *Ester* Connectivity

Ringsdorf et al.[43a] reported the single pot preparation of liquid crystalline dendrimers with terminal chiral groups. Preparation of the hyperbranched macromolecule **28** (Scheme 6.9) utilized two different building blocks; (1) an AB$_2$-type monomer **29** possessing a mixed anhydride and two mesogenic biphenylacetate moieties, and (2) a terminal unit **30**, possessing a 3,4-disubstituted benzoyl chloride. Both of the building blocks were synthesized from methyl 3,4-dihydroxybenzoate (**31**). Dendrimer synthesis was effected by the treatment of diacetate monomer **29** with *p*-toluenesulfonic acid and heat (240 °C, 8 h), followed by the addition of more acid catalyst and the chiral capping agent **30**. Thermal analysis (200 °C, 1 h) of the CHCl$_3$-soluble macromolecule **28** by polarization microscopy, DSC, and X-ray scattering were included in the characterization of this asymmetric polymer; polarimetric experiments substantiated the macromolecular chirality by recording an optical rotation of $[\alpha]_D = +6.8°$ in CH$_2$Cl$_2$.

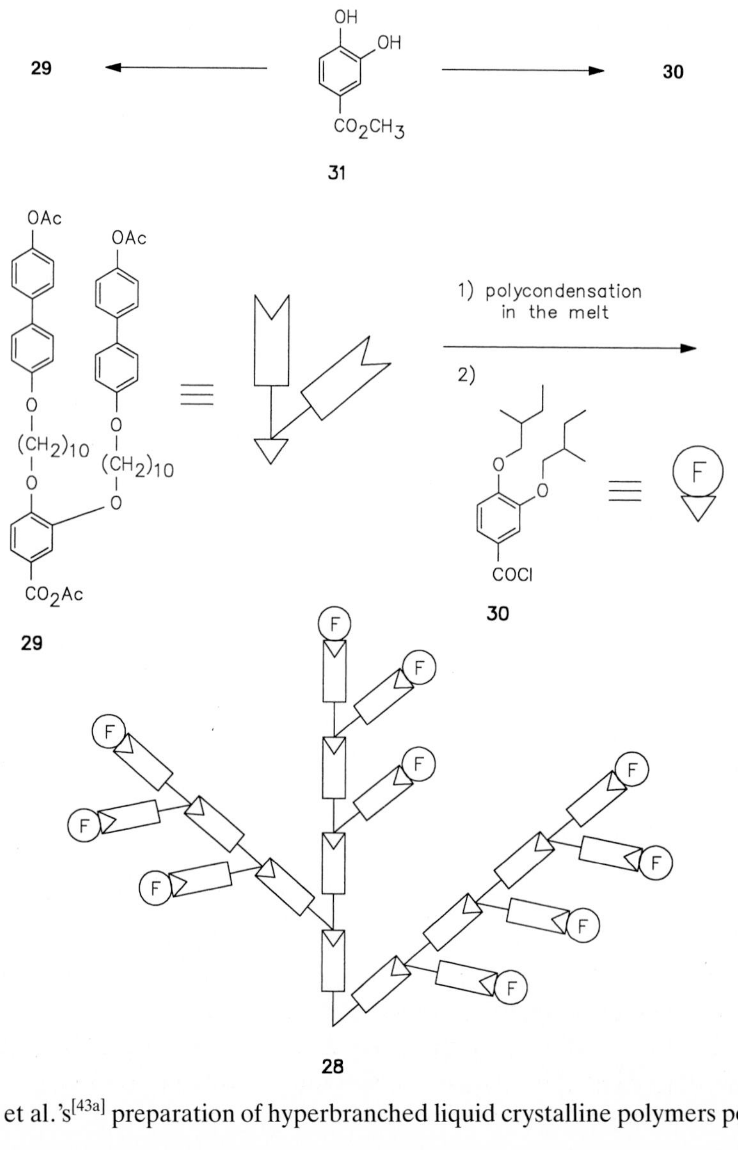

Scheme 6.9. Ringsdorf et al.'s[43a] preparation of hyperbranched liquid crystalline polymers possessing chiral capping units.

6.2.9 1 → 2 *Aryl*-Branched, *S*-Connectivity

Kakimoto et al.[43b] have reported the preparation of hyperbranched engineering plastic, poly(phenylene sulfide)(PPS) dendrimers via the polymerization of methyl 3,5-bis-(phenylthio)-phenyl sulfoxide. The degree of polymerization and weight average molecular weight were determined to be 80 and 25,700 amu, respectively T_g values were determined to increase (102 to 124 °C) proportionately relative to molecular weight.

6.3 1 → 2 *C*-Branched

6.3.1 1 → 2 *C*-Branched, *Ester* Connectivity

Kambouris and Hawker[44] described the one-step polymerization of methyl 4,4-bis(4'-hydroxyphenyl)pentanoate (**32**) and the structural elucidation of the resultant hyper-

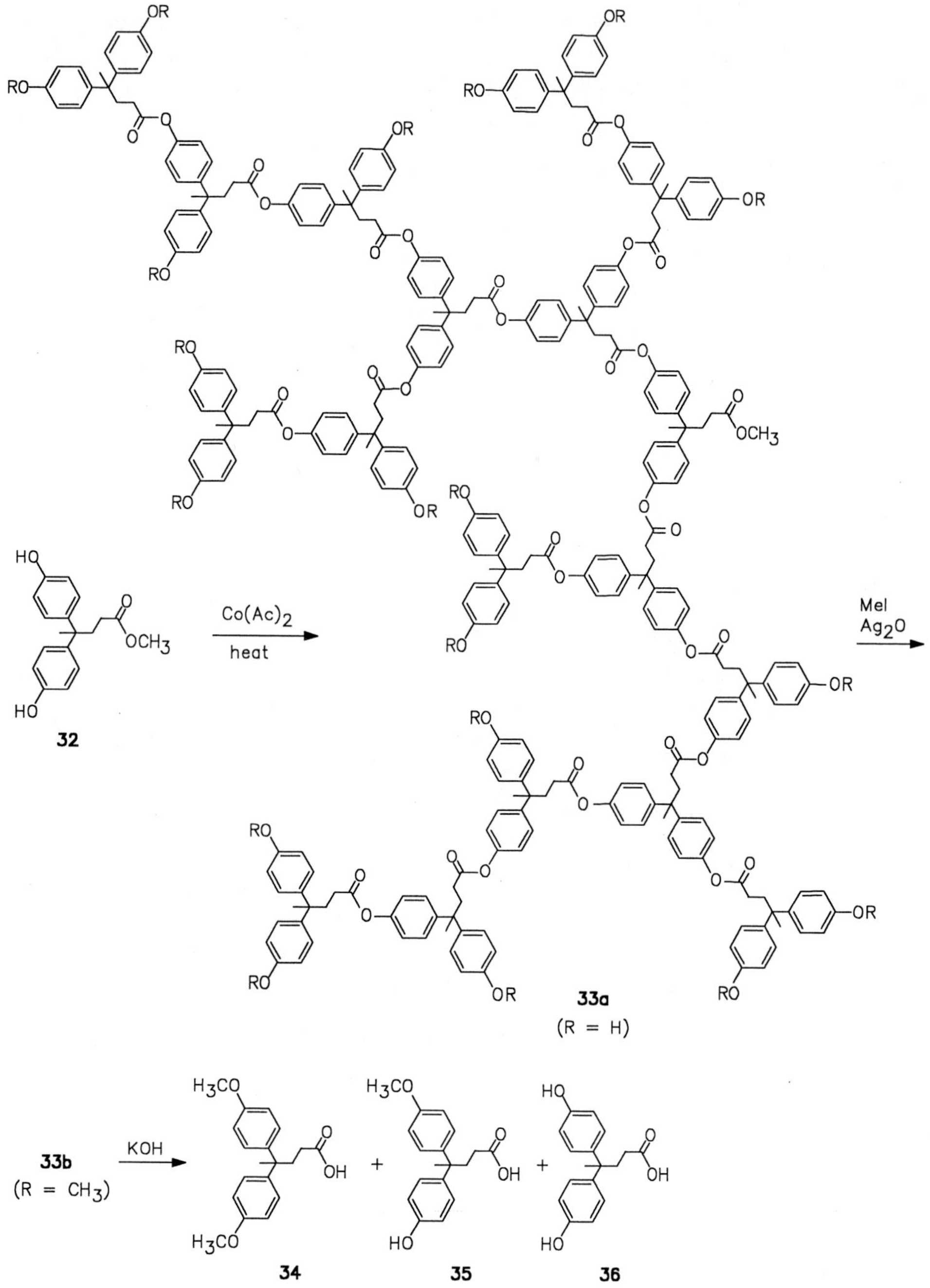

Scheme 6.10. Hawker et al.'s[44] cobalt catalyzed polymerization affording polyester macromolecules. The degree of branching was determined to be ca. 49 % based on degradative product distribution assessment.

branched polymer **33** (Scheme 6.10). Neat methyl ester monomer **32** was melted (120 °C) in a vacuum and in the presence of a catalytic amount of cobalt acetate giving (90 %) the phenolic-terminated, hyperbranched polyester possessing M_w of 47,000 amu.

Application of a previously described method[22] was not helpful in ascertaining the degree of branching; therefore, a new degradative technique was devised and demonstrated.[44] Treatment of phenolic polymer **33 a** (R = H) with methyl iodide and silver oxide gave the terminated methyl ether **33 b** (R = Me), which was formed in high yield with no evidence of side reactions. This ethereal ester was then saponified (KOH, THF, H_2O) to afford only three building block products in 24 % (**34**), 51 % (**35**), and 25 % yield (**36**), as determined by capillary gas chromatography and HPLC. Thus, from the product distribution, the degree of branching was determined to be 49 %.

A hyperbranched aliphatic polyester (**39**) based on 2,2-bis(hydroxymethyl)propionic acid (**37**) as the AB$_2$ monomer and 2-ethyl-2-(hydroxymethyl)-1,3-propanediol (**38**) as the core has been reported.[45,46] In general, the one-step esterification (Scheme 6.11) was performed in the bulk using an acid catalyst, involved no purification steps, and resulted in theoretical molar masses of 1200–44,300 amu. The degree of branching was determined to be ca. 80 %, suggesting that the polymers are highly branched. These materials also exhibited good thermal stability when analyzed with TGA in an inert atmosphere. Details concerning the relaxation processes in these hyperbranched macromolecules have been reported.[47]

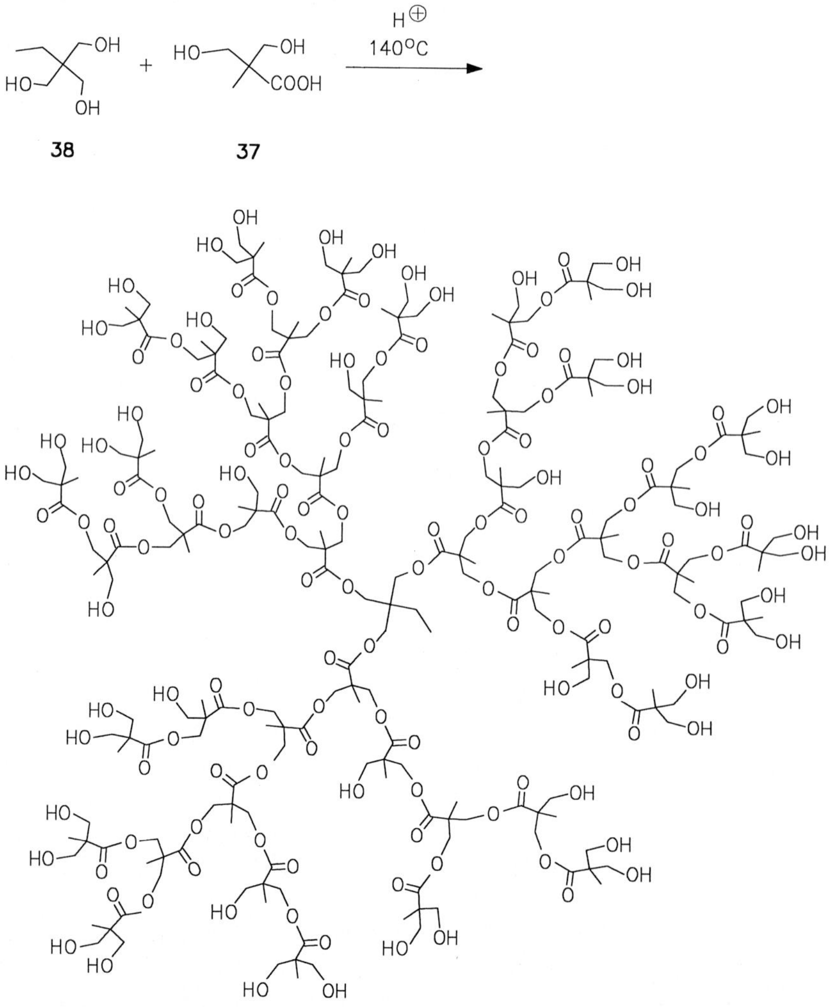

Scheme 6.11. Preparation of hyperbranched aliphatic polyesters.

6.3.2   1 → 2 *C*-Branched, *Ether* Connectivity

Percec and Kawasumi[48,49] initially reported a preliminary insight into the preparation of a fascinating series of thermotropic, liquid crystalline polymers possessing tertiary *C*-branching centers. Percec et al.[50–52] later published more complete summaries of their branching systems. Since dendritic design was initially derived from the branching architecture of trees, the authors compared the synthetic approach to these novel cascades to that of the 'willow' tree due to conformationally flexible branching centers inherent in the monomers. Under conditions where the monomer units adopt 'gauche' configurations, the dendritic arms will be relatively extended, thus importing typical cascade morphology. Conditions favoring monomer 'anti' configurations will cause a minimization of the interior void volume by the formation of a layered, nematic mesophase.

"Willow-like" cascade construction was facilitated by the preparation of the AB$_2$-type building blocks, e.g., 6-bromo-1-(4-hydroxy-4'-biphenylyl)-2-(4-hydroxyphenyl)hexane (**40**; Scheme 6.12), 13-bromo-1-(4-hydroxyphenyl)-2-[4-(6-hydroxy-2-naphthalenylyl)-phenyl]tridecane (**41**), and 13-bromo-1-(4-hydroxyphenyl)-2-(4-hydroxy-4"-*p*-terphenylyl)tridecane (**42**). Scheme 6.12 illustrates the monomer preparative strategy for the incorporation of the monomeric mesogenic moieties predicated on conformational (*gauche* versus *anti*) isomerism. Thus, one equivalent of 4-hydroxybiphenyl was allowed to react with 1,4-dibromobutane to yield the monobromide **43**, which was converted under Finkelstein conditions to the corresponding iodide **44**. Treatment of the latter with ketone **45** under phase transfer conditions afforded the α-substituted ketone **46**. Reduc-

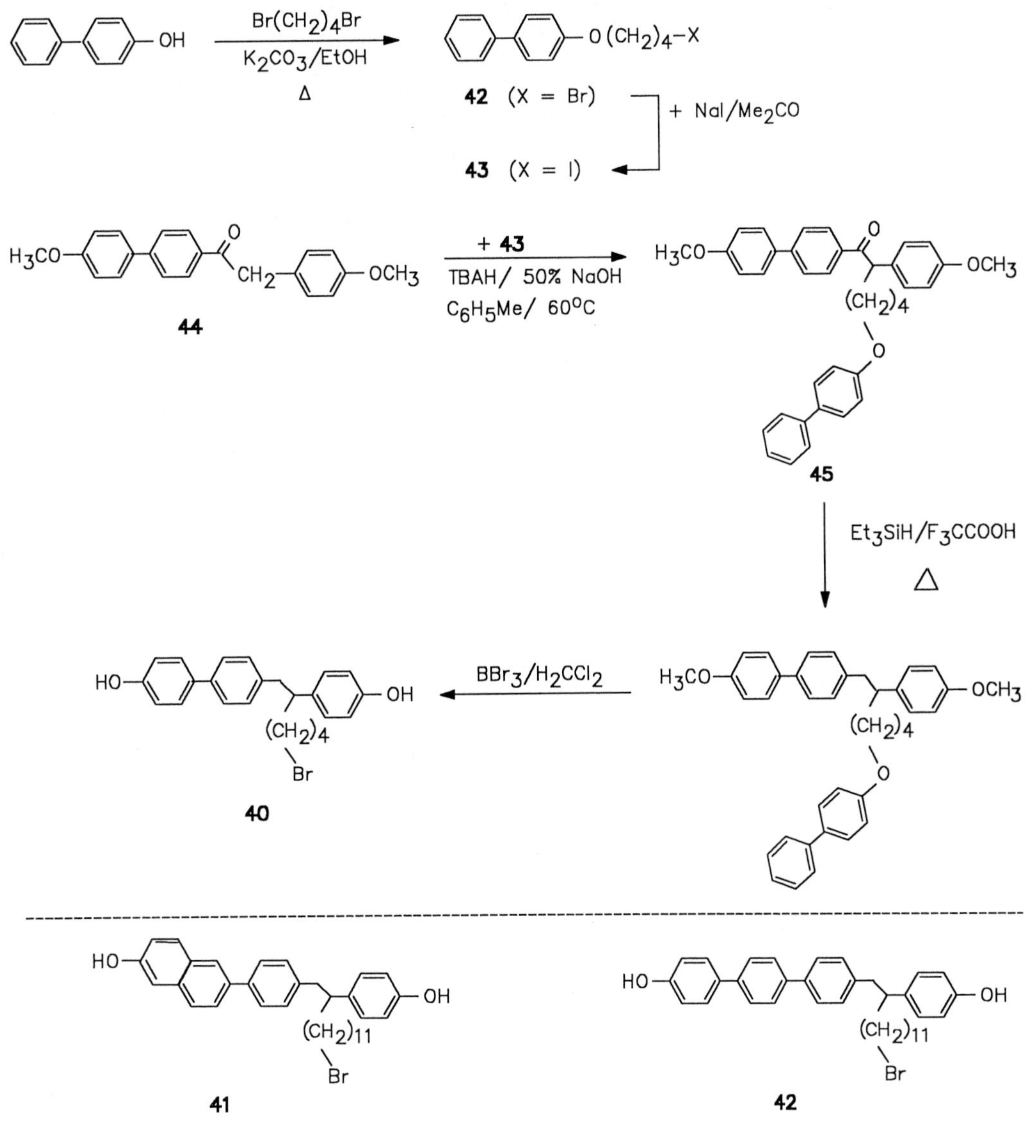

Scheme 6.12. Monomer preparation for the construction of liquid crystalline polymers.

tion of the carbonyl moiety, followed by concomitant demethylation and bromination, gave the desired monomer **40**.

Synthesis of the phenolic bromide building block **40** was similar to the that of previously reported monomer[48] but the alkyl chain spacer unit was shortened in an attempt to increase the nematic mesophase isotropization temperature.[50] Construction of the related building blocks **41** and **42** followed similar pathways; the latter were prepared to create hyperbranched macromolecules with a broader mesophase.[50]

One-pot construction[53] of the polyether dendrimers (**47**; Scheme 6.13) utilized standard phenoxide alkylation methods (Bu$_4$NHSO$_4$, NaOH(10M)). Capping of the phenolic terminal groups with either alkyl or benzyl groups gave hyperbranched macromolecule **48**. Scheme 6.13 also shows idealized representations of the isotropic (**47a**) and nematic (**47b**) states. These polymers were all observed to be soluble in *o*-dichlorobenzene and other more typical solvents when the counterion to the phenolic termini was tetrabutyl-ammonium ion. Polymers possessing alkali-metal phenolate termini were water-soluble. Extensive discussion of the characteristic ^{1}H NMR spectra of the materials was presented as well as comments on the diverse side reactions, such as alkyl halide hydrolysis and cyclizations. Other uses for these novel building blocks, such as the preparation of macrocyclic liquid crystalline oligopolyethers,[54-56a] and convergently generated, non-spherical dendrimers that also exhibit thermotropic liquid crystalline phases,[56b] have been reported.

6.3.3   1 → 2 *C*-Branched, *Amide*-Connectivity

Bergbreiter, Crooks, et al.[56c] have described the synthesis of hyperbranched polymer films grafted onto self-assembled monomers. Branching within the polymeric superstructure was effected by the use of an α,ω-diamino-terminated poly(*tert*-butyl acrylate) building block. Advantages to this procedure include (1) the production of thick polymer layers even when reactions proceed in low yield and (2) a resultant polymer film possessing a high density of modifiable functional groups.

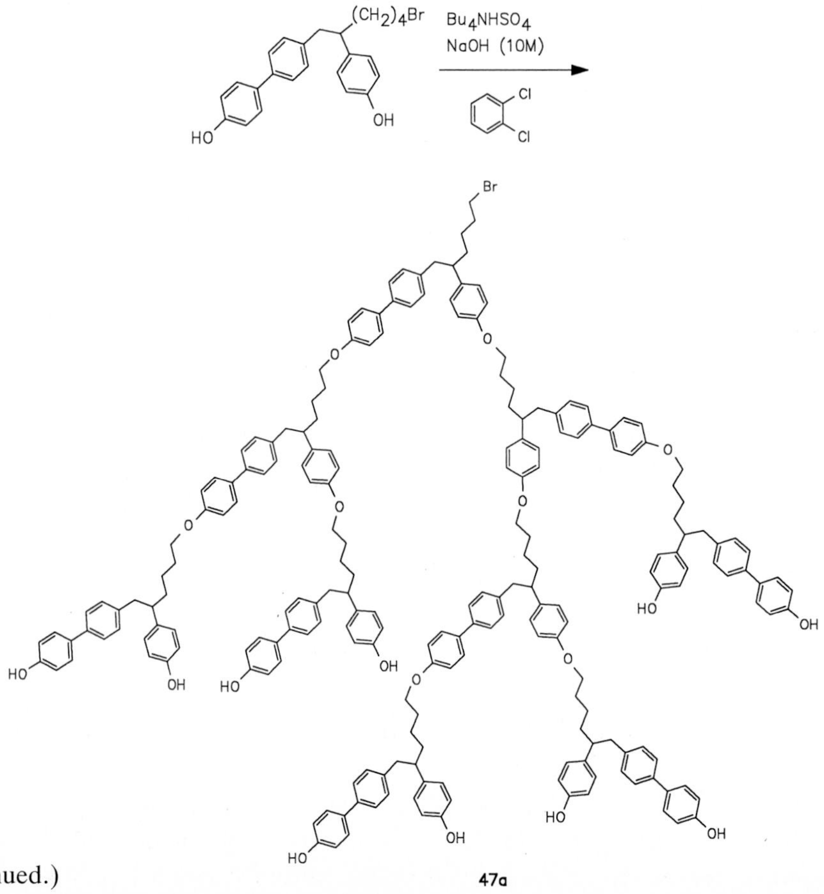

Scheme 6.13. (Continued.)   **47a**

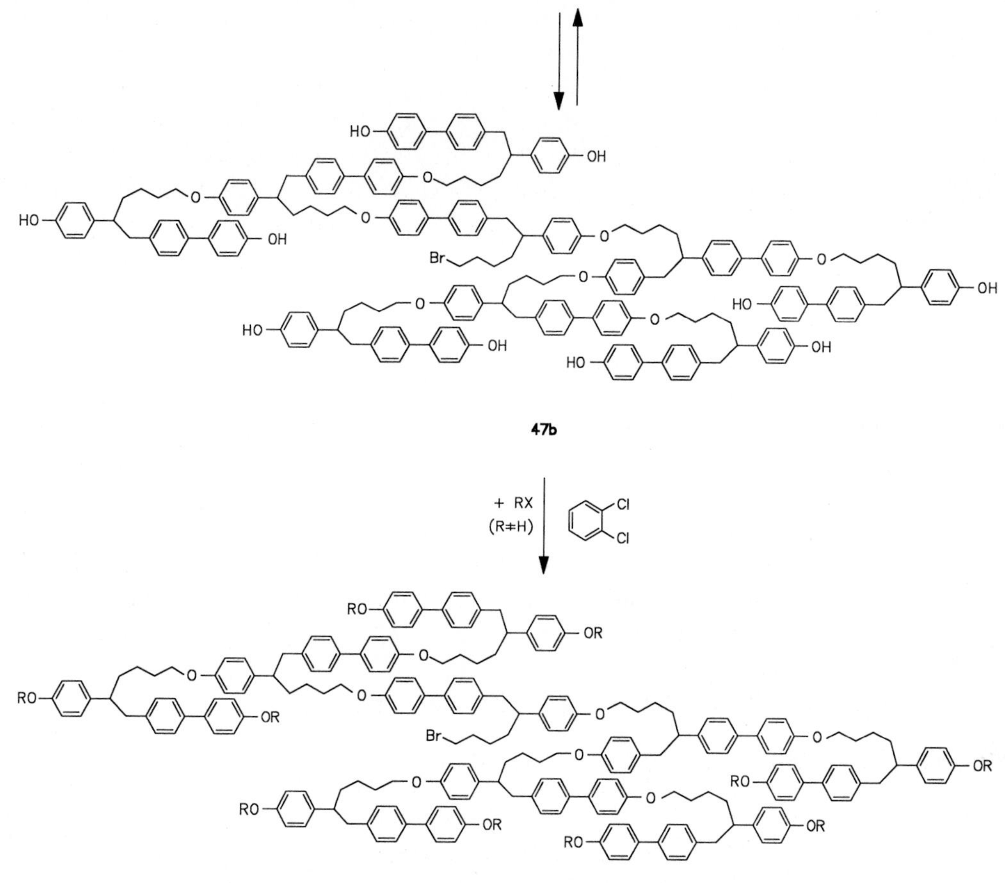

47b

48

Scheme 6.13. Percec et al.'s[53] application of mesogenic monomers for the preparation of hyperbranched polymers that form isotropic and nematic phases based on monomeric gauche and anti conformations.

6.3.4 1 → 2 *Aryl*-Branched, *C*-Connectivity

Fréchet et al.[56d, e] have reported the preparation of dendritic material employing a self-condensing vinyl polymerization of 3-(1-chloroethyl)ethenylbenzene. The preparative reaction for dendrimer growth is essentially a "living" vinyl polymerization resulting in increasing numbers of reactive sites and branches on the growing chains. Polymers synthesized via this method possess molecular weights greater than 100,000 amu. A similar, hyperbranched polystyrene has been reported.[56f]

6.3.5 1 → 2 *N*-Branched and Connectivity

Suzuki et al.[57] explored the Pd-catalyzed ring-opening of the monomer 5,5-dimethyl-6-ethenylperhydro-1,3-oxazin-2-one to give a hyperbranched polyamine. The polymerization was conducted at ambient temperature in THF and catalyzed with $Pd_2(dba)_3 \cdot 2$ dppe, affording after the evolution of carbon dioxide, the desired hyperbranched polyamine. NMR spectroscopy confirmed the high yield conversion of the monomer. The degree of branching was ascertained (from NMR data) as the ratio of tertiary amine units to the total of secondary and tertiary moieties.

6.4 1 → 3 *Ge*-Branched and Connectivity

The anionic polymerization of tris(pentafluorophenyl)germanium halide (**49**) was reported by Bochkarev et al.[58] to give *Ge*-based polymers (**50**; Scheme 6.14), a 9-Cascade:germane[3]:(2,3,5,6-tetrafluorogermanylidyne):pentafluorobenzene. These poly-

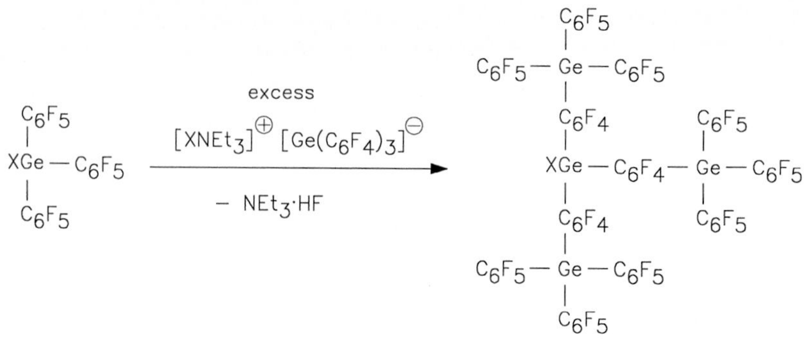

Scheme 6.14. Synthesis of siloxane hyperbranched polymers effected by Pt-catalysis.

mers possessed molecular masses in the range 100,000–170,000 amu. Polymerization was attributed to the formation of a metal anion, leading to aromatic substitution of the *p*-fluoro substituent.

6.5   1 → 3(2) *Si*-Branched and Connectivity

Mathias and Carothers[59–62] reported the classical polymerization (Scheme 6.15) of a A₃B-type, *Si*-based monomer (**51**) employing Pt-catalyzed alkene hydrosilylation to afford the hyperbranched poly(siloxysilanes), e.g., **52**. ¹H NMR integration suggested polymer growth continued through the third or fourth generation. The addition of more catalyst to the reaction did not increase the molecular weight (ca. 19,000 amu), as determined by a single, narrow SEC peak, which is probably due to the onset of dense packing. Gradual coupling of the unreacted Si−H moieties to form Si−O−Si bonds was circumvented by termination with allyl phenyl ether and allyl-terminated oxyethylene oligomer.

Mathias et al.[63] also reported the preparation of a rigid four-directional core for the construction of dendrimers possessing "sterically dispersed" functionality. Pathways for

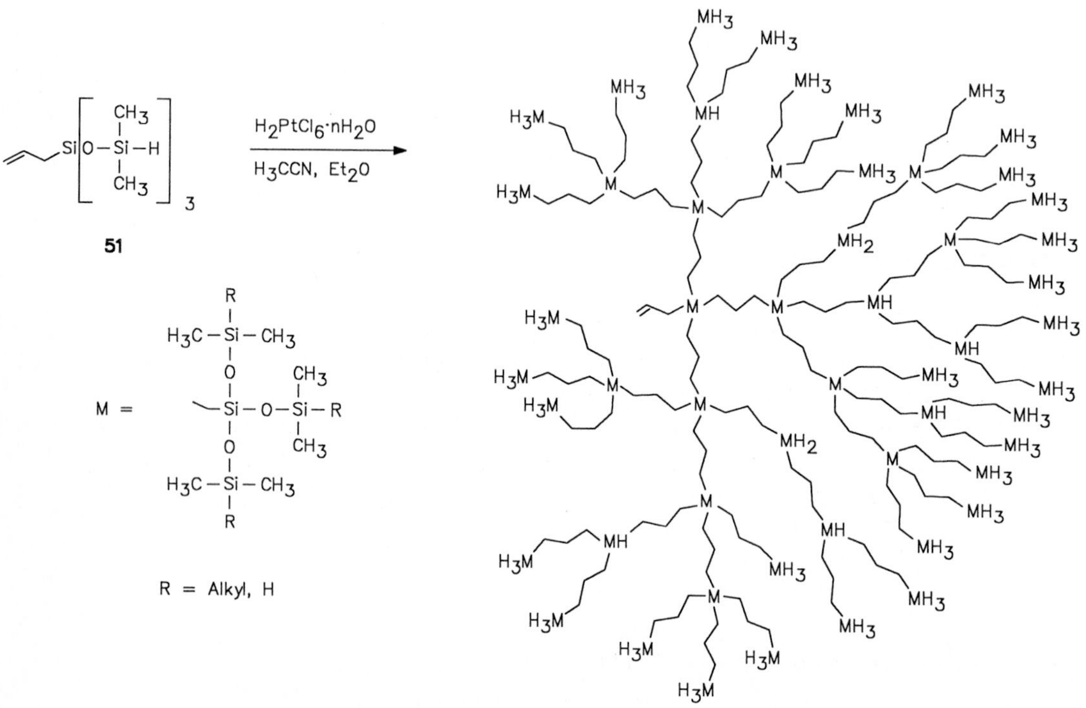

Scheme 6.15. Hyperbranched poly(siloxysilanes) via polymerization of *Si*-based monomer.

divergent cascade synthesis beginning with a 1,3,5,7-tetrakis(4-iodophenyl)adamantane core[64] are subsequently described. Conventional polymerization of a 1 → 3 branched, polysiloxane monomer by hydrosilylation polyaddition was also reported. Terminal functionalization of the siloxane-based polymers and potential applications of these novel materials, such as controlled interphases for composites, artificial blood substitutes, and "microreactors" possessing terminally bound catalytic moieties were noted.

Muzafarov, Golly and Möller[64b] have prepared similar poly(alkoxysilanes) that are easily hydrolyzed under acidic conditions and are thus biodegradable. Additional branched organosilicon polymers of interest include those of Ishikawa et al.[64c]

6.6 References

[1]   P. J. Flory, *J. Am. Chem. Soc.* **1952**, *74*, 2718.

[2]   P. J. Flory, *Principles of Polymer Chemistry*, Cornell University Press, Ithaca, New York, **1953**.

[3]   a) M. Gauthier, M. Möller, *Macromolecules* **1991**, *24*, 4548. b) M. Gauthier, L. Tichagwa, J. S. Downey, S. Goo, *Macromolecules* **1996**, *29*, 519. c) D. A. Tomalia, D. M. Hedstrand, M. S. Ferritto, *Macromolecules* **1991**, *24*, 1435. d) Y. Rouault, O. V. Borisov, *Macromolecules* **1996**, *29*, 2605.

[4]   Y. H. Kim, O. W. Webster, *J. Am. Chem. Soc.* **1990**, *112*, 4592.

[5]   Y H. Kim, *Polym. Preprints* **1993**, *34*(1), 56.

[6]   Y. H. Kim, O. W. Webster, *Macromolecules* **1992**, *26*, 5561.

[7]   Y. H. Kim, *Adv. Mater.* **1992**, *4*, 764.

[8]   N. Miyaura, T. Yanagai, A. Suzuki, *Syn. Commun.* **1981**, *11*, 513.

[9]   Y. H. Kim, *Macromol. Symp.* **1994**, *77*, 21.

[10] Y. H. Kim, O. W. Webster, *Polym. Preprints* **1988**, *29*(2), 310.

[11] Y.-H. Lai, *Synthesis* **1981**, 585.

[12] Y. H. Kim, R. Beckerbauer, *Macromolecules* **1994**, *27*, 1968.

[13] K. L. Wooley, C. J. Hawker, J. M. J. Fréchet, *J. Am. Chem. Soc.* **1991**, *113*, 4252.

[14] a) O. W. Webster, Y. H. Kim, F. P. Gentry, R. D. Farlee, B. E. Smart, *Polym. Preprints* **1992**, *33*(1), 186. b) A. Gügel, P. Belik, M. Walter, A. Kraus, E. Harth, M. Wagner, J. Spickermann, K. Müller, *Tetrahedron* **1996**, *52*, 5007.

[15] C. J. Hawker, R. Lee, J. M. J. Fréchet, *J. Am. Chem. Soc.* **1991**, *113*, 4583.

[16] K. L. Wooley, C. J. Hawker, R. Lee, J. M. J. Fréchet, *Polym. J.* **1994**, *26*, 187.

[17] K. L. Wooley, J. M. J. Fréchet, C. J. Hawker, *Polymer* **1994**, *35*, 4489.

[18] B. I. Voit, S. R. Turner, *Polym. Preprints* **1992**, *33*(1), 184.

[19] S. R. Turner, B. I. Voit, T. H. Mourey, *Macromolecules* **1993**, *26*, 4617.

[20] S. R. Turner, F. Walter, B. I. Voit, T. H. Mourey, *Macromolecules* **1994**, *27*, 1611.

[21] F. Walter, R. Turner, B. I. Voit, *Polym. Preprints* **1993**, *34*(1), 79.

[22] C. J. Hawker, R. Lee, J. M. J. Fréchet, *J. Am. Chem. Soc.* **1991**, *113*, 4583.

[23] D. J. Massa, K. A. Shriner, S. R. Turner, B. I. Voit, *Macromolecules* **1995**, *28*, 3214.

[24] H. R. Kricheldorf, O. Stöber, D. Lübbers, *Macromolecules* **1995**, *28*, 2118.

[25] H. R. Kricheldorf, G. Schwarz, F. Rusher, *Macromolecules* **1991**, *24*, 3485.

[26] a) H. R. Kricheldorf, D. Lübbers, *Makromol. Chem. Rapid Commun.* **1991**, *12*, 691. b) Kumar and Ramakrishnan (*Macromolecules* **1996**, *29*, 2524) have prepared hyperbranched polyesters using 3,5-dihydroxybenzoic acid and its derivatives via standard self-condensation conditions (i.e., transesterification). Spacer length between branching centers was varied using mesogenic segments and the resulting dendrimers were studied using DSC. These hyperbranched materials were determined to be amorphous with no liquid crystalline phases. This was postulated to arise via "random distribution of the mesogenic segments."

[27] K. E. Uhrich, C. J. Hawker, J. M. J. Fréchet, S. R. Turner, *Macromolecules* **1992**, *25*, 4583.

[28] C. J. Hawker, J. M. J. Fréchet, *J. Am. Chem. Soc.* **1990**, *112*, 7638.

[29] T. M. Miller, T. X. Neenan, E. W. Kwock, S. M. Stein, *Polym. Preprints* **1993**, *34*(1), 58.

[30] T. M. Miller, T. X. Neenan, E. W. Kwock, S. M. Stein, *J. Am. Chem. Soc.* **1993**, *115*, 356.

[31] T. M. Miller, T. X. Neenan, E. W. Kwock, S. M. Stein, *Macromol. Symp.* **1994**, *77*, 35.

[32] C. J Hawker, F. Chu, P. Kambouris, *Polym. Preprints* **1995**, *35*(1), 747.

[33] F. Chu, C. J. Hawker, *Polym. Bull.* **1993**, *30*, 265.

[34] Y. H. Kim, *J. Am. Chem. Soc.* **1992**, *114*, 4947.

[35] V. R. Reichert, L. J. Mathias, *Macromolecules* **1994**, *27*, 7024.

[36] K. L. Wooley, C. J. Hawker, J. M. Pocham, J. M. J. Fréchet, *Macromolecules* **1993**, *26*, 1514.

[37] V. R. Reichert, L. J. Mathias, *Macromolecules* **1994**, *27*, 7015.

[38] K. E. Uhrich, S. Boegeman, J. M. J. Fréchet, S. R. Turner, *Polym. Bull.* **1991**, *25*, 551.

[39] R. Spindler, J. M. J. Fréchet, *Macromolecules* **1993**, *26*, 4809.

[40] Z. W. Wicks, *Prog. Org. Coat.* **1975**, *3*, 73.

[41] J. H. Saunders, K. C. Frisch, *Polyurethanes: Chemistry and Technology,* Wiley, New York, **1962**, pp. 118–121.

[42] A. Kumar and S. Ramakrishnan, *J. Chem. Soc., Chem. Commun.* **1993**, 1453.

[43] a) S. Bauer, H. Fischer, H. Ringsdorf, *Angew. Chem.* **1993**, *105*, 1658; *Angew. Chem., Int. Ed. Engl.* **1993**, *32*, 1589. b) M. Jikei, Z. Hu, M.-a. Kakimoto, Y. Imai, *Macromolecules* **1996**, *29*, 1062.

[44] P. Kambouris, C. J. Hawker, *J. Chem. Soc., Perkin Trans. 1* **1993**, 2717.

[45] E. Malmström, M. Johansson, A. Hult, *Macromolecules* **1995**, *28*, 1698.

[46] M. Johansson, E. Malmström, A. Hult, *J. Polym. Sci. A, Polym. Chem.* **1993**, *31*, 619.

[47] E. Malmström, F. Liu, R. H. Boyd, A. Hult, U. W. Gedde, *Polym. Bull.* **1994**, *32*, 679.

[48] V. Percec, M. Kawasumi, *Polym. Preprints* **1992**, *33*, 221.

[49] V. Percec, M. Kawasumi, *Macromolecules* **1992**, *25*, 3843.

[50] V. Percec, P. Chu, M. Kawasumi, *Macromolecules* **1994**, *27*, 4441.

[51] V. Percec, J. Heck, G. Johansson, D. Tomazos, M. Kawasumi, *J. Macromol. Sci., Pure Appl. Chem.* **1994**, *A31*, 1031.

[52] V. Percec, J. Heck, G. Johansson, D. Tomazos, M. Kawasumi, P. Chu, G. Ungar, *J. Macromol. Sci., Pure Appl. Chem.* **1994**, *A31*, 1719.

[53] V. Percec, M. Kawasumi, *Polym. Preprints* **1993**, *34*(2), 158.

[54] V. Percec, M. Kawasumi, *Polym. Preprints* **1993**, *34*(1), 154.

[55] V. Percec, M. Kawasumi, *Polym. Preprints* **1993**, *34*(1), 156.

[56] a) V. Percec, P. Chu, *Polym. Preprints* **1995**, *36*(1), 743. b) V. Percec, P. Chu, G. Ungar, J. Zhou, *J. Am. Chem. Soc.* **1995**, *117*, 11441. c) Y. Zhou, M. L. Bruening, D. E. Bergbreiter, R. M. Crooks, M. Wells, *J. Am. Chem. Soc.* **1996**, *118*, 3773. d) J. M. J. Fréchet, M. Henmi, I. Gitsov, S. Aoshima, M. R. Leduc, R. B. Grubbs, *Science* **1995**, *269*, 1080. e) M. Freemantle, *Chem. & Eng. News* **1995**, Aug. 28th, 7. f) S. C. Gaynor, S. Edelman, K. Matyjaszewski, *Macromolecules* **1996**, *29, 1079*.

[57] M. Suzuki, A. Ii, T. Saegusa, *Macromolecules* **1992**, *25*, 7071.

[58] M. N. Bochkarev, V. B. Cilkin, L. P. Mayorova, G. A. Razuvaev, U. D. Cemchkov, V. E. Sherstyanux, *J. Organometal. Chem. USSR* **1988**, *1*, 196.

[59] T. W. Carothers, L. J. Mathias, *Polym. Preprints* **1993**, *34*(1), 503.

[60] L. J. Mathias, T. W. Carothers, *J. Am. Chem. Soc* **1991**, *113*, 4043.

[61] L. J. Mathias, R. M. Bozen, *Polym. Preprints* **1992**, *33*(2), 146.

[62] L. J. Mathias, T. W. Carothers, *Polym. Prepints* **1991**, *32*(3), 633.

[63] L. J. Mathias, V. R. Reichert, T. W. Carothers, R. M. Bozen, *Polym. Preprints* **1993**, *34*(1), 77.

[64] a) L. J. Mathias, V. R. Reichert, A. V. G. Muir, *Polym. Preprints* **1992**, *33*(2), 144. b) A. M. Muzafarov, M. Golly, Möller, *Macromolecules* **1995**, *28*, 844. c) A. Kunai, E. Toyoda, I. Nagamoto, T. Horio, M. Ishikawa, *Organometallics* **1996**, *15*, 75.

7 Chiral Dendritic Macromolecules

7.1 Divergent Procedures to Chiral Dendrimers

As noted in Section 4.2, Denkewalter et al.[1-3] were the first to report a divergent procedure to high molecular weight cascade polymers which utilized a chiral carboxylic acid protected, naturally occurring amino acid building block, N,N'-bis($tert$-butoxycarbonyl)-L-lysine. These poly(α,ε-L-lysine) dendrimers with purported molecular weights up to 233,600 amu (at the 10th generation), were described as having utility as "surface modifying agents, metal chelating agents, and substrates for the preparation of pharmaceutical dosages".[1] Discussion of the dendrimers' chirality or lack thereof, in the cited patents, as well as in a related physical properties study,[4] was excluded. More recent reports using related lysine building blocks for dendritic scaffolding construction include the preparation of (1) a Lipid-Core-Peptide system,[5] (2) multiple antigen peptides,[6-8] (3) octameric synthetic HIV-1 antigen,[9] (4) dendritic sialoside inhibitors,[10] and (5) dendritic block copolymer surfactants of "hydraamphiphiles".[11a]

Avnir et al.[11b] have examined the classical definitions and terminology of chirality and subsequently determined that they are too restrictive to describe complex objects such as large random supermolecular structures and spiral *d*iffusion-*l*imited *a*ggregates (DLAs). Architecturally, these structures resemble chiral (and fractal) dendrimers; therefore, new insights into chiral concepts and nomenclature are introduced that have a direct bearing on the nature of dendritic macromolecular assemblies, for example, "continuous chirality measure" and "virtual enantiomers."

7.1.1 1 → 3 C-Branched, *Ether* and *Amide* Connectivity

In 1991, Newkome and Lin[12] reported a series of acid terminated poly(ether amido) dendrimers (e.g., **3, 4**), generated from the tetraacid core (**1**) and sequential use of the amino acid building block **2** (Scheme 7.1). These acid-terminated dendrimers were readily transformed[13] into the related polytryptophane analogues (**5–7**) by treatment with tryptophane methylester hydrochloride employing a common peptide coupling procedure (DCC; 1-HBT in DMF).[14] The resultant chiral cascade series was examined via ORD/CD and the preliminary data indicate a linear relationship between optical rotation and the number of surface tryptophane moieties.

7.1.2 1 → 2 C-Branched

Seebach et al.[15] reported the utilization of their chiral tris(hydroxymethyl)methane derivatives[16] as building blocks for divergent dendrimer construction (Scheme 7.2). Analogues of polyol **8** were constructed by means of an aldol addition of the enolate of (2R,6R)-2-*tert*-butyl-6-methyl-1,3-dioxan-4-one with various aldehydes, followed by reduction with lithium aluminum hydride. Triol **8**, possessing three chiral centers, was derivatized with functionalized halides, possessing allyl, 4-(silyloxy)but-2-en-1-yl or 4-substituted benzyl moieties as well as pent-4-enoic and 3,5-dinitrobenzoic acid chlorides. Homologation of alcohol **8** was accomplished by treatment with allyl bromide (NaH, THF) to give triene **9**, which was followed by hydroboration (BH₃, THF) to afford triol **10**. Conversion of the extended triol **10** to the corresponding trimesylate **11** allowed the formation of nonaester **12** via reaction with triethyl sodiomethanetricarboxylate[17a] at elevated temperature (110 °C); ester **12**, however, proved to be difficult to purify. The asymmetric triol **8** was also subjected to a Michael-type addition to acrylonitrile affording, albeit in low yield, trinitrile **13**. Hydrolysis and esterification of **13** smoothly provided the tris(methyl ester) **14**. Murer and Seebach[17b] have extended the utility of their asymmetric building blocks to include the preparation of dendrimers (through tier 3) with doubly and triply branched architecture. Chirality was incorporated into the core as well as the superstructure.

Tam and coworkers reported the synthesis of a *multiple antigen peptide* (MAP) [6,18,19] system utilizing divergently constructed lysine-based scaffolding similar to that of Denkewalter's non-draining spheres.[1–4] The dendritic scaffolding preparation was accomplished using a Merrifield-type, solid-phase method, which eliminated the traditional step of peptide to carrier conjugation. Six different antigen peptide sequences were attached to the cascade imine periphery via triglycyl extenders.

Rao and Tam[19] reported the connection of a 24-residue peptide to a small, lysine-based dendrimer (Scheme 7.3). The peptide connection was effected via treatment of the octameric, glyoxylyl-terminated dendrimer **15** with the 1,2-aminothiol polypeptide **16** to afford the thiazolidinyl peptide dendrimer **17**. The thiazolidine linkage also introduces a new asymmetric center.

Tam and Shao[20a] later extended the usefulness of the terminal glyoxylyl moieties to the preparation (Scheme 7.4) of peptide dendrimers possessing additional modes of pep-

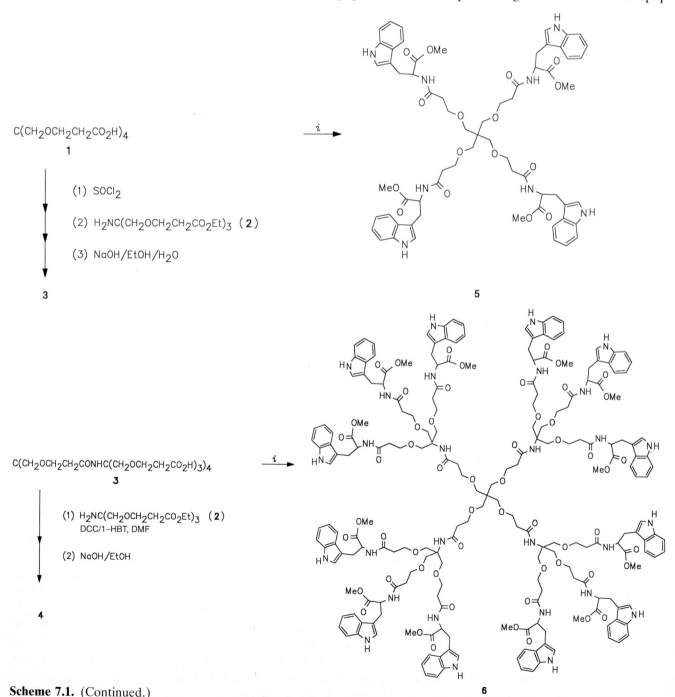

Scheme 7.1. (Continued.)

6

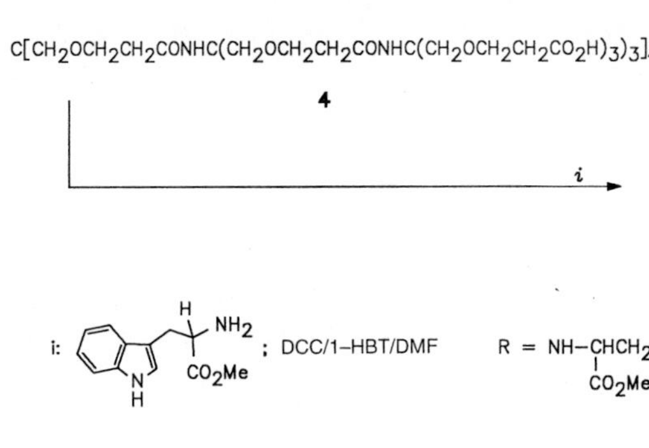

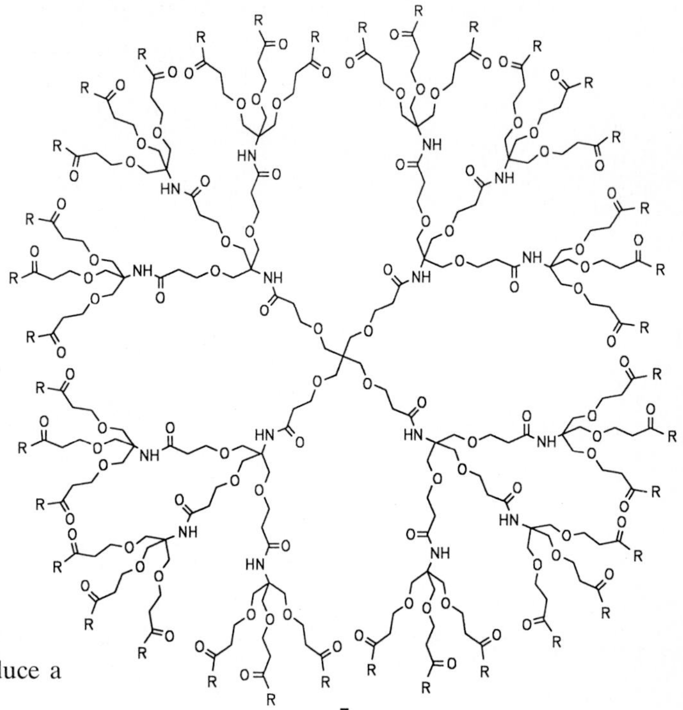

C[CH₂OCH₂CH₂CONHC(CH₂OCH₂CH₂CONHC(CH₂OCH₂CH₂CO₂H)₃)₃]₄

4

i →

i: [structure] ; DCC–1-HBT/DMF   R = [structure]

Scheme 7.1. Reaction of tryptophane methyl ester with a series of carboxylic acid terminated dendrimers to produce a family of aminoacid coated cascades.

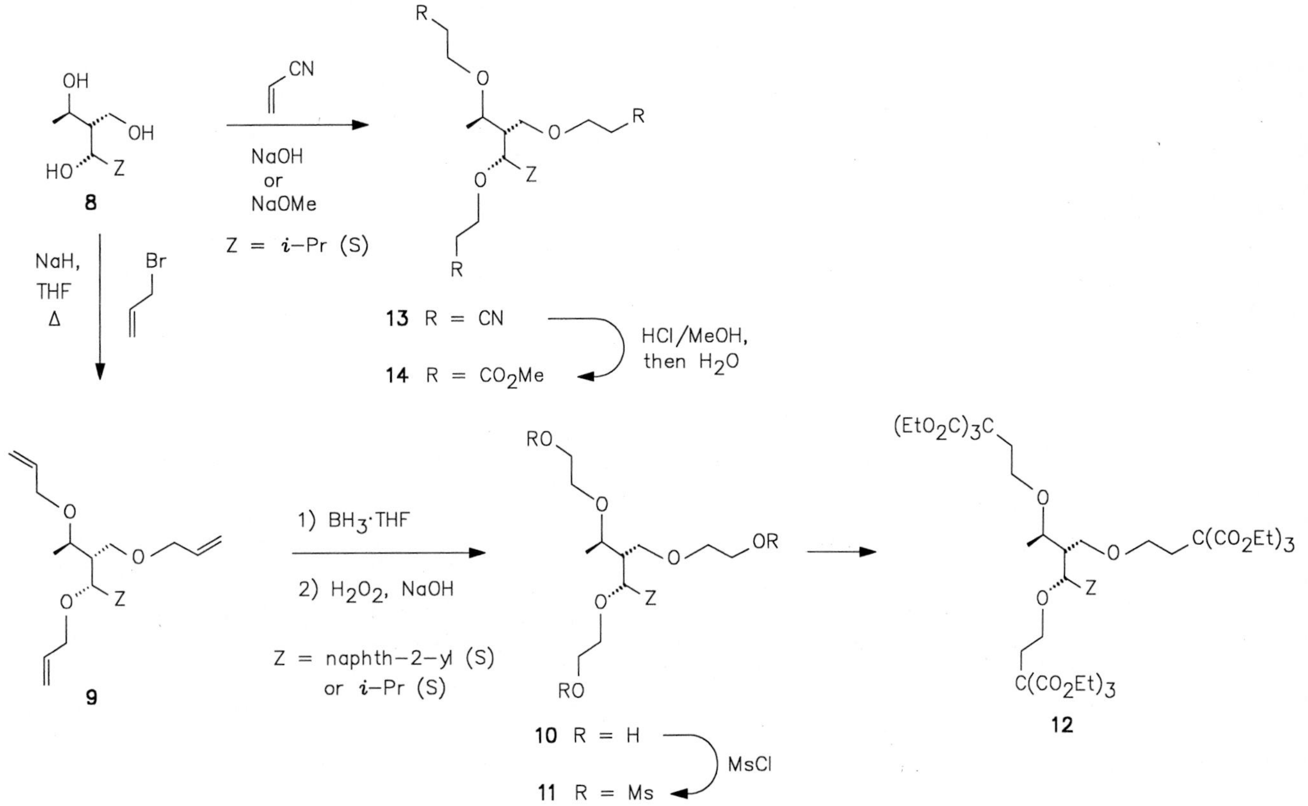

13 R = CN
14 R = CO₂Me   HCl/MeOH, then H₂O

Z = *i*-Pr (S)

8

NaH, THF, Δ   Br

9

1) BH₃·THF
2) H₂O₂, NaOH

Z = naphth-2-yl (S) or *i*-Pr (S)

10 R = H
11 R = Ms   MsCl

12

(EtO₂C)₃C

C(CO₂Et)₃

C(CO₂Et)₃

Scheme 7.2. Seebach et al.'s[15] chiral core preparation.

tide ligation (i.e., terminal peptide connectivity). Thus, the tetrakisglyoxylyl cascade **19**, prepared from the tetrakisacetal **18**, was reacted with peptides containing terminal hydrazide, 1,2-aminothiol, and oxime moieties to give the hydrazone, thiazolidine, and oxime peptide ligated dendrimers **20, 21**, and **22**, respectively.

Sharpless et al.[20b] have constructed chiral dendrimers by employing the "double-exponential" synthetic method.[20c] Chirality was induced via the use of aryl acetonide monomers prepared by asymmetric dihydroxylation (AD) of the corresponding prochiral alkenes. Each monomeric unit possessed two asymmetric carbons. Examples described are C_3-symmetric and contain up to 45 chiral building blocks with 24 acetonide termini.

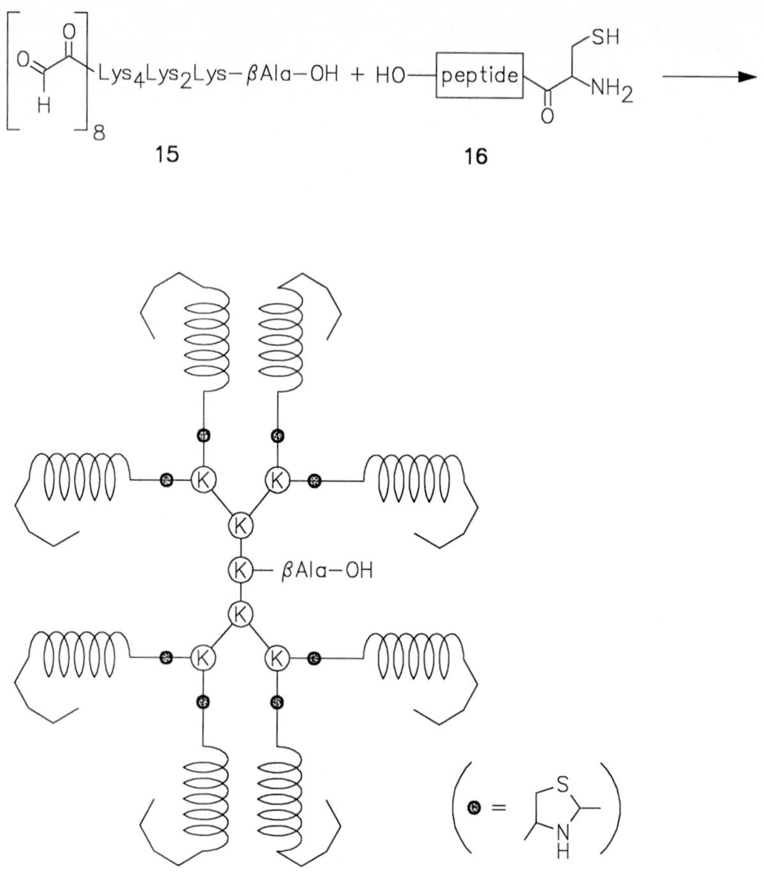

Scheme 7.3. Use of a lysine-based scaffolding for the ordering of 24-residue peptides.

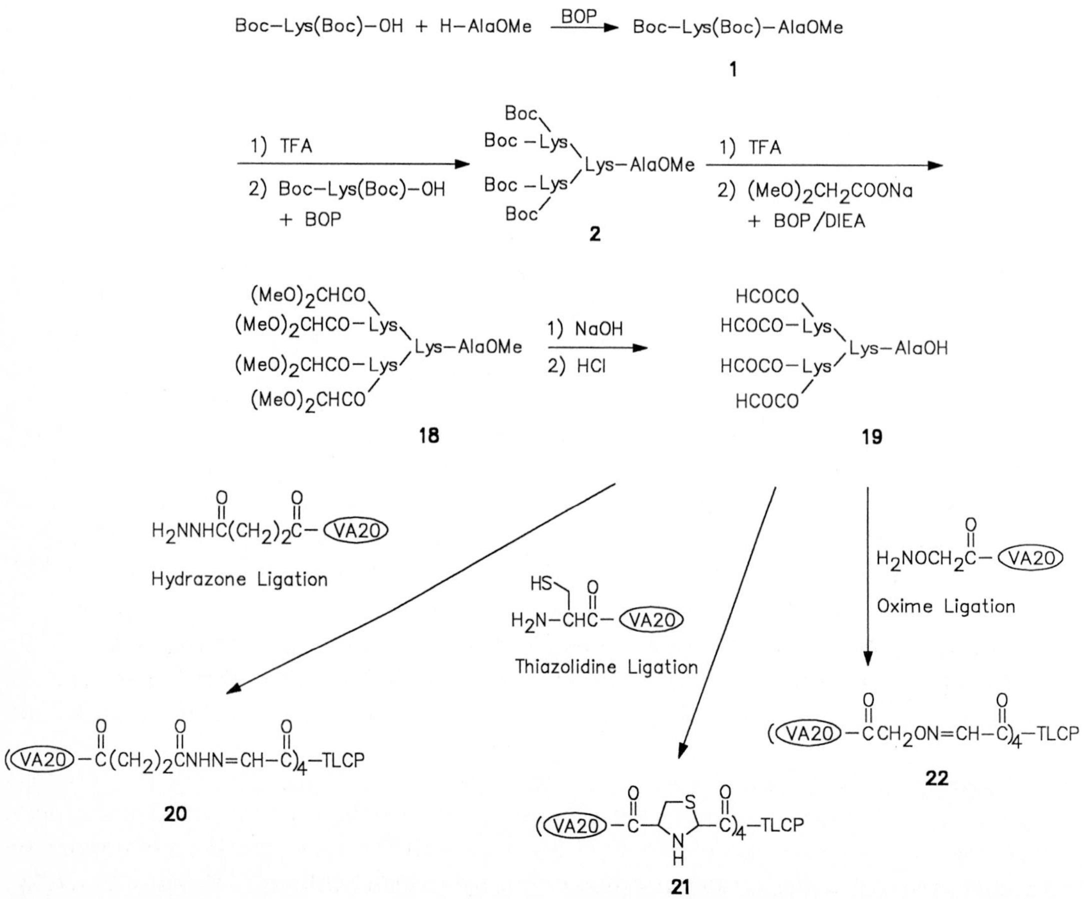

Scheme 7.4. Employment of terminal glyoxylyl groups to effect differing modes of peptide attachment.

7.1.3   1 → 2 *Aryl*-Branched, *Ester* and *Amide* Connectivity

Seebach et al.[16,21] prepared a series of chiral aryl dendrimers from related chiral triol cores (e.g., **8**) by treatment with 3,5-dinitrobenzoyl chloride to afford the first generation, hexanitro cascade **23**, which was then catalytically reduced to give the hexaamine **24** (Scheme 7.5). Reaction of this hexaamine with the same aroyl chloride (**26**) gave rise to the dodecanitro dendrimer **25** in 83 % yield (when R = isopropyl). A notable change in the molar rotation was observed in proceeding from the first tier, hexanitro dendrimer **23** $[\Phi_D^n-124]$ to the second tier, dodecanitro dendrimer **25** $[\Phi_D^n + 155]$, suggesting that there are significant optical activity contributions due to chromaphoric conformational asymmetry at the dendritic surface. These amino and nitro-terminated dendrimers were further shown to act as hosts in the formation of host–guest complexes via generation of inclusion complexes with ethyl acetate, methanol, and acetone.

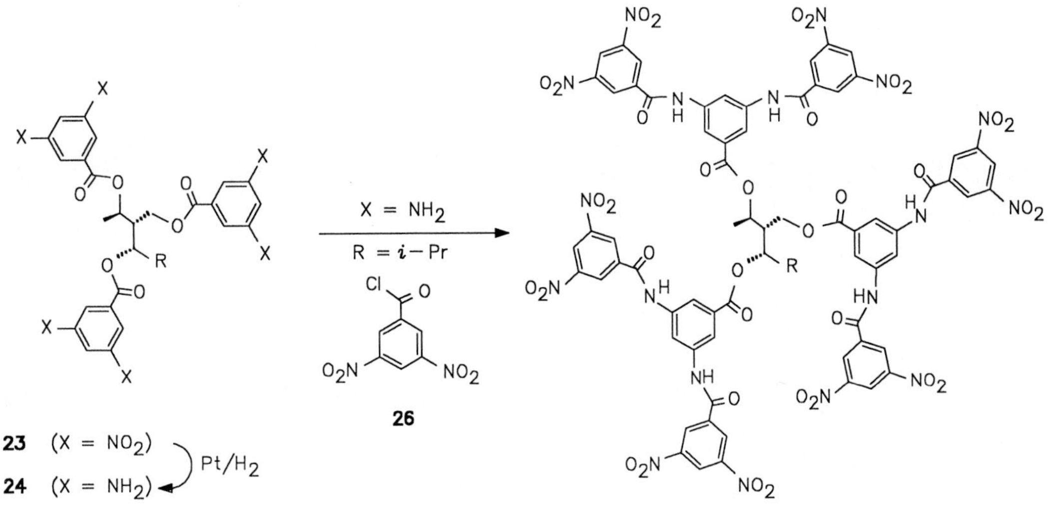

Scheme 7.5. Divergent construction of polyaramide dendrimers possessing a chiral, ester-based core.

7.1.4   1 → 2 *Aryl*-Branched, *Ether* and *Ester* Connectivity

Ringsdorf et al.[22] reported the single-pot preparation of liquid crystalline hyperbranched polymers possessing terminal chiral groups. Preparation of the branched polymer **27** (Scheme 7.6) utilized two different building blocks: an AB$_2$-type monomer **28**, containing a mixed anhydride and two mesogenic biphenylacetate moieties and an end-capping 3,4-disubstituted benzoyl chloride (**29**). Both the AB$_2$-type monomer (**28**) and the end-capping monomer (**29**) were prepared in five and three steps, respectively, from methyl 3,4-dihydroxybenzoate (see Section 6.2.7).

The chiral polymer showed good solubility in typical organic solvents, such as chloroform, toluene, and THF; it was further determined by gel permeation chromatography to possess an intermediate degree of branching ($P_n = 6$). Optical rotation, as ascertained by polarographic measurements in CH$_2$Cl$_2$, was established at $[\alpha]_D^{20} = +6.8°$. Wide-angle X-ray scattering patterns indicated at 4.4 Å average mesogen–mesogen distance, while small-angle reflections suggested 28 Å spacing between cholesteric phase layering.

7.1.5   1 → 2 *N*-Branched and Connectivity

Meijer et al.[23] constructed a series of novel "dendritic boxes" by surface modification of divergently prepared, amine-terminated poly(propyleneimine) dendrimers[24] (e.g., **30**; Scheme 7.7) by attachment of *N*-protected chiral amino acid caps (see Section 4.3.1).

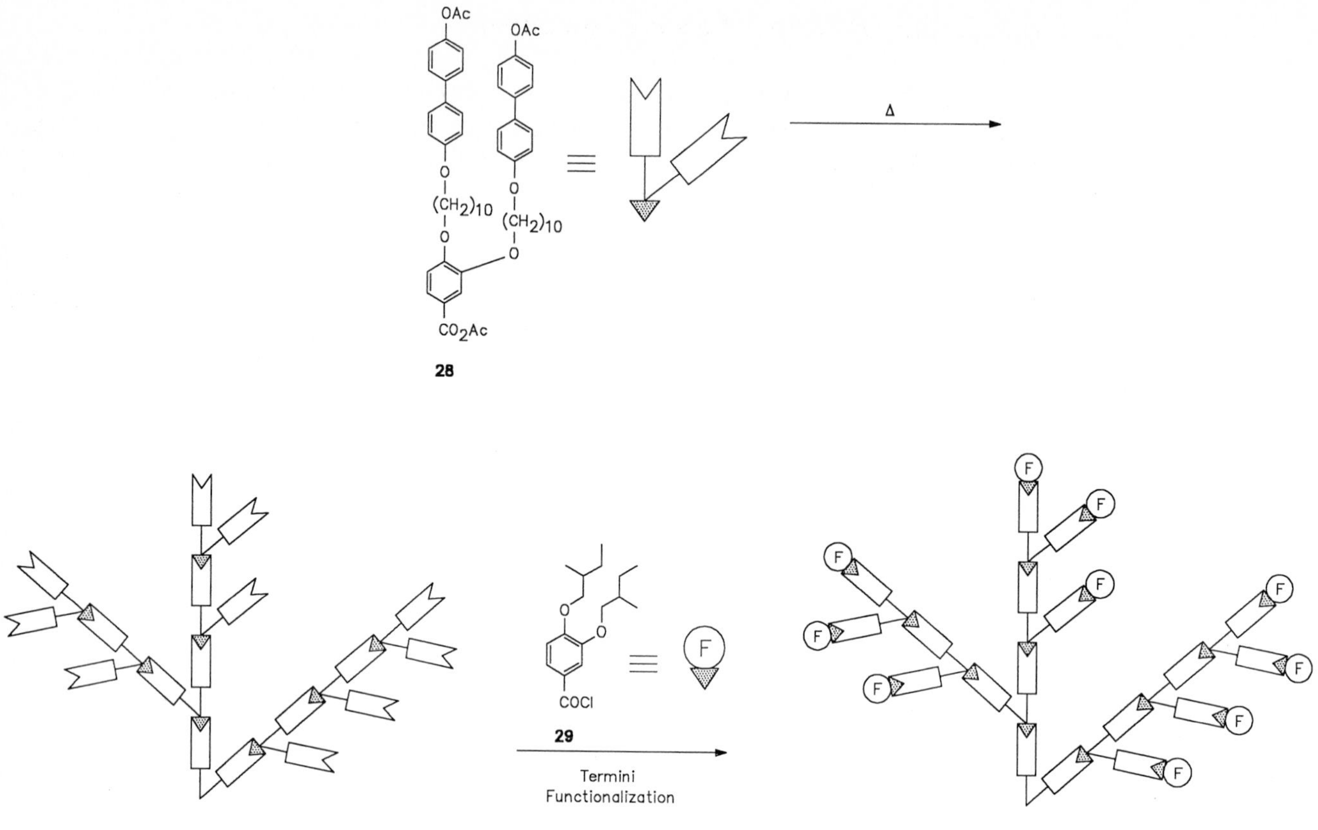

Scheme 7.6. Preparation of a hyperbranched liquid crystalline polymer possessing mesogenic building blocks and chiral termini.

Thus, treatment of polyamine **30**, possessing 64 terminal amino groups, with the *N*-hydroxysuccinimide activated ester of *N*-(*tert*-butoxycarbonyl)phenylalanine **31** afforded the capped dendrimer **32**. Using this procedure, the amino-terminated cascade family from generations 1 through 5 were converted to the corresponding protected amino acid derivatives, e.g., L-alanine, L-leucine, L- and D-phenylalanines, L-tyrosine, and L-tryptophane. The fifth generation dendrimer possessing 64 amino groups was also coated with other chiral amino acid moieties (e.g., L-*t*-Bu serine, L-tyr-cysteine, L-*t*-Bu aspartic ester) by similar technology.[25]

Restricted rotation, as surface conjestion increases, was postulated as being responsible for the observed optical rotations as the generations increase,[25] the possibility of racemization during amino acid acylation was also investigated. Amino acid isolation and high-performance liquid chromatography (HPLC) analysis following acidic hydrolysis of the asymmetric dendrimers revealed an enantiomeric excess > 96 %.

Other aspects of these macromolecules that were discussed included glass transition temperature (T_g), spin−spin (T_2) and spin−lattice (T_1) relaxation measurements, and molecular inclusion via dense packed entrapment of guest molecules.

7.1.6   1 → 2 *N*-Branched, *Amide*-Connectivity

Okada et al.[26a] have reported the persubstitution of PAMAM dendrimers with either lactose or maltose derivatives. Treatment of PAMAM (e.g., **33**; Scheme 7.8) with an excess of either *O*-β-D-galactopyranosyl-(1 → 4)-D-glucono-1,5-lactone (**34**), or the related *O*-α-D-glucopyranosyl-(1 → 4)-D-glucono-1,5-lactone, in DMSO at 27–40 °C under nitrogen afforded the surface coated cascade **35**, which was ascertained to be a single component by SEC and characterized by IR and ^{13}C NMR data. By ^{1}H NMR and vapor pressure osmometry, the surface of the PAMAM was almost quantitatively coated by the sugar residue. These "sugar balls" were shown to be water soluble but insoluble in etha-

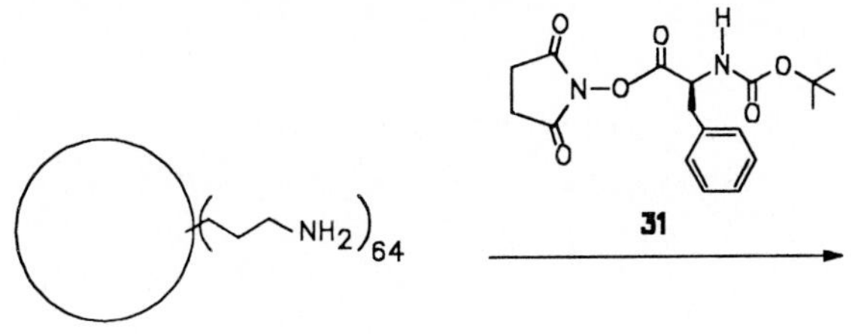

30

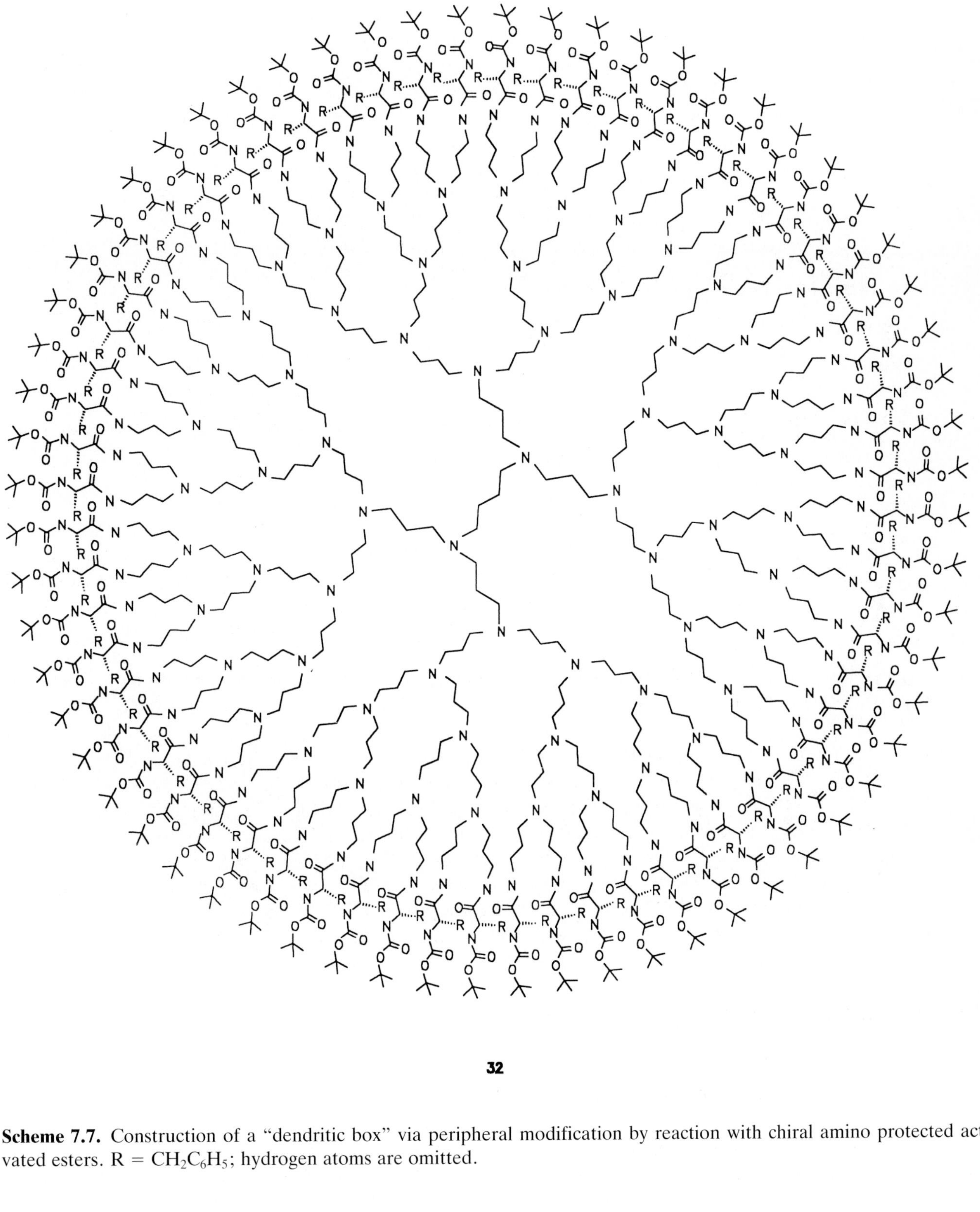

32

Scheme 7.7. Construction of a "dendritic box" via peripheral modification by reaction with chiral amino protected activated esters. R = $CH_2C_6H_5$; hydrogen atoms are omitted.

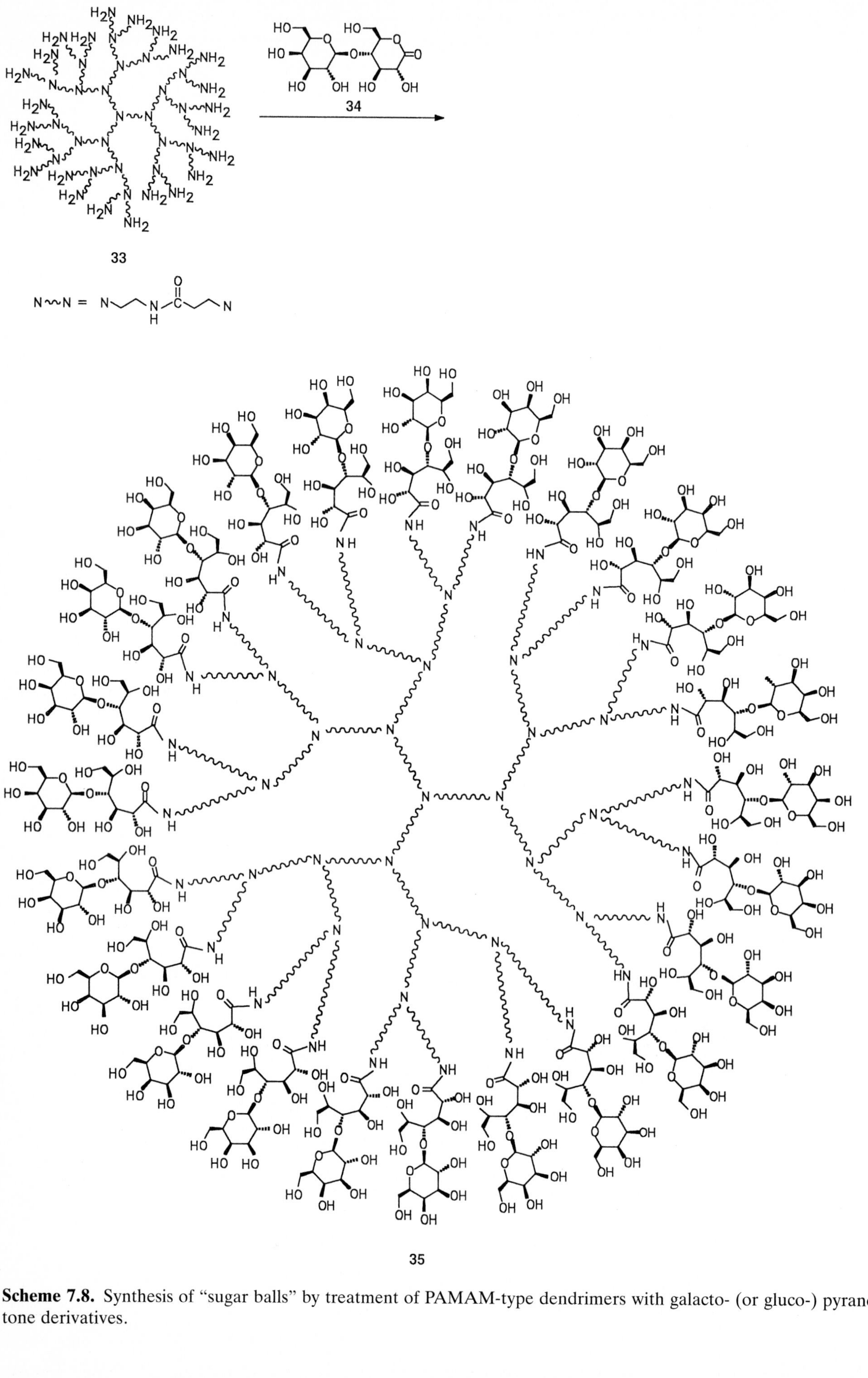

Scheme 7.8. Synthesis of "sugar balls" by treatment of PAMAM-type dendrimers with galacto- (or gluco-) pyranosyl lactone derivatives.

nol and chloroform. The molecular recognition of the sugar groups was also investigated; concanavalin A showed a strong interaction with the gluco-coated sugar ball and little or no interaction with the galaco-coated surface. Apparent aggregation of the lactose-sugar ball with peanut agglutinin (*Arachis hypogaea*) was demonstrated, whereas with the maltose derivative, no interaction was shown.

Roy et al. have reported the synthesis of hyperbranched dedritic lactosides[26b] and glycosides.[26c] Additional sugar terminated, dendritic structures of interest include high mannose content nonamannan residue,[26d] clustered glycopolymers,[26e] and tris(sialyl Lewis N-glycopeptides).[26f] It should be mentioned here that Hanessian et al.[26g] were the first to report (1985) the preparation of polysugar, "megacaloric cluster compounds". Preconstructed glucopyranoside-based arms were attached to tetra- and penta-valent cores. Binder and Schmid[26h] have also reported the synthesis of small carbohydrate bearing molecules.

Antibodies have been coupled[27] to starburst (PAMAM) dendrimers; these reagents offer an attractive approach to the development of immunoassays. These dendritic reagents are stable and retain full immunological activity – both in solution as well as when bound to a solid substrate; their analytical sensitivity is equal to or is better than that of the established methods. The dendritic approach to radial partition immunoassays has been shown to possess the best traits of both homogeneous and heterogeneous immunoassay formats.

Haensler and Szoka[28] reported the use of polyamidoamine polymers for the "transfection" of, or the introduction of DNA into, cultured cells. Dendrimer-mediated transfection thus has ramifications in the area of hereditary disease treatment via gene therapy. Transfection was determined to depend on dendrimer to DNA ratio and dendritic diameter with a 6:1 terminal amine to phosphate charge ratio and a cascade diameter of 68 Å maximizing transfection of firefly luciferase, which was used as an expression vector. When membrane disrupting amphipathic peptides (GALA[29]) were attached to the periphery of the cascade (e.g., an average of 13 GALA residues fifth tier, with the 96 amine dendrimer), DNA transfections increased significantly using a 1:1 dendrimer/DNA complex.

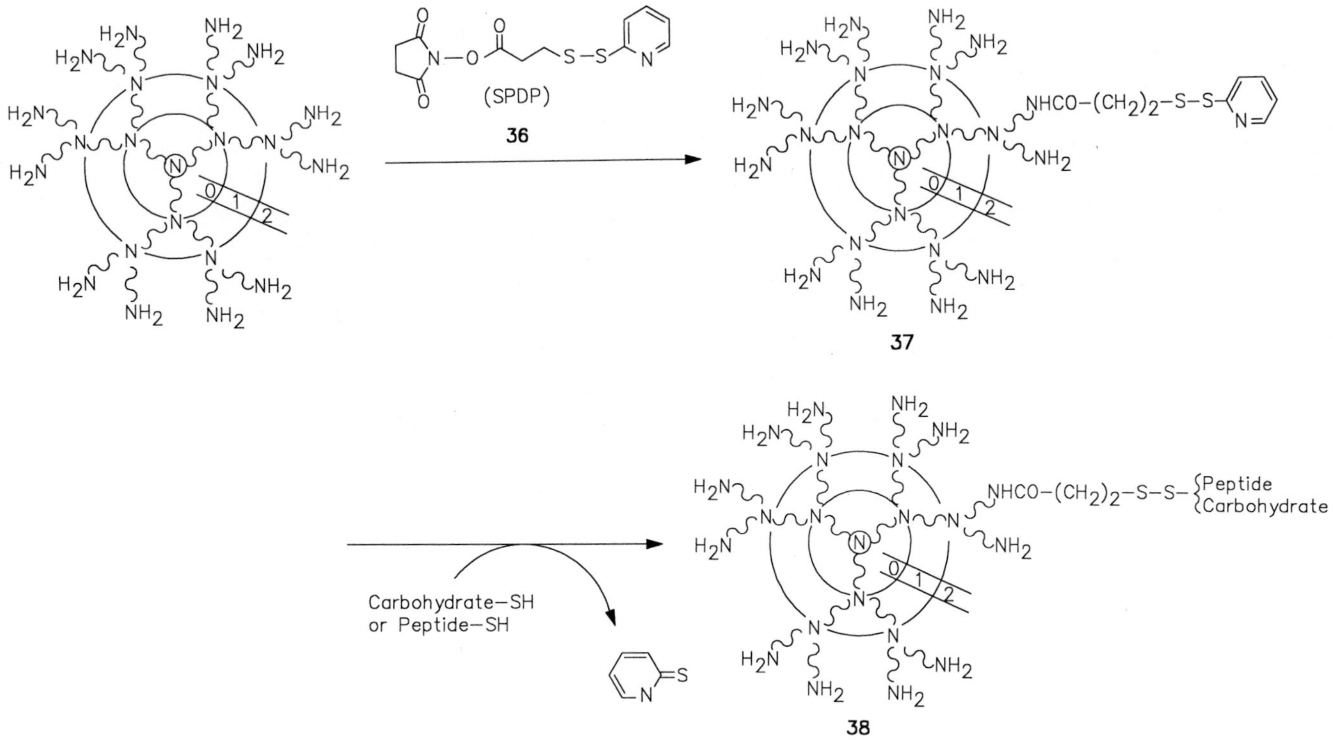

Scheme 7.9. Use of a disulfide linker for the non-site specific connection of amphipathic peptides and carbohydrates to amine terminated dendrimers.

Preparation of the amphipathic peptide modified cascade (**38**, Scheme 7.9) was facilitated by reaction of the polyamine dendrimer with *N*-succinimidyl 3-(2-pyridinyldithio)propionate (SPDP: **36**) to give the polypyridinyldisulfide (**37**). Subsequent treatment of the pyridinyldisulfide moieties with the GALA peptide possessing a cysteine amino acid afforded the modified cascades (**38**).

Barth et al.[30] have reported the preparation of boronated starburst dendrimer–monoclonal antibodies immunoconjugates as a potential delivery system for boron neutron capture therapy.[31,32] Starburst dendrimers have also been employed as "linker molecules" for the covalent connection of synthetic porphyrins to antibodies.[33]

7.2 Convergent Procedures to Chiral Dendrimers

7.2.1   1 → 2 *Aryl*-Branched, *Ether* Connectivity

Seebach and coworkers[34] successfully utilized a convergent approach to the preparation of chiral dendrimers. They wished to address three basic questions: (1) Will the incorporation of chiral cores impact asymmetry to dendritic structures? (2) Can enantioselective host-guest interactions occur near the core? (3) Can chiral recognition be translated to the dendritic surface?

Employing the above asymmetric tris(hydroxymethyl)methane derivatives **8** and **10** (Scheme 7.2) as cores, cascades **39** and **40** were prepared (Scheme 7.10) by the attachment of benzyl ether dendrons, or wedges (e.g., **41**), constructed using the method of Fréchet et al.[35] The resultant optical activity exhibited by these dendrimers was diminished {**39**, $[\alpha]_D^{25} = +3.7$; **40** $[\alpha]_D^{25} = -0.2$} with respect to that of the cores {**8**, $[\alpha]_D^{25} = +12.6$; **10**, $[\alpha]_D^{25} = -14.2$} suggesting a chiral dilution effect. The authors also noted that the optical rotation of the tris(allyl ether) precursors {$[\alpha]_D^{25} = +10.8$} to the homologated triol core **10** reverses in sign. Polyether **39** was shown to form a clathrate with CCl_4; extrusion of CCl_4 was possible only by heating at 100 °C (0.5 mm) for several hours.

A novel approach to chiral dendrimer synthesis has been reported by Kremer and Meijer.[36] Conceptionally, this unique cascade can be envisioned as a tetrasubstituted pentaerythritol whereby each core oxygen is substituted by a different generation dendritic wedge, i.e. bromides **43**, **44**, and **45** corresponding to the first through third tier wedges, respectively. Construction of the poly(benzyl ether) **42** was slightly more problematic than its conception (Scheme 7.11). Beginning with diethyl (ethoxymethylene) malonate, diol **46** was prepared in five steps. Bissilylation of diol **46** followed by aldehyde liberation, reduction, and reaction with benzyl bromide gave the bis(*tert*-butyldimethylsilane) (**47**). Key to the introduction of chirality via core manipulation was the removal ($ZnBr_2$, CH_2Cl_2) of a *single tert*-butyldimethylsilyl (TBDMS) group from polyether **47** in the presence of a similar TBDMS group as well as a methoxymethyl (MEM) ether moiety to afford monoalcohol **48**. Subsequent reaction with the first tier bromide **43**, desilylation (Bu₄NF), and treatment with the second tier bromide **44** afforded the MEM-protected alcohol **49**. MEM removal (*β*-chlorocatechoborane) and final attachment of the third tier wedge **45** gave the desired racemic, asymmetric ether **42**.

The use of chiral auxiliaries to induce enantioselectivity was unsuccessful, as was attempted racemic separation via chiral HPLC stationary phase. It should be noted, however, that this is an excellent example of the utility of the convergent process coupled with controlled deprotection of a core unit.

7.2.2   1 → 2 *C*-Branched, *Amide* Connectivity

Mitchell et al. reported[37] the synthesis of chiral dendrimers using L-glutamic acid as the monomer. Construction (Scheme 7.12) was accomplished convergently beginning with *N*-benzyloxycarbonyl protected L-glutamic acid, which was converted to the bis(activated ester) (**50**) by reaction with *N*-hydroxysuccinimide, DCC, and a catalytic amount

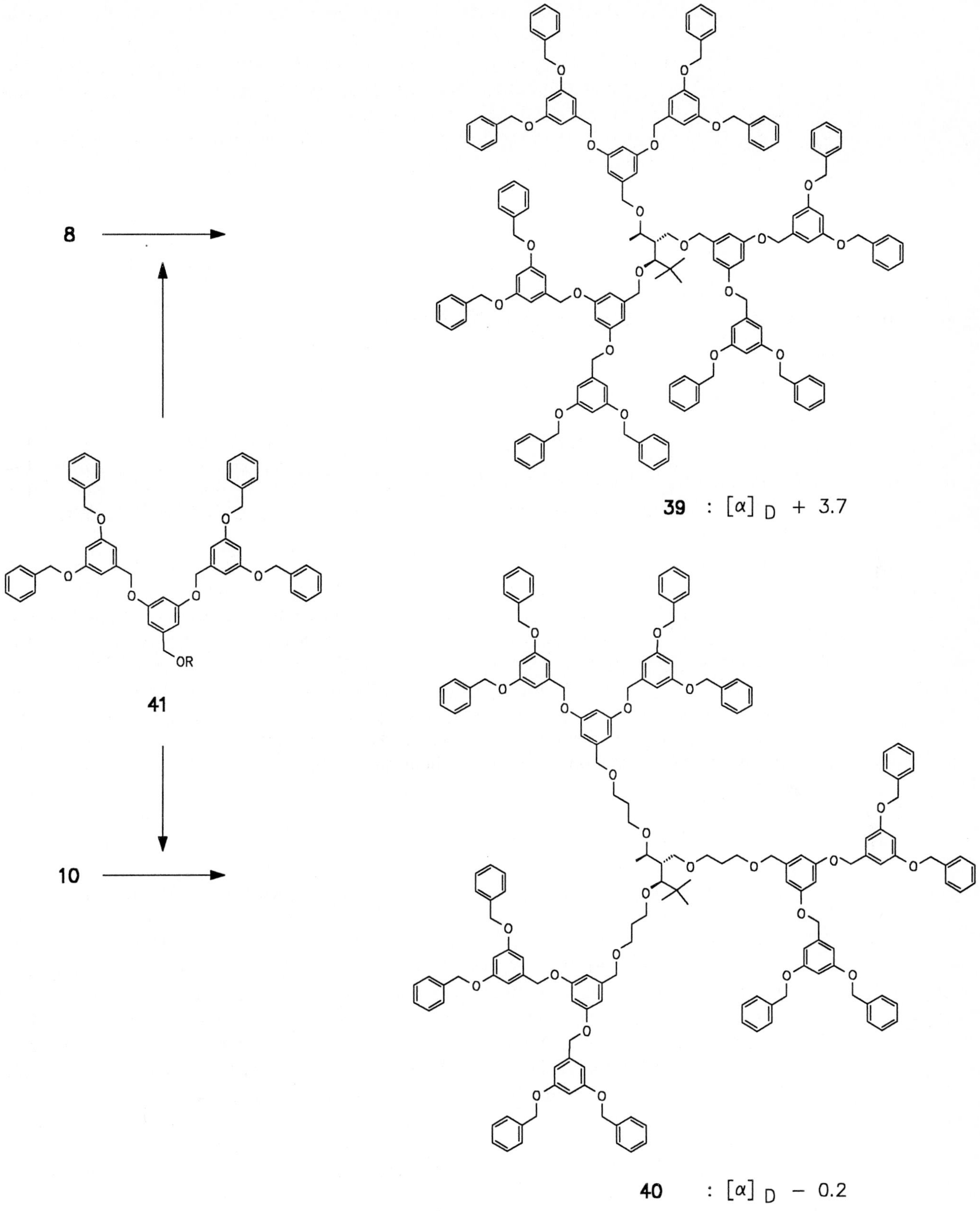

8 ⟶

41

10 ⟶

39 : [α] D + 3.7

40 : [α] D − 0.2

Scheme 7.10. Preparation of chiral dendrimers via the attachment of benzyl ether dendrons to asymmetric, trigonal cores.

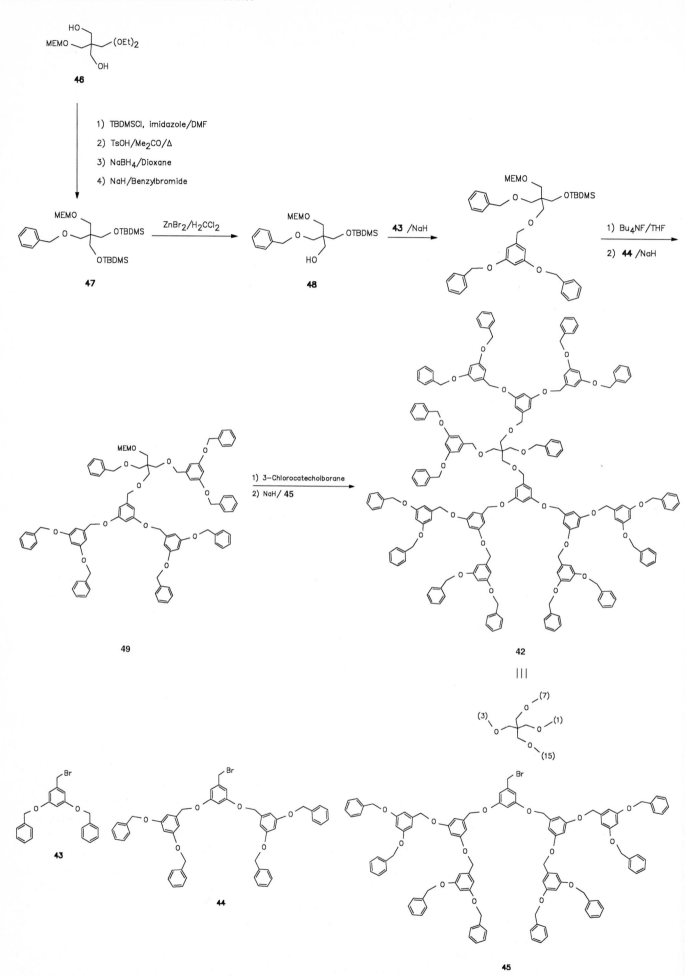

Scheme 7.11. Synthesis of a chiral dendrimer based on the attachment of different generation wedges to a symmetric pentaerythrityl-based core.

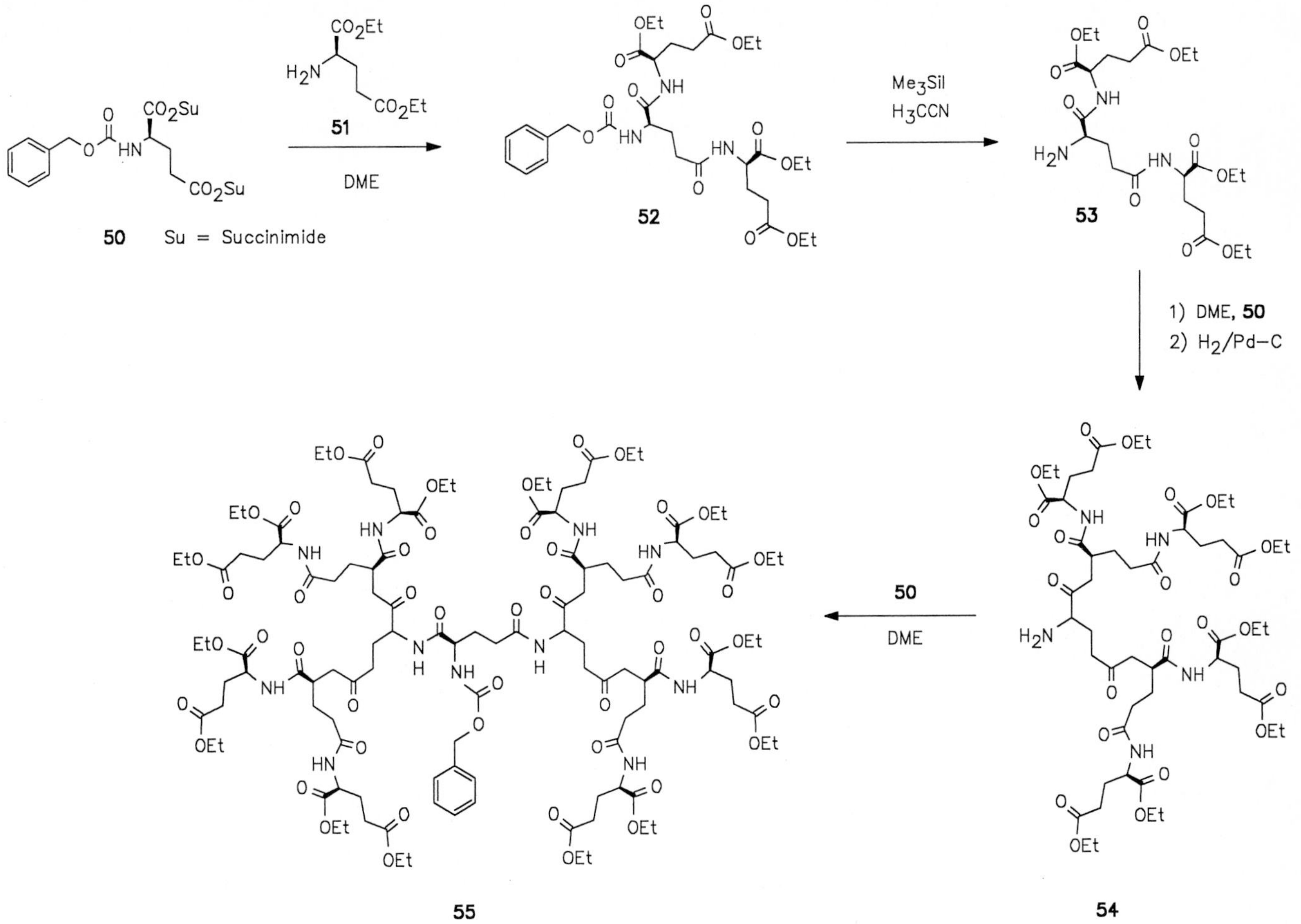

Scheme 7.12. Employment of *L*-glutamic acid building blocks for the construction of chiral superstructures.

of *N,N*-dimethylaminopyridine (DMAP). Treatment of the activated ester **50** with L-glutamic acid diethyl ester (**51**) yielded the tetraethyl ester **52**, which was debenzylated (ISiMe₃) to give the free amine **53**. Acylation of two equivalents of amino tetraester **53** with one equivalent of bis(succinimide ester) (**50**) resulted in the second generation, i.e., the octaethyl ester, that was subsequently catalytically deprotected (H₂/Pd−C) to give the free amine **54**. This alternative method was employed when attempted cleavage with iodotrimethylsilane failed, presumably due to the increased steric hindrance caused by the growing dendrimer.

The octaester **54** was then treated with additional activated ester **50** to afford an *N*-protected dendrimer **55** possessing 16 terminal ester moieties and 15 chiral centers, all with the identical L-configuration. No optical activity data were reported.

Hudson and Damha described[38] the synthesis of dendritic polymers based on nucleic acids using an automated DNA synthesizer and employing thymidine and adenosine as building blocks. Convergent construction began by anchoring a thymidine moiety (Scheme 7.13) to a long-chain alkylamine-controlled-pore glass (LCAA−CPG) support possessing long alkyl spacers (18 Å) that enhance bound reagent accessibility and a 500 Å pore size that facilitates high molecular weight oligonucleotide synthesis. Repetitive treatment of the thymidine bound units (**56**) with deoxythymidine phosphoramidite in a "chain extension" procedure afforded pentathymidine chains (**57**). The polymer-bound chains were then "branched" or coupled, via reaction with a tetrazole activated adenosine 2′,3′-bis(phosphoramidito) reagent. Continued "chain extension" and "branching" allowed preparation of an 87-mer (**58**), which was simultaneously cleaved from the support and deprotected employing NH₄OH.

Due presumably to increased distances between reactive sites, coupling of two equivalents of 38-mer (**59**) was unsuccessful. Further chain extension to give the 43-mer (**60**) subsequently allowed the preparation of the two-directional dendrimer (**58**) via standard coupling.

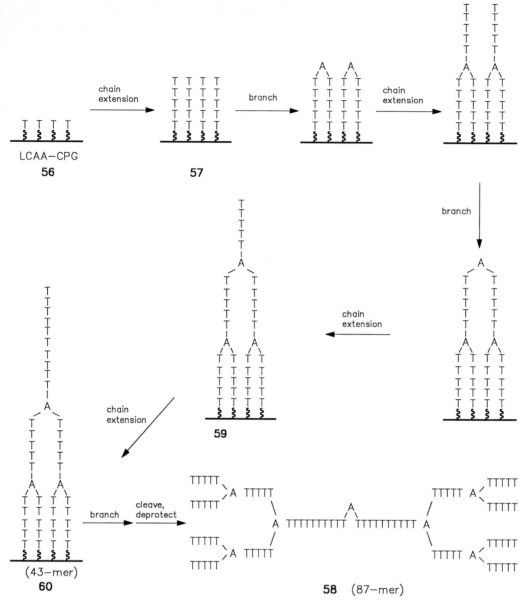

Scheme 7.13. Synthesis of thymidine and adenosine based dendrimers using a DNA synthesizer.

Characterization of the dendrimers included composition analysis by snake venom phosphodiesterase and alkaline phosphatase (SVPDE/AP) enzymatic digestion to afford inosine derived from the deamination of adenosine and thymidine in the expected quantities based on HPLC. Polyacrylamide gel electrophoresis was also employed. The authors noted that increasing quantities of branch defects, described as "default" oligomers, suggested that divergent methodology might be better suited for the construction of macromolecules using this technique.

The construction of an asymmetric diamino acid has also been reported[39] that was envisioned as a replacement for the *C*-terminal half of a class of DNA binding proteins known as the Leucine Zipper class of proteins.

7.2.3   1 → 2 *Aryl*-Branched, *Ether* Connectivity

Chow and coworkers[40,41] reported the synthesis and characterization of homochiral dendrimers through the first generation by using (2*R*, 3*R*)-tartaric acid as the chiral building block. The mode of preparation is based on three parts: the three-directional core, 1,3,5-trihydroxybenzene (**61**), the chiral connector, 1-*O*-tosyl-4-*tert*-butyl-phenoxy-2,3-*O*-isopropylidene-L-threitol (**62**), and a branching unit, 1-benzyloxy-3,5-dihydroxybenzene (**63**).

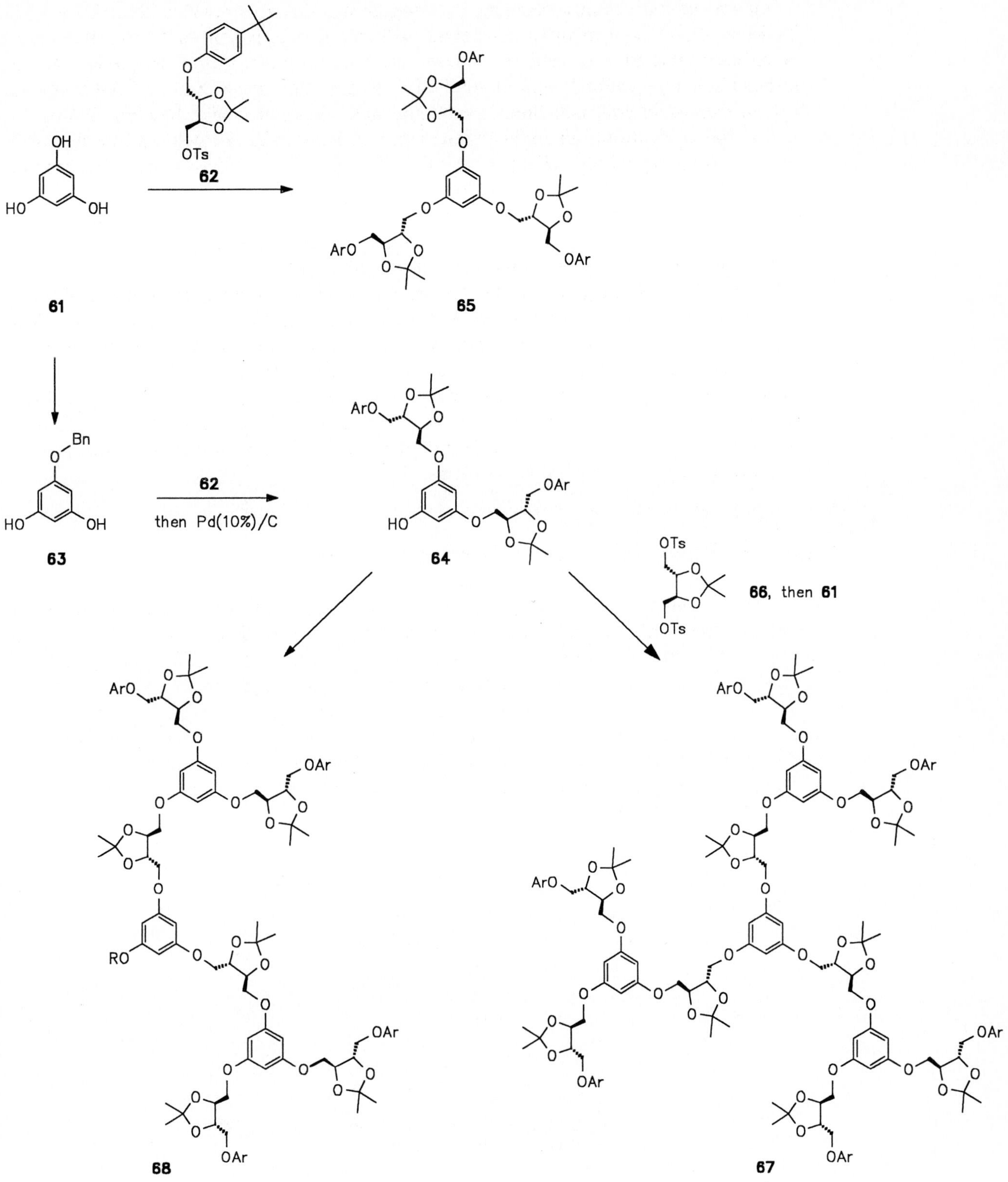

Scheme 7.14. Construction of homochiral dendrimers effected using (2*R*,3*R*)-tartaric acid-based building blocks.

Convergent construction (Scheme 7.14) began with the treatment of (2*S*, 3*S*)-(−)-1,4-di-*O*-tosyl-2,3-*O*-isopropylidene-L-threitol with one half equivalent of 4-*tert*-butylphenol under basic conditions to yield the mono-*O*-arylation product (**62**). Subsequent reaction of the chiral aryl ether **62** with one-half equivalent of dihydroxybenzene[42] **63**, followed by hydrogenolysis with palladium on charcoal gave the dendron **64**. Attempted preparation of the disubstituted phenol **64** via reaction of the tosylate **62** with one third equivalent core **61** resulted in low yields (ca. 26 %). Poor solubility of the tris-phenolic trianion was suggested as a rationale for the low conversion.

The 0th cascade **65** was prepared from core **61** and chiral block **62**, whereas the first generation dendrimer **67** was synthesized from the extension of dendron **64** with the bis-tosylate **66**, followed by the addition of the same core **61**.

The specific rotation {[*α*]D} of the 0th and first generation dendrimers (**65** and **67**) was recorded at − 59.60 and − 69.70, respectively; molar rotation was reported as − 5690 (**65**) and − 17690 (**67**), and the molar rotation per tartaric acid unit was determined to be − 1900 (**65**) and − 1970 (**67**), respectively. The use of bis(aryl ether) **64** allowed construction of tetrakis(aryl ether) **68**, which, when treated with the bistosylate **66**, afforded the corresponding dendron to be used for the preparation of the second generation dendrimers; subsequent reaction with trisphenol **61** failed to give the desired material presumably due to steric and solubility problems.

7.2.4   1 → 3 *P*- and *Aryl*-Branched, *P*- and *Ether*-Connectivity

Brunner and Fürst[43] reported the synthesis of a series of optically active, extended chelate phosphines, one of which was derived from the reaction of 5-bromo-1,3-di(borneoxymethyl)benzene (**69**) and bis(dichlorophosphino)methane (**70**). Technically, the product (−)-1,1-bis{3′,5′-di(borneoxymethyl)phenyl]phosphino}methane (**71**) is both a first tier *P*- and 1,3-aryl-branched dendrimer. Excluding distortions, the maximum distance between the *P* atom and the most distant *H* atom(s) is 11.9 Å for the two-layer ligand **71**. The Rh-catalyzed {[Rh(cod)Cl]₂} hydrogenation of (*α*)-*N*-acetamidocinnamic acid in the presence of **71** afforded a "disappointingly low" [5.2 % *ee* (*R*)] optical induction.

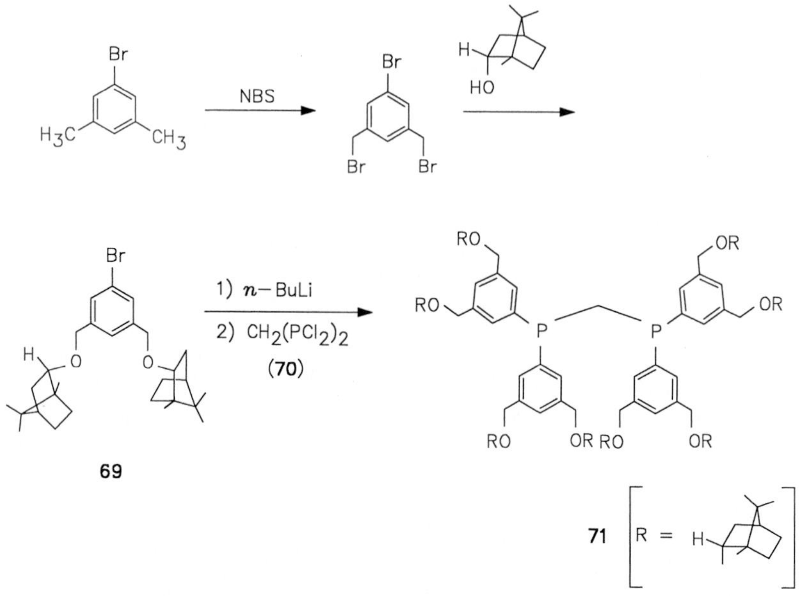

Scheme 7.15. First generation phosphine-core dendrimers possessing optically active (−)-borneol terminated groups.

7.3 References

[1] R. G. Denkewalter, J. F. Kolc, W. J. Lukasavage, U. S. Pat. 4,410,688, (**1979**).

[2] R. G. Denkewalter, J. F. Kolc, W. J. Lukasavage, U. S. Pat. 4,360,646 (**1982**).

[3] R. G. Denkewalter, J. F. Kolc, W. J. Lukasavage, U. S. Pat. 4,289,872 (**1981**).

[4] S. M. Aharoni, C. R. Crosby III, E. K. Walsh, *Macromolecules* **1982**, *15*, 1093.

[5] I. Toth, M. Danton, N. Flinn, W. A. Gibbons, *Tetrahedron Lett.* **1993**, *34*, 3925.

[6] J. P. Tam, *Proc. Natl. Acad. Sci. USA* **1988**, *85*, 5409.

[7] A. Pessi, E. Bianchi, F. Bonelli, L. Chiappinelli, *J. Chem. Soc. Chem. Commun.* **1990**, 8.

[8] C. Rao, J. P. Tam, *J. Am. Chem. Soc.* **1994**, *116*, 6975.

[9] C. Y. Wang, D. J. Looney, M. L. Li, A. M. Walfield, J. Ye, B. Hosein, J. P. Tam, F. Wong-Staal, *Science* **1991**, *254*, 285.

[10] R. Roy, D. Zanini, S. J. Meunier, R. Romanowska, *J. Chem. Soc., Chem. Commun.* **1993**, 1869.

[11] a) T. M. Chapman, G. L. Hillyer, E. J. Mahan, K. A. Shaffer, *J. Am. Chem. Soc.* **1994**, *116*, 11195. b) O. Katzenelson, H. Z. Hel-Or, D. Avnir, *Chem. Eur. J.* **1996**, *2*, 174

[12] G. R. Newkome, X. Lin, *Macromolecules* **1991**, *24*, 1443.

[13] G. R. Newkome, X. Lin, C. D. Weis, *Tetrahedron: Asymmetry* **1991**, *2*, 957.

[14] Y. S. Klausner, M. Bodansky, *Synthesis* **1972**, 453.

[15] J.-M. Lapierre, K. Skobrides, D. Seebach, *Helv. Chim. Acta* **1993**, *76*, 2419.

[16] D. Seebach, J.-M. Lapierre, W. Jaworek, P. Seiler, *Helv. Chim. Acta* **1993**, *76*, 459.

[17] a) G. R. Baker, G. R. Newkome, *Org. Prep. Proc. Int.* **1986**, *18*, 117. b) P. Murer, D. Seebach, *Angew. Chem. Int. Ed. Engl.* **1995**, *34*, 2116.

[18] K. J. A. Chang, W. Pugh, S. G. Blanchard, J. McDermed, J. P. Tam, *Proc. Natl. Acad. Sci. U. S.* **1988**, *85*, 4929; D. N. Posnett, H. McGrath, J. P. Tam, *J. Biol. Chem.* **1988**, *263*, 1719.

[19] C. Rao, J. P. Tam, *J. Am. Chem. Soc.* **1994**, *116*, 6975.

[20] a) J. Shao, J. P. Tam, *J. Am. Chem. Soc.* **1995**, *117*, 3893. b) H.-T. Chang, C.-T. Chen, T. Kondo, G. Siuzdak, K. B. Sharpless, *Angew. Chem. Int. Ed. Engl.* **1996**, *35*, 182. c) T. Kawaguchi, K. L. Walker, C. L. Wilkins, J. S. Moore, *J. Am. Chem. Soc.* **1995**, *117*, 2159.

[21] D. Seebach, J. M. Lapierre, G. Greiveldinger, K. Skobridis, *Helv. Chim. Acta* **1994**, *77*, 1673.

[22] S. Bauer, H. Fischer, H. Ringsdorf, *Angew. Chem.* **1993**, *105*, 1658; *Angew. Chem. Int. Ed. Engl.* **1993**, *32*, 1589.

[23] J. F. G. A. Jansen, E. M. M. de Brabander-van den Berg, E. W. Meijer, *Science* **1994**, *266*, 1226.

[24] E. M. M. de Brabander-van den Berg, E. W. Meijer, *Angew. Chem.* **1993**, *105*, 1370; *Angew. Chem., Int. Ed. Engl.* **1993**, *32*, 1308.

[25] J. F. G. A. Jansen, E. W. Meijer, E. M. M. de Brabander - van den Berg, *J. Am. Chem. Soc.* **1995**, *117*, 4417. J. F. G. A. Jansen, H. W. I. Peerlings, E. M. M. de Brabander – van den Berg, E. W. Meijer, *Angew. Chem.* **1995**, *107*, 1321; *Angew. Chem. Int. Ed. Engl.* **1995**, *34*, 1206.

[26] a) K. Aoi, K. Itoh, M. Okada, *Macromolecules* **1995**, *28*, 5391. b) R. Roy, W. K. C. Park, Q. Wu, S.-N. Wang, *Tetrahedron Lett.* **1995**, *36*, 4377. c) D. Zanini, W. K. C. Park, R. Roy, *Tetrahedron Lett.* **1995**, *36*, 7383. d) P. Grice, S. V. Ley, J. Pietruszka, H. W. M. Priepke, *Angew. Chem. Int. Ed. Engl.* **1996**, *35*, 197. e) T. Furuike, N. Nishi, S. Tokura, S.-I. Nishimura, *Chem. Lett.* **1995**, 823. f) U. Sprengard, M. Schudok, W. Schmidt, G. Kretzschmar, H. Kung, *Angew. Chem. Int. Ed. Engl.* **1996**, *35*, 321. g) S. Hanessian, C. Hoornaert, A. G. Vernet, A. M. Nadzan, *Carbohydr. Res.* **1985**, *137*, C14. h) W. H. Binder, W. Schmid, *Monatsh. Chem.* **1995**, *126*, 923.

[27] P. Singh, F. Moll, III, S. H. Lin, C. Ferzli, K. S. Yu, R. K. Koski, R. G. Saul, P. Cronin, *Clin. Chem.* **1994**, *40*, 1845.

[28] J. Haensler, F. C. Szoka, Jr., *Bioconjugate Chem.* **1993**, *4*, 372.

[29] N. K. Subbarao, R. A. Parente, F. C. Szoka, L. Nadasdi, K. Pongracz, *J. Biol. Chem.* **1987**, *26*, 2964.

[30] R. F. Barth, D. M. Adams, A. H. Soloway, F. Alam, M. V. Darby, *Bioconjugate Chem.* **1994**, *5*, 58.

[31] R. F. Barth, A. H. Soloway, R. G. Fairchild, R. M. Baggert, *Cancer* **1992**, *70*, 2955.

[32] R. F. Barth, A. H. Soloway, R. G. Fairchild, *Cancer Res.* **1990**, *50*, 1061.

[33] J. C. Roberts, Y. E. Adams, D. Tomalia, J. A. Mercer-Smith, D. K. Lavallee, *Bioconjugate Chem.* **1990**, *1*, 305.

[34] D. Seebach, J.-M. Lapierre, K. Skobridis, G. Greiveldinger, *Angew. Chem.* **1994**, *106*, 457; *Angew. Chem., Int. Ed. Engl.* **1994**, *33*, 440.

[35] C. Hawker, J. M. J. Fréchet, *J. Chem. Soc., Chem. Commun.* **1990**, 1010.

[36] J. A. Kremers, E. W. Meijer, *J. Org. Chem.* **1994**, *59*, 4262. J. A. Kremers, E. W. Meijer, *Macromol. Symp.* **1995**, *98*, 491.

[37] L. J. Twyman, A. E. Beezer, J. C. Mitchell, *Tetrahedron Lett.* **1994**, *35*, 4423.
[38] R. H. E. Hudson, M. J. Damha, *J. Am. Chem. Soc.* **1993**, *115*, 2119.
[39] F. Bambino, R. T. C. Brownlee, F. C. K. Chiu, *Tetrahedron Lett.* **1994**, *35*, 4619.
[40] H.-F. Chow, L. F. Fok, C. C. Mak, *Tetrahedron Lett.* **1994**, *35*, 3547.
[41] H.-F. Chow, C. C. Mak, *J. Chem. Soc., Perkin Trans. 1* **1994**, 2223.
[42] W. D. Curtis, J. F. Stoddart, G. H. Jones, *J. Chem. Soc., Perkin Trans. 1* **1977**, 785.
[43] H. Brunner, J. Fürst, *Tetrahedron* **1994**, *50*, 4303.

8 Dendrimers Containing Metal Sites

The introduction of metal centers on the surface or within dendrimers has been a recent trend reflecting a shift from the initial synthetic directions to a more applied emphasis. In general, surface modification is the easier mode of attachment than the incorporation of metal centers within the interior of the dendrimers or hyperbranched polymers due to a lack of appropriately designed building blocks or the internal availability of binding loci. The organometallic dendrimers afford attractive advantages over their polymeric counterparts in that they possess a precise molecular architecture as well as a predetermined chemical composition. Metal centers have been shown to act as connectors, branching, or terminal (surface) centers; also, metal sites can be located at either specific or random loci within the dendritic infrastructure.

8.1 Metals as Branching Centers

Balzani et al.[1–3] reported the syntheses of homo- and heterometallic (ruthenium and osmium) polypyridine cascade complexes, in which their 'complexes as metals' and 'complexes as ligands' strategy is tantamount to a *divergent−convergent procedure*.[4] The decanuclear polypyridine complex **1** was prepared (Scheme 8.1) from a core, $M(BL)_3^{2+}$ (**2**), consisting of a metal [M, e.g., Ru(II) or Os(II)] coordinated to three 2,4-bis(2-pyridyl)pyrazine (BL) ligands[5] and the preconstructed arms, $Ru[(BL)M(L)_2]_2Cl_2^{4+}(PF_6)_4$ (**3**), where L is bipyridine or biquinoline[6] to afford $M[(BL)M[(BL)M(L)_2]_2]_3^{20+}$ (**1**). In a typical procedure,[1] the attachment of the arms with three-directional hexacoordinate branching centers to a core [M(BL)₃] was accomplished in very good yield in refluxing mixture of methanol−water−ethyleneglycol−silver nitrate. Utilizing this general procedure, Balzani et al.[7] reported the preparation of cascade molecules derived from luminescent and redox-active transition metal complexes.[8–10] This procedure allows the preparation of dendrimers possessing two or more

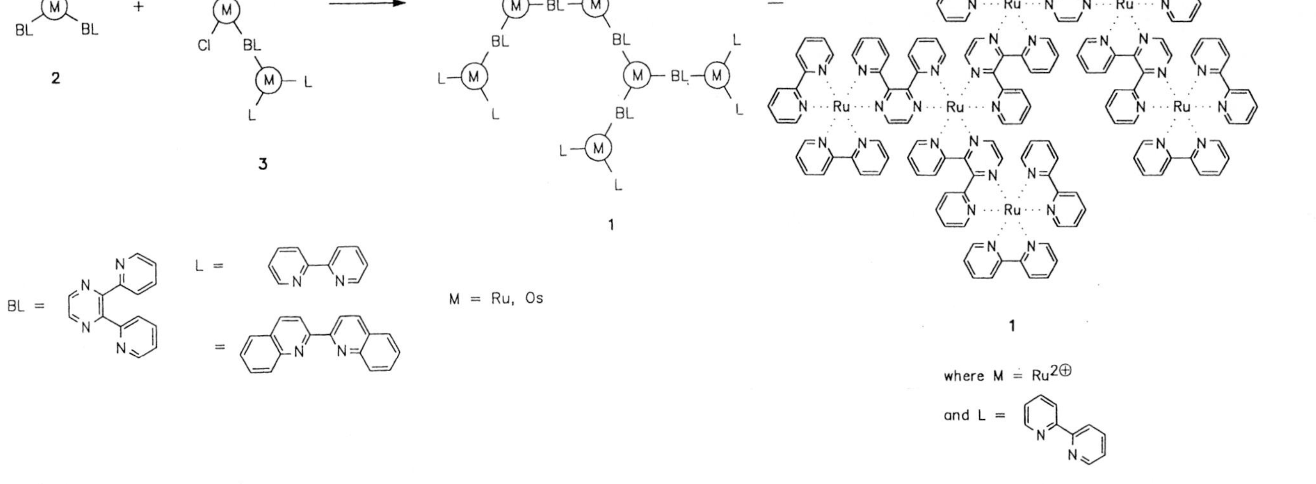

Scheme 8.1. Balzani et al.'s[1–3] pyridyl-based building blocks have afforded entrance into novel polymetallic dendrimers.

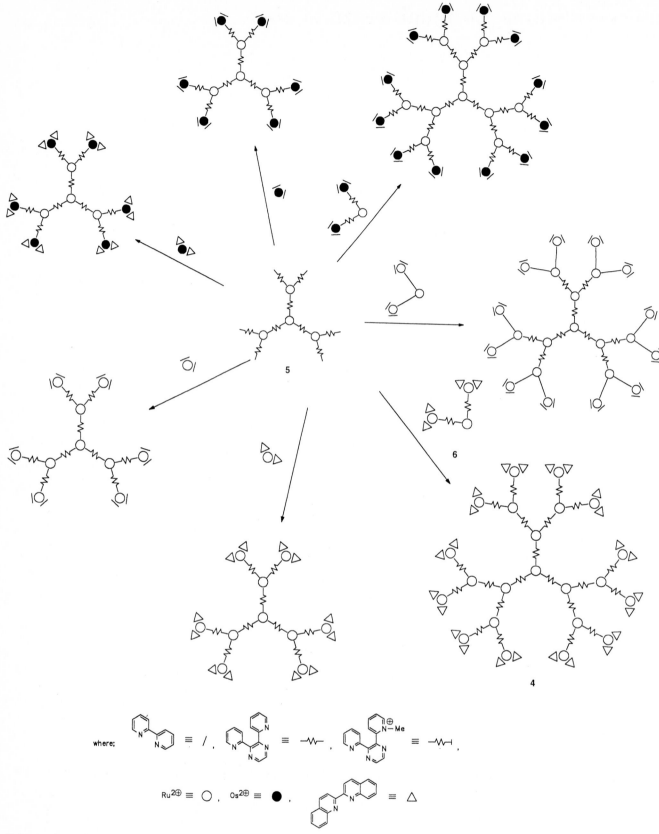

Scheme 8.2. Absorption and luminescence characteristics of this family of Ru^{+2} and Os^{+2} dendrimers have been studied (e.g., **4:** $\lambda^{298K}_{max(abs)} = 542$ nm and $\lambda^{298K}_{lum} = 786$ nm)

different concentric metal spheres as each generation is added; this concept of construction utilizing transition metals within the infrastructure instills specific properties, such as luminescence or electrochemical effects, which can be easily modified; application of this technology could potentially lead to light harvesting devices.[11,12]

More recently, Campagna, Denti, Balzani and coworkers[4, 13] have extended their method to prepare larger family members of this series (Scheme 8.2). The largest member (**4**), reported to date, possesses 22 metal centers and was prepared by treatment of the tetrakisRu(II) core **5** with the trisRu(II) dendron **6**. Oxidation and reduction potentials have been tabulated for this series as well as the absorption and emission properties. Some electrochemical interactions between neighboring units have been observed; however, each metal-based building block in the polymetallic dendrimers possesses its own absorption properties. The absorption and luminescence coefficients are thus correspondingly large (e.g., $\lambda^{298K}_{max(absorbance)} = 542$ nm and $\lambda^{298K}_{max(luminescence)} = 786$ nm for **4**).

Using similar metal−ligand connectivity as that of Balzani and coworkers,[13] Brewer et al.[14] reported photoinitiated electron collection wherein two covalently coupled ruthenium chromophores function in the same mixed metal (Ru and Ir) tricomplex (**7**; Scheme 8.3) to effect electron collection; while this example is not highly dendritic, it can be considered to be a first generation. Photolysis of $\{[(bpy)_2Ru(dpb)]_2IrCl_2\}^{+5}$ [py = 2,2′-bipyridine; dpb = 2,3-bis(2-pyridyl)benzoquinoxaline] in the presence of dimethylaniline generates the doubly reduced species $\{[(bpy)_2Ru(dpb^-)]_2IrCl_2\}^{3+}$, possessing the $[Ir(dpb)_2Cl_2]^{1+}$ moiety, which was previously demonstrated[15a] to deliver electrons stored on the bridging ligands to a substrate. Phenanthroline cores have also been attached to Fréchet-type dendrons (Section 5.4.2) and subsequently complexes to Ruthenium.[15b]

Bochkarev et al.[16] prepared hyperbranched dendrimers via polymerization of tris(pentafluorophenyl)germanium anions resulting in branched macromolecules with Ge-branching centers. Building block propagation was attributed to metal anion, aromatic substitution of a phenyl, *para*-fluoro-moiety. Thus, treatment of Ge-monomer **8** with excess Ge-anion afforded hyperbranched macromolecules with an internal framework resembling tetrakisGedendrimer **9** (Scheme 8.4). Subsequently, Sn and Si were employed for hyperbranched dendrimer construction.

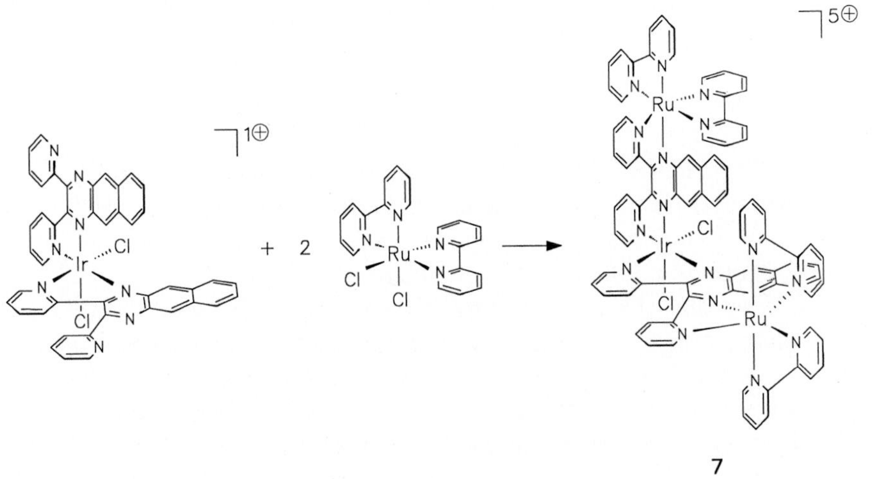

7

Scheme 8.3. Construction of a mixed valence branched assembly capable of photoinitiated electron collection.

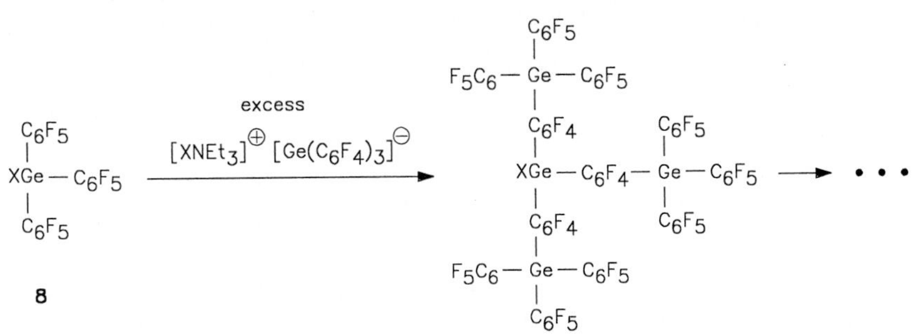

8

9

Scheme 8.4. Preparation of polygermanium hyperbranched dendrimers was reported[16] in 1988.

8.2   Metals as Building Block Connectors

The general use of the tridentate ligand 2,2′:6′,2″-terpyridine (tpy) has been reported[17–19] to control the assembly of coordination oligomers and polymers. Newkome et al.[20] reported the preparation of dendrimers incorporating novel ligand−metal−ligand connectivity for building block attachment (Scheme 8.5). The resultant dendrimer (**15**) possessed twelve pseudooctahedral Ru(II) centers,[21] each coordinated by two orthogonal, 4′-substituted 2,2′:6′,2″-terpyridines. Synthesis of the dendritic building blocks **10** and **11** (derived from the previously reported[22] 1 → 3 C-branching precursors (e.g., **12**) was predicated on the facile alkoxylation of 4′-chloroterpyridine.[23] Dendritic construction was then readily accomplished by coupling of the known tetracarboxylic acid[24] **13** with amine **10** to afford the uncomplexed, dodecaterpyridine dendrimer **14**, which, when treated with twelve equivalents of paramagnetic Ru(III) building block **11** in the presence of N-ethylmorpholine, generated the heteroleptic dodecaruthenium complex **15**.

This tpy−ruthenium−tpy (<Ru>) connectivity has been applied[25] to the development of complexes designed to explore the construction of "ordered, three-dimensional matrices" comprised of specifically positioned multiple dendrimer assemblies or networks[26,27] (see Chapter 9). Bis-dendrimer **16** was subsequently prepared (Scheme 8.6) via the single, Ru(II)-based connection of two independently prepared dendrimers each possessing a tpy unit. Due to the proximity of the incorporated terpyridine units to the core branching centers, these participating dendrimers were described as a "key" (**17**) and a "lock" (**18**). Both the lower generation key **17** and the lock **18** were constructed in an analogous manner starting with the alkoxylation of 4′-chloroterpyridine by an ω-hydroxyacid to provide the requisite functionalized cores. The tpy-carboxylic acids were then each treated[24a] sequentially with aminotriester **19** and formic acid to afford the desired generations of key and lock combinations, e.g., **16**. Electrochemical studies of the "complex" series revealed that as the steric hindrance around the Ru−tpy redox-active center increases, the reversible redox behavior exhibited by the cyclic voltammograms is adversely affected. Such a redox-driven, macromolecular "unlocking" forms the basis for a reversible "lock−unlocking" mechanism. Also Constable and Harverson[24b] have developed a convergent stategy for the construction of terpyridin-Ru-based branched and linear macromolecules.

Dendrimers possessing multiple internally incorporated 3,3′-substituted bipyridine ligands have also been reported.[28] Cascade construction was effected via amine homologation of the aminotriester building block **19** (see Scheme 8.6) allowing the insertion of various diamino-based functional groups near the dendritic core. Other polynucleating pyridinyl-based molecules that have been reported include 1,3,5-tris(4′-terpyridinyl)benzene,[29] hexakis(4-pyridyl)benzene,[30] and a series[31] of poly(4-vinylpyridino)benzenes and poly(4-ethynylpyridino)benzenes.

Organoplatinum dendrimers have been successfully created by Achar and Puddephatt,[32] in which construction of the cascades was conducted by a convergent procedure using two reactions: (1) oxidative−addition of PtMe$_2$(t-Bu$_2$bpy) via CH$_2$Br insertion generating stable Pt(IV) centers and (2) SMe$_2$ ligand displacement from Pt$_2$Me$_4$(μ-SMe$_2$)$_2$ by a free bipyridine (Scheme 8.7). Thus, two equivalents of [PtMe$_2$(t-bu$_2$bpy)] (**20**) were treated with one equivalent of 4,4′-bis(bromomethyl)-2,2′-bipyridine (**21**) to give the 1st generation cascade (**22**) containing two [Pt(IV)] centers where Pt oxidative-addition serves as the mode of building block connectivity. Treatment of the bis Pt-complex **22** with the dimeric, platinum dimethylsulfide complex afforded the mixed valence, tris-platinum complex **23**; subsequent oxidation−addition and Pt(II) metalation gave the heptaplatinum dendrimer **24**. Repetition of the sequence smoothly afforded the third generation Pt$_{14}$ cluster. Attempts to generate the next tier were only moderately successful, presumably due to steric constraints. This example of cascade construction constituted the first report of *alkyl* transition metal dendrimers. The authors later reported[33] the oxidative−addition reaction of four equivalents of each Pt-complex **20**, **23**, and **24** to 1,2,4,5-tetrakis(bromomethyl)benzene to form poly-Pt cascades possessing 4, 12 (**25**), and 28 (**26**) metal centers, respectively.

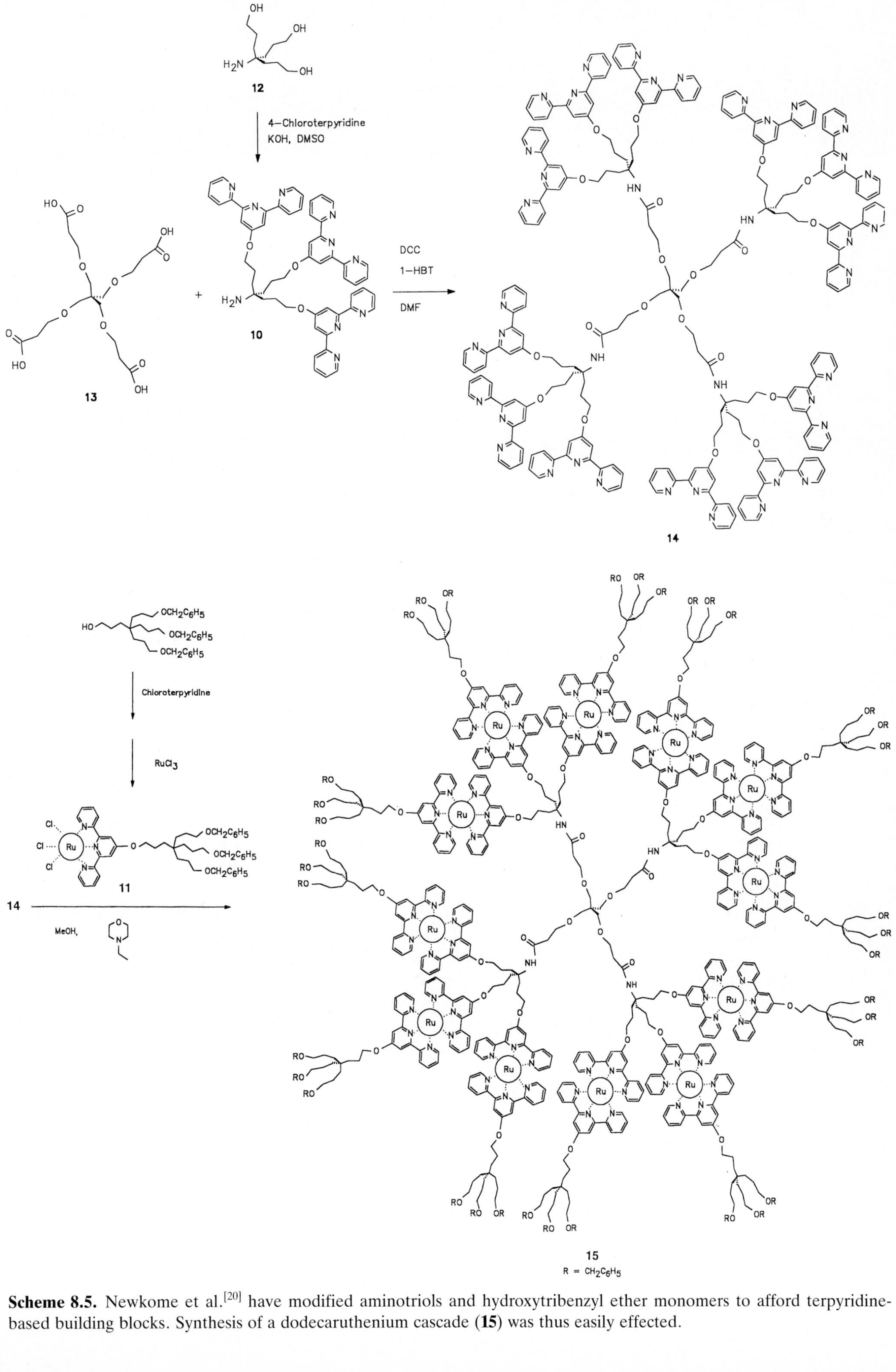

Scheme 8.5. Newkome et al.[20] have modified aminotriols and hydroxytribenzyl ether monomers to afford terpyridine-based building blocks. Synthesis of a dodecaruthenium cascade (**15**) was thus easily effected.

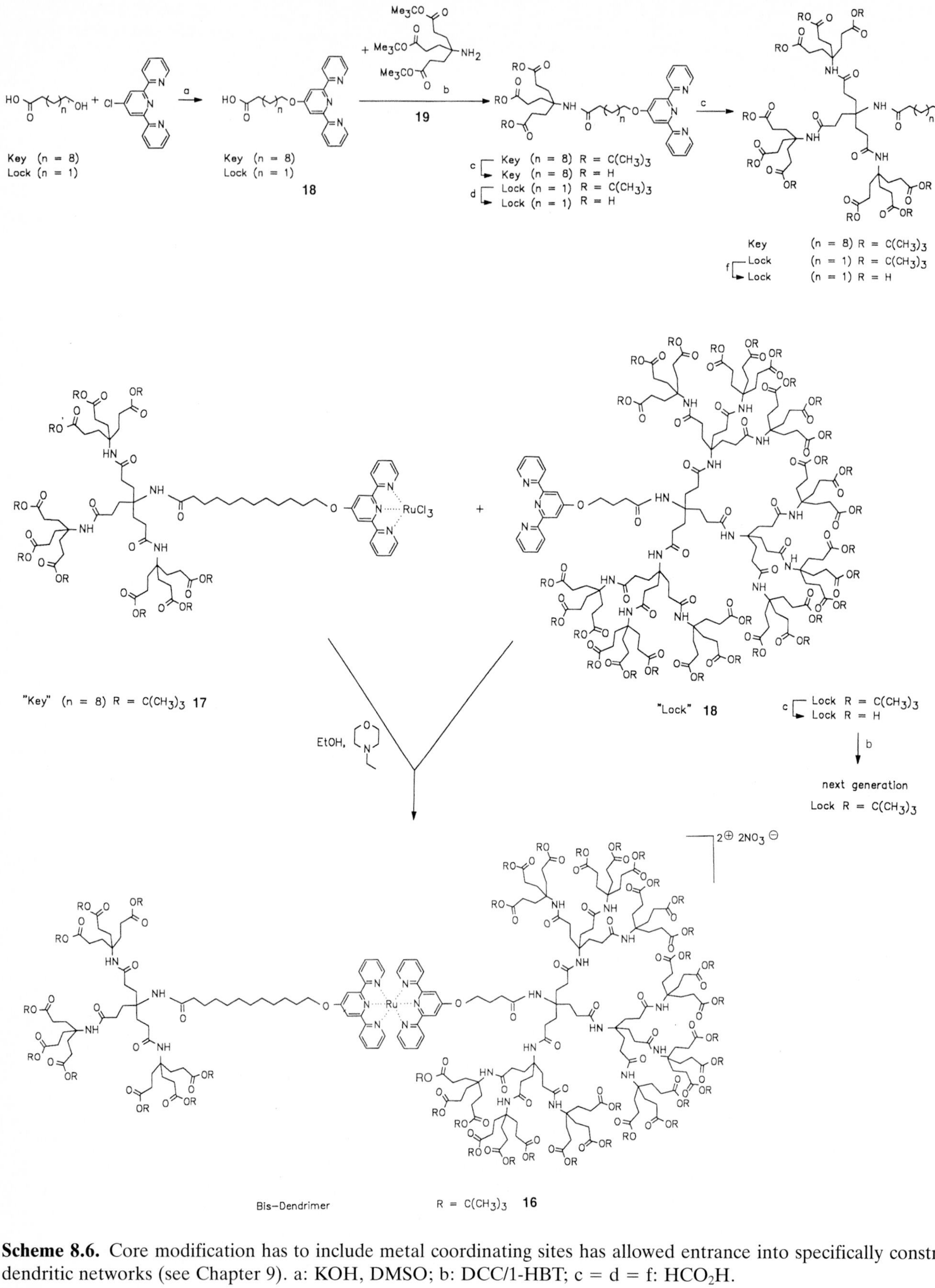

Scheme 8.6. Core modification has to include metal coordinating sites has allowed entrance into specifically constructed dendritic networks (see Chapter 9). a: KOH, DMSO; b: DCC/1-HBT; c = d = f: HCO₂H.

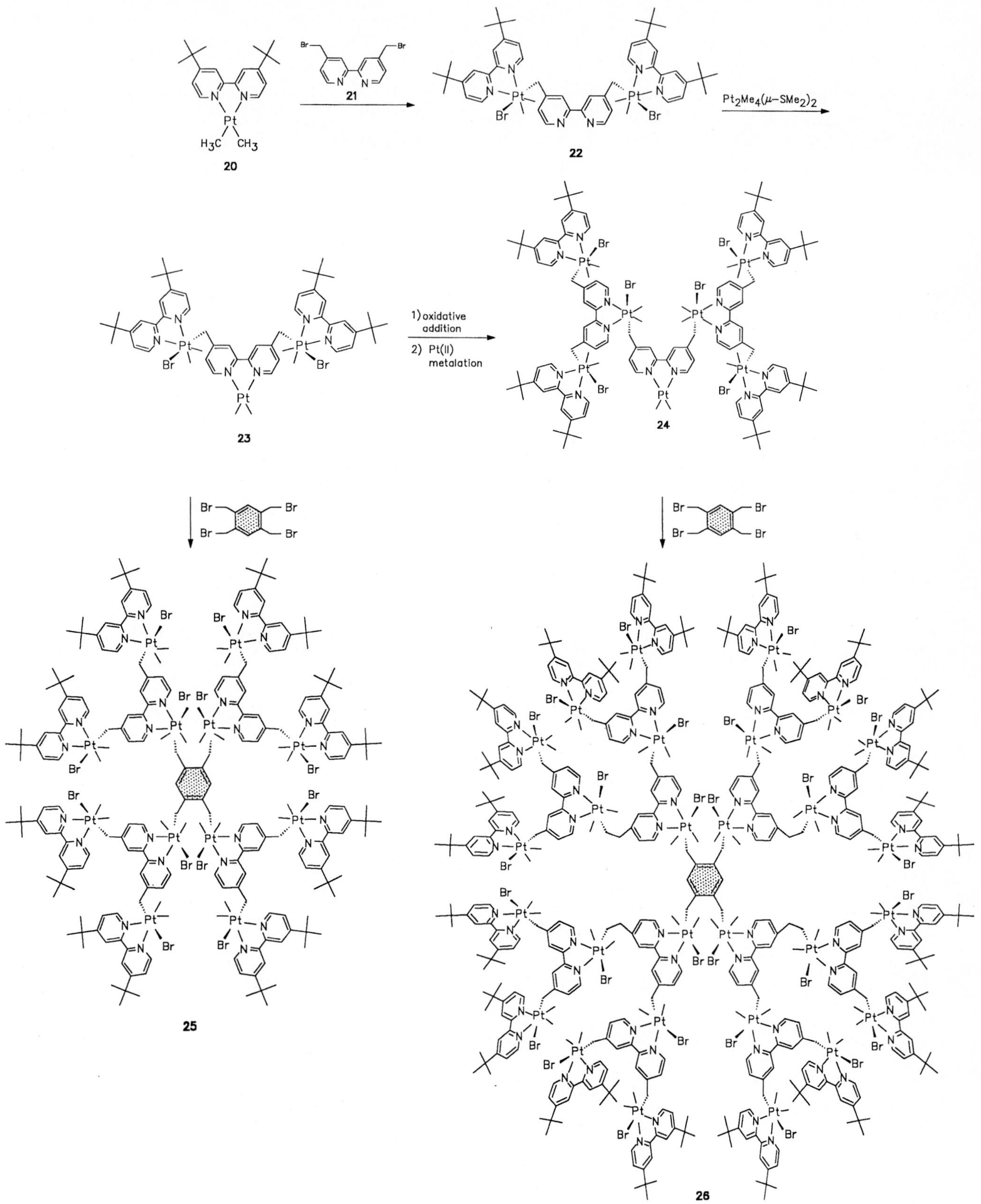

Scheme 8.7. Convergent procedures have been employed for the preparation of high molecular weight organoplatinum dendrimers.

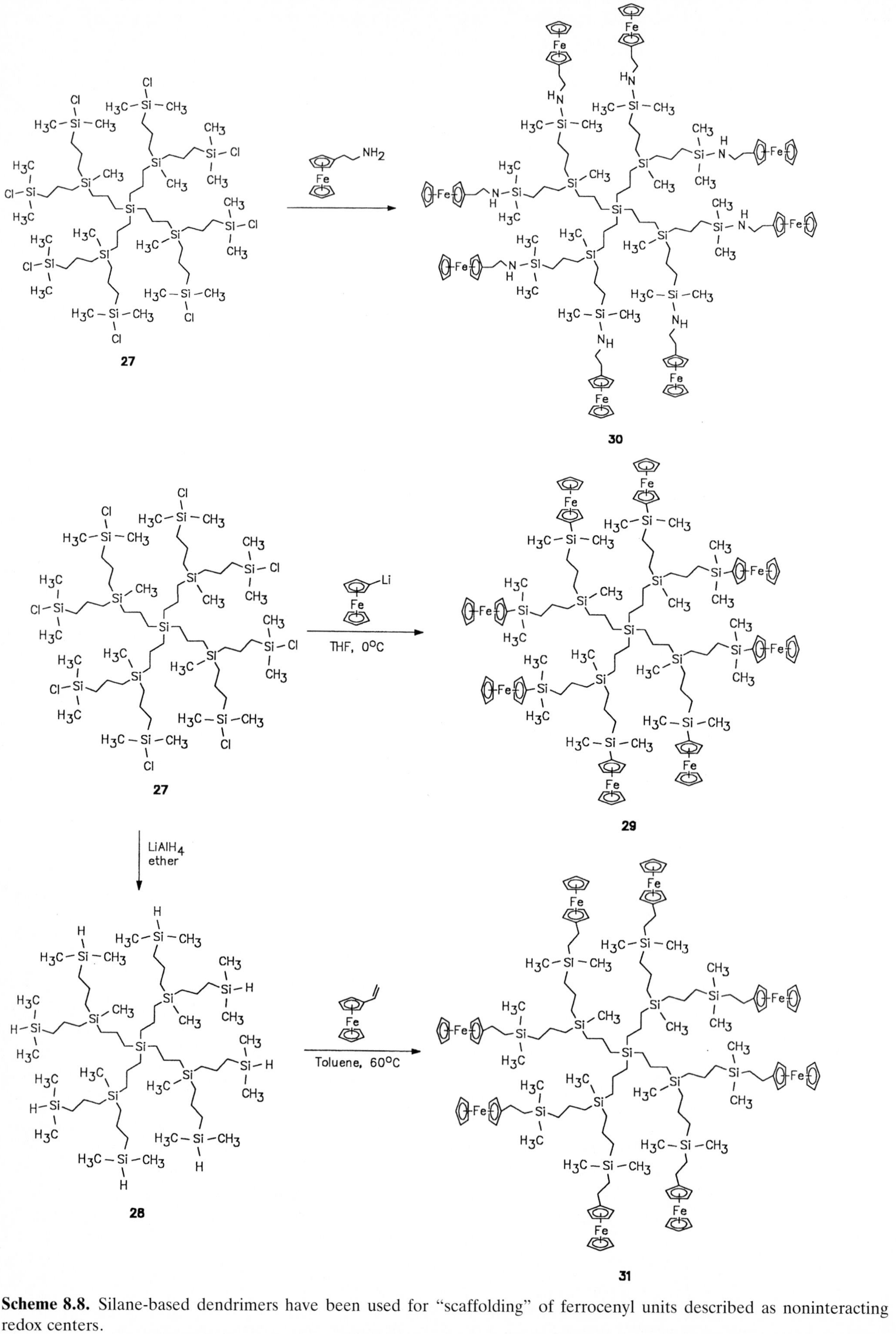

Scheme 8.8. Silane-based dendrimers have been used for "scaffolding" of ferrocenyl units described as noninteracting redox centers.

8.3 Metal Centers as Termination Groups (Surface Functionalization)

Ferrocenyl-terminated dendrimers have been created by Morán, Cuadrado and co-workers, based on their[34] and others[35–37] organosilicon molecular infrastructures. Morán's studies initially focused on the terminal functionalization of Si−Cl and Si−H surface moieties, described in Chapter 4.10, as represented by dendrimers **27** and **28**, respectively (Scheme 8.8). Thus, treatment of octachlorosilane **27** with monolithioferrocene $[(\eta^5\text{-}C_5H_4Li)Fe(\eta^5\text{-}C_5H_5)]$ in THF at 0 °C afforded the ferrocenyl-substituted dendrimer **29**.[34] An alternative procedure utilized (β-aminoethyl)ferrocene $[(\eta^5\text{-}C_5H_4CH=CH_2)Fe(\eta^5\text{-}C_5H_5)]$ as the terminal transform for reaction with chlorosilane **27** to generate octaferrocene **30**; whereas, hydrosilylation of vinylferrocene[38,39] $[(\eta^5\text{-}C_5H_4CH=CH_2)Fe(\eta^5\text{-}C_5H_5)]$ with hydridosilane **28** in toluene at 60 °C in the presence of a catalytic amount of Karstedt catalyst afforded octaferrocene **31**. These ferrocenyl derivatives were fully characterized and, from computer-generated molecular models, were demonstrated to possess an approximate 3 nm diameter at the second generation. Chemical oxidation of the ferrocenyl moieties was accomplished[40] upon treatment with $NOPF_6$ in CH_2Cl_2; the resulting polynuclear dendritic cations were characterized by EPR spectra in which the "polyferrocenium" species have the unpaired electrons essentially localized on the *Si*-substituted cyclopentadienyl fragments. The cyclic voltammograms of these tetra- and octanuclear ferrocenyl macromolecules show a single reversible oxidation process representing a simultaneous multielectron transfer as expected for an independent reversible one-electron process at the same potential.

Since these organometallic dendrimers have *non-interacting* ferrocenyl redox centers, Morán and coworkers[40] successfully modified electrode surfaces by electrodeposition of these dendritic materials in their oxidized forms. Electrodeposition was accomplished by controlled potential electrolysis or by repetitive anodic and cathodic cycling; the amount of electroactive dendrimer deposited could thus be regulated.

Morán et al.[41] attached the $Cr(CO)_3$ moiety to tetrakis(phenylsilane) **32** (prepared by the hydrosilylation of tetraallylsilane with four equivalents of dimethylphenylsilane) by treatment with excess $Cr(CO)_6$ in dibutyl ether−THF at 140 °C affording the air-stable, crystalline tetrakis(chromium carbonyl) dendrimer **33**, which was also prepared by reaction of tetraallylsilane with $\{[\eta^6\text{-}C_6H_5Si(Me)_2H]Cr(CO)_3\}$ (Scheme 8.9).[42a] Reaction of the corresponding eight phenyl-terminated analogues afforded the partially metalated silane dendrimer **34** as the major product, even with an excess of $Cr(CO)_6$.

Other surface functionalization (Scheme 8.10) was achieved by treatment of the first tier, tetrachlorosilane core (**35**) with sodium cyclopentadienide followed by $Co_2(CO)_8$ in refluxing CH_2Cl_2 to afford the tetracobalt cascade **36**; whereas reaction with $\{Na[\eta^5\text{-}C_5H_5Fe(CO)_2]\}$ on the same core (**35**) afforded dendrimer **37** containing four silicon−iron σ-bonds. The reaction of the 0th generation Si−H core **38** with $Co_2(CO)_8$ gave rise to the related Si−Co molecule **39**. Seyferth et al.[42b] reported the construction of carbosilane dendrimers bearing peripheral dicobalt hexacarbonyl units via the reaction of alkyne terminated dendrimers with Co_2CO_8.

Morán et al. have also reported the preparation of hyperbranched ferrocenyl *Si*-based polymers (Figure 8.1). The construction of ferrocenyl−silicon polymers **40** and **41** was effected by the reaction of dilithioferrocene·TMEDA with the tetrachlorosilane **35** (see Scheme 8.10) and the Pt-mediated hydrosilylation of 1,1'-divinylferrocene with tetrasilylhydride **38**. The 3-dimensional motif exhibited by hyperbranched polymers **40** and **41** is analogous to that depicted by the ferrocenyl−silicon network structures[38,43] **42** and **43** (Figure 8.2).

Van Koten et al.[44] functionalized the surface of *Si*-dendrimers by the elegant treatment of the dodecachlorosilane **44** with the aryl halide reagent **45**, which possesses a spacer sufficient to afford surface mobility and allow generation of dodecaarylbromide **46**. Subsequent reaction of aryl bromide **46** with the zero valent nickel reagent $[Ni(PPh_3)_4]$ in THF at 60–70 °C generated the desired aryl nickel(II) dendrimer **47** with twelve nickel centers (Scheme 8.11). These dendritic organometallic macromolecules have been shown to be effective homogeneous catalysts for the Kharasch addition reaction of polyhaloalkanes to olefins. These large, "dendritic metals" are amenable to removal from solution via filtration, affording recyclable homogeneous catalysts.

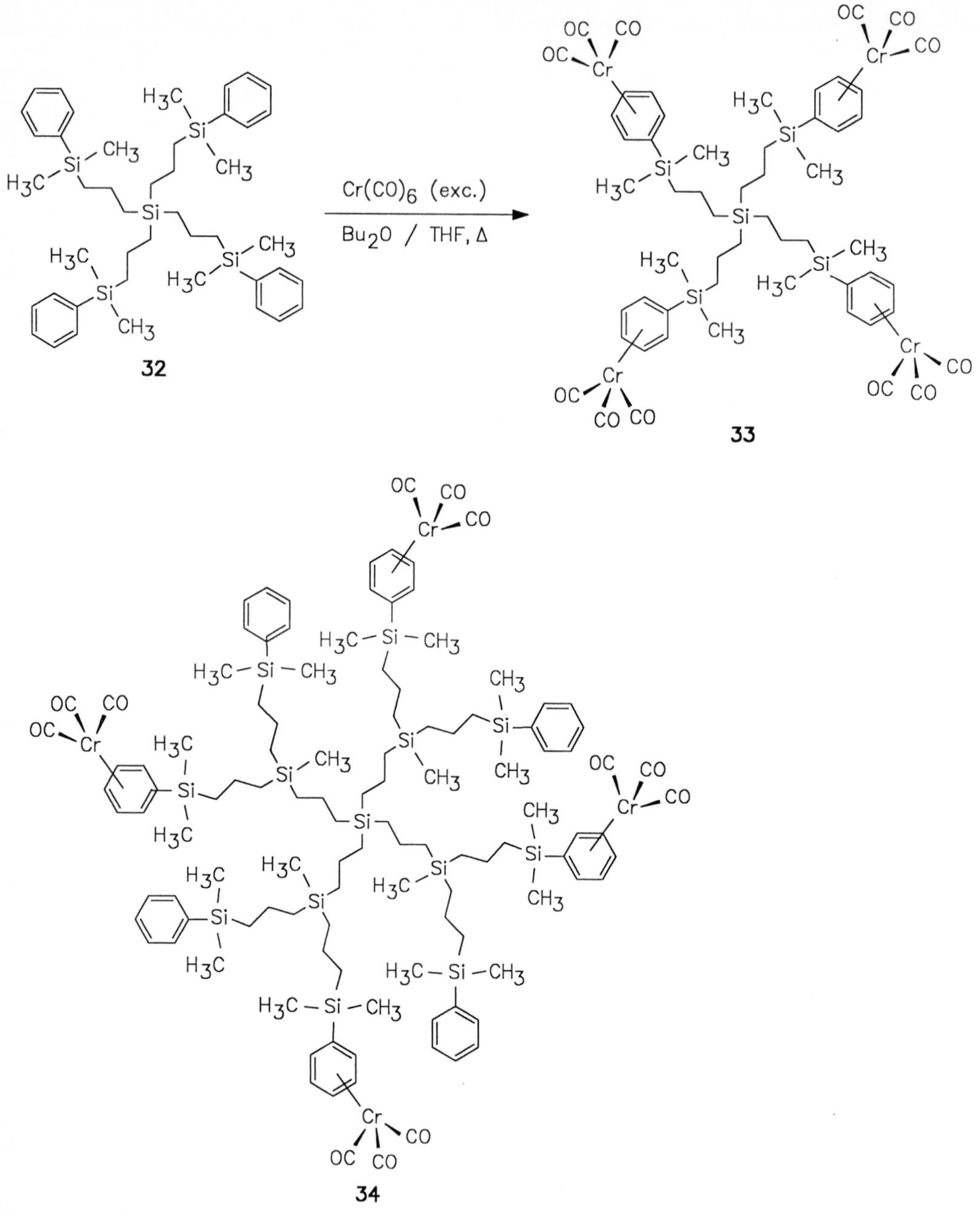

Scheme 8.9. Cr(CO)$_3$ groups have been attached at the termini of small dendrimers.

DuBois et al.[45] have reported the preparation of a tetrakis(tri-*P*-ligand) **48**, generated by the free-radical addition of bis[(diethylphosphino)ethyl]phosphine to tetravinylsilane. Reaction of *P*-based cascade **48** with [Pd(MeCN)$_4$(BF$_4$)$_2$] gave the tetrapalladium complex **49** (Scheme 8.12). Interestingly, the electrochemical studies of poly-Pd-complex **49** in acidic DMF and in the presence of CO$_2$ showed that CO$_2$ is reduced to CO. An analogous P$_{15}$ dendrimer **50** was constructed[45] via sequential radical addition of diethyl vinylphosphate to primary phosphines and LiAlH$_4$ reduction. Treatment of dendrimer **50** with the Pd(II) reagent afforded, in this case, the penta-Pd-complex **51**. These square planar Pd-complexes (**49** and **50**) are initially comprised of a triphosphine and an acetonitrile ligand, which can be readily exchanged, e.g., with triethylphosphine, affording the related tetra *P*-complexes. It was noted that there are other possibilities; however, the structures derived from the alkylphosphines appear to dominate. Complexes derived from terminal arylphosphine ligands (not shown) appeared to be more complicated than expected based on NMR data, although ^{31}P NMR firmly establish the connectivity of the ligands. When the acetonitrile-coordinated complexes were subjected to electrochemical studies in the presence of carbon dioxide, an enhanced current was observed; this was attributed to catalytic reduction of CO$_2$ to CO. The initial study of the chemistry of these metallated dendrimers demonstrates their potential for performing catalytic reactions.

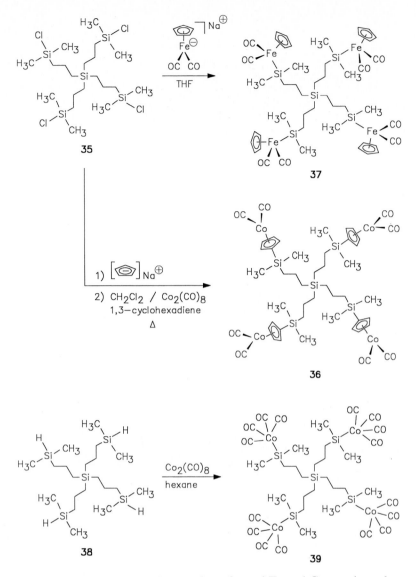

Scheme 8.10. Silicon-based chemistry has led to the creation of novel Fe and Co terminated structures.

Astruc et al.[46–51] have utilized metal-activated, methyl aromatic sandwich complexes to generate organometallic molecular trees. Treatment of {[(C₅H₅]Fe(η^6-mesitylene)](PF₆)} (**52**) with allyl bromide in the presence of base gave the nonaene **53**, which was demetalated by photolysis in the presence of triphenylphosphine in MeCN to give the free arene **54**. Hydroboration of polyene **54** with diisamylborane followed by H₂O₂ oxidation afforded the desired nonaol; subsequent nonaarylation, achieved by treatment of the nonaol with [(C₅H₅)Fe(η^6-*p*-MeC₆H₄F)(PF₆)] (**55**), gave the nonairon sandwich complex **56** (Scheme 8.13). Since the external aryl groups are complexed, the methyl substituents are activated for further iteration by a similar branching procedure.

Astruc and his coworkers[52,53] have further employed the hexahapto coordination of aromatics by the [(η-C₅H₅)Fe⁺] moiety (e.g., **57**) thus activating benzyl positions and allowing hexafunctionalization of permethylbenzene (Scheme 8.14). However, even under forcing conditions, allylation stops at the rigid dodeca-allylated complex to give polyene **58** after demetallation. Cobaltocene benzyl activation has been similarly exploited[47] to give the decabenzyl cobalt complex **59**.

Liao and Moss[54] prepared by a convergent procedure[55,56] dendrimers containing 48 ruthenium atoms at the periphery of the macromolecule (i.e., **60** (not drawn); Scheme 8.15). Preparation of the requisite dendrons, such as **61** and **62**, was effected starting with the reaction of {(η^5-C₅H₅)Ru(CO)₂[(CH₂)₃Br]} (**63**) and 3,5-dihydroxybenzyl alcohol. Subsequent convergent "generational layering" and core attachment was conducted via Fréchet's methodology[55] (see Chapter 5.2.2).

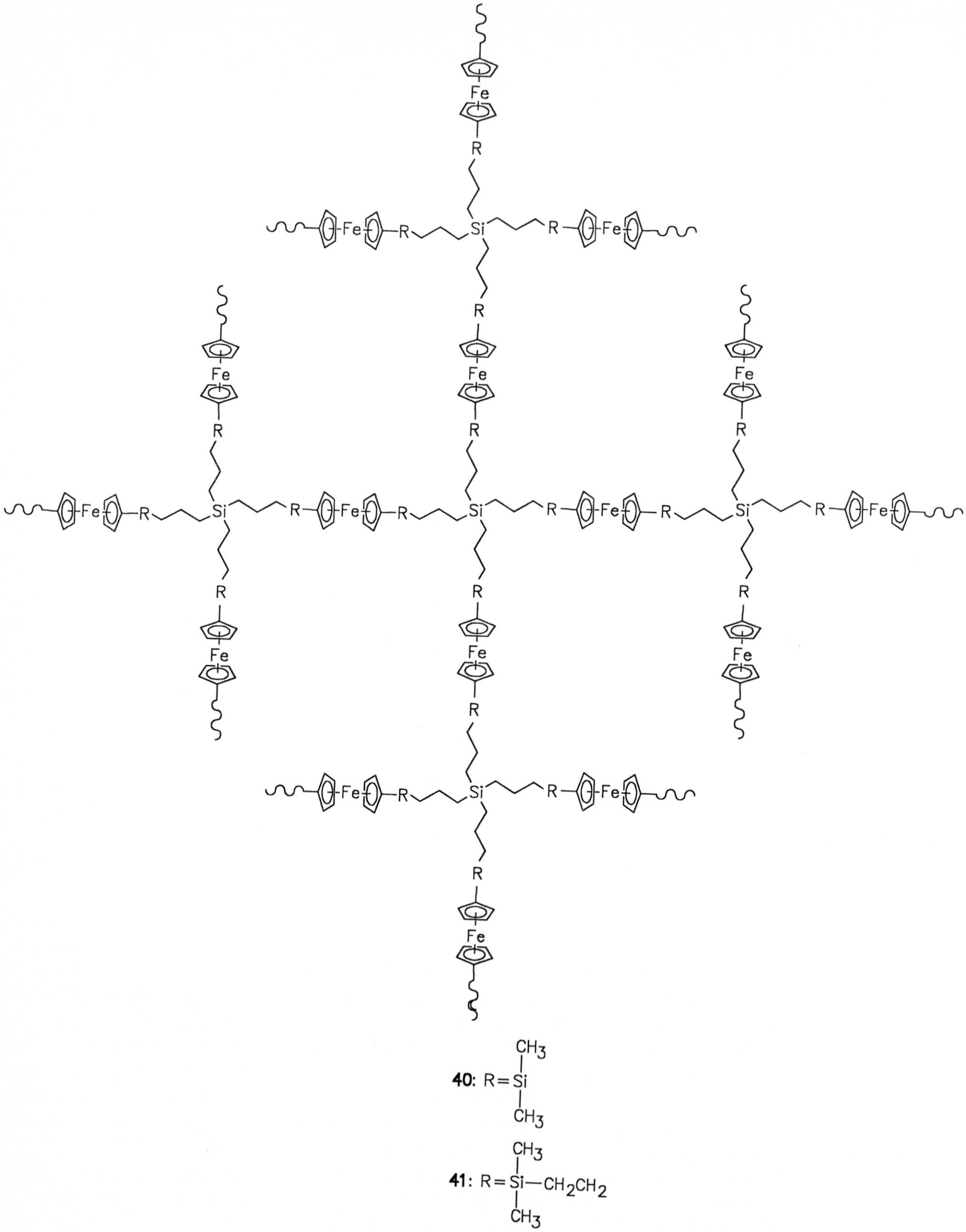

40: R = $Si(CH_3)_2$

41: R = $Si(CH_3)_2$—CH_2CH_2

Figure 8.1. Four-directional Si and ferrocenyl connectors have been incorporated into hyperbranched polymers.

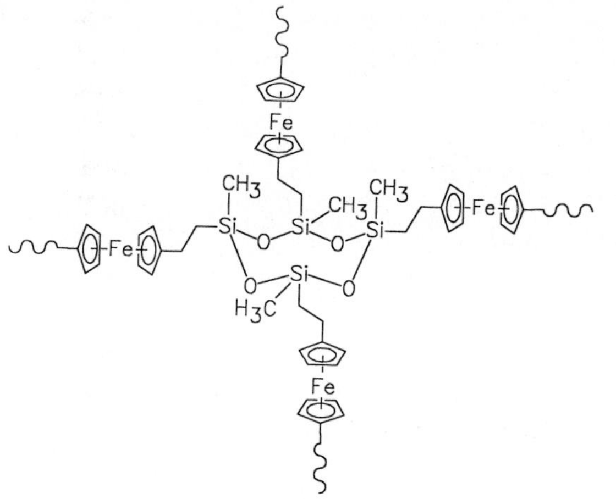

42

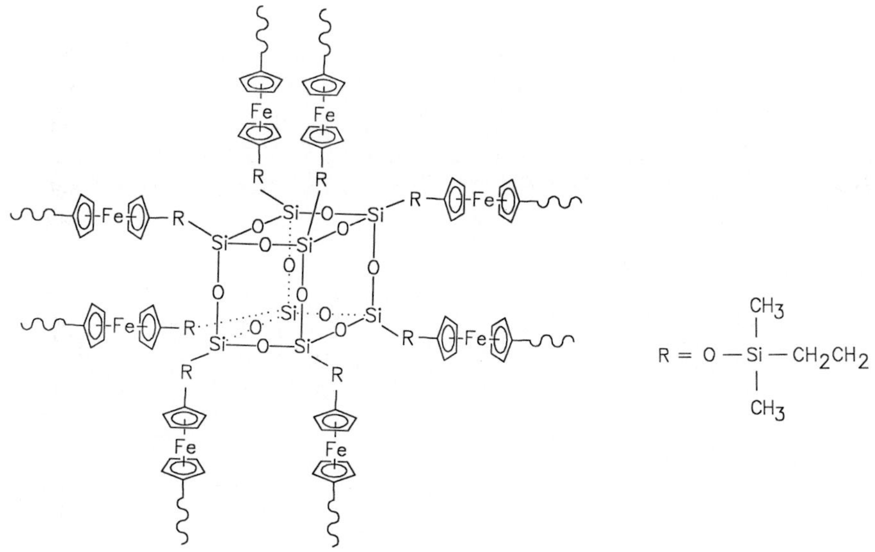

43

$$R = O-Si-CH_2CH_2$$

Figure 8.2. Ferrocenyl-siloxane-based "network structures" prepared by Moran et al.[38,43]

Dodecaferrocene **64** (Figure 8.3) was prepared[57] for the study of multiple redox centers. The calixarene core was constructed by the acid-catalyzed condensation of ferrocenecarboxaldehyde with resorcinol. Beer et al.[58] have also constructed poly Na^+ salt **65** by the connection of crown ethers to bipyridine ligands, which were subsequently organized by the complexation to a Ru(II) metal core.

While investigating the potential to improve on the original method of cascade preparation[59], Moors and Vögtle[60a] constructed a hexaimine dendrimer, which was readily converted to the tris-Co complex (**66**) in quantitative yield. Lhotak and Shankai[60b] have addressed the potential for creation of calix[4]arene-based dendrimers and examined some metal-binding properties for the corresponding oligomers.

Engel et al.[61] reported the high yield coordination of Au(I) with the free phosphine core of a phosphonium cascade (see Chapter 4.14). As a prelude to the fascinating chemistry yet to be explored by the integration of transition metals with dendrimers, Majoral et al.[62] have reported the creation of a mixed valence *P*-cascade (see Scheme 4.27) possessing 3072 terminal AuCl moieties at the 10th generation (theoretical MW 1,020,302)! The authors noted that structural defects could not be ruled out in the limits of ^{31}P NMR accuracy (1%); the number of terminal groups (e.g., AuCl) was thus expected to be close to 3000. The Au-dendrimers were further characterized by high-resolution transmission electron micrographs affording measured diameters of 150 ± 5 Å.

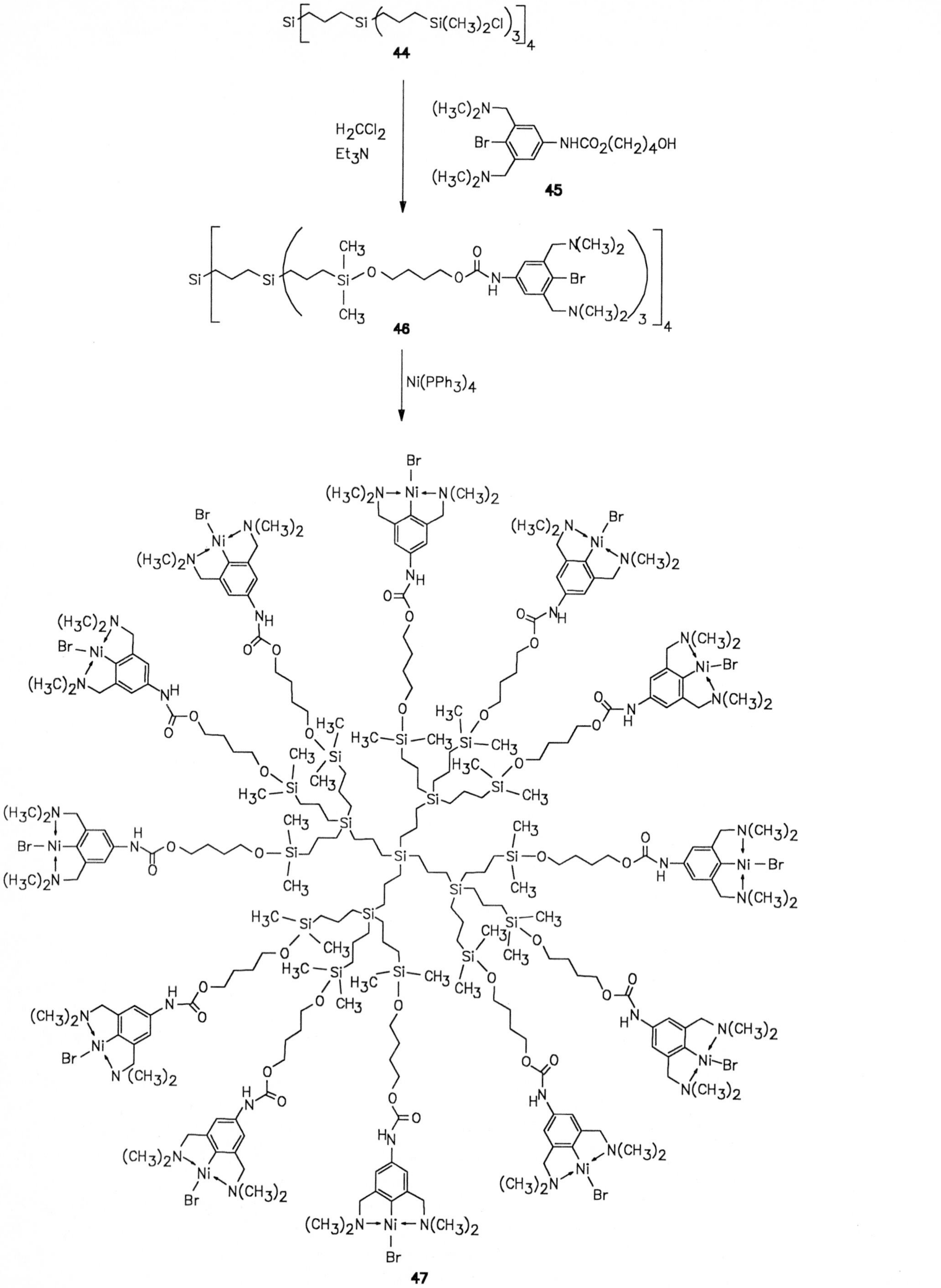

Scheme 8.11. Aryl Ni(II) surface modified dendrimers were demonstrated[44] to be effective at catalyzing the Kharasch addition reactions.

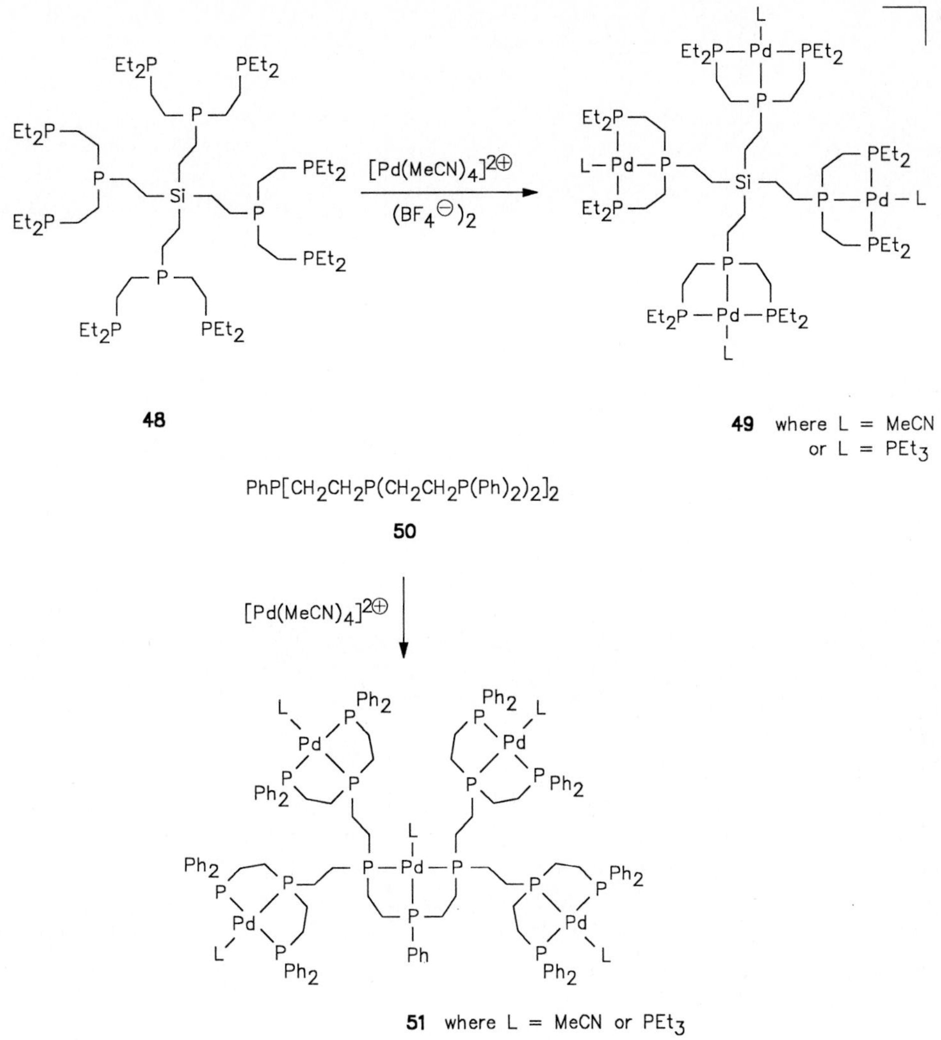

48

49 where L = MeCN
or L = PEt$_3$

PhP[CH$_2$CH$_2$P(CH$_2$CH$_2$P(Ph)$_2$)$_2$]$_2$

50

[Pd(MeCN)$_4$]$^{2\oplus}$

51 where L = MeCN or PEt$_3$

Scheme 8.12. DuBois et al.[45] have prepared *P*-based poly(palladium complexes) capable of reducing CO$_2$ to CO.

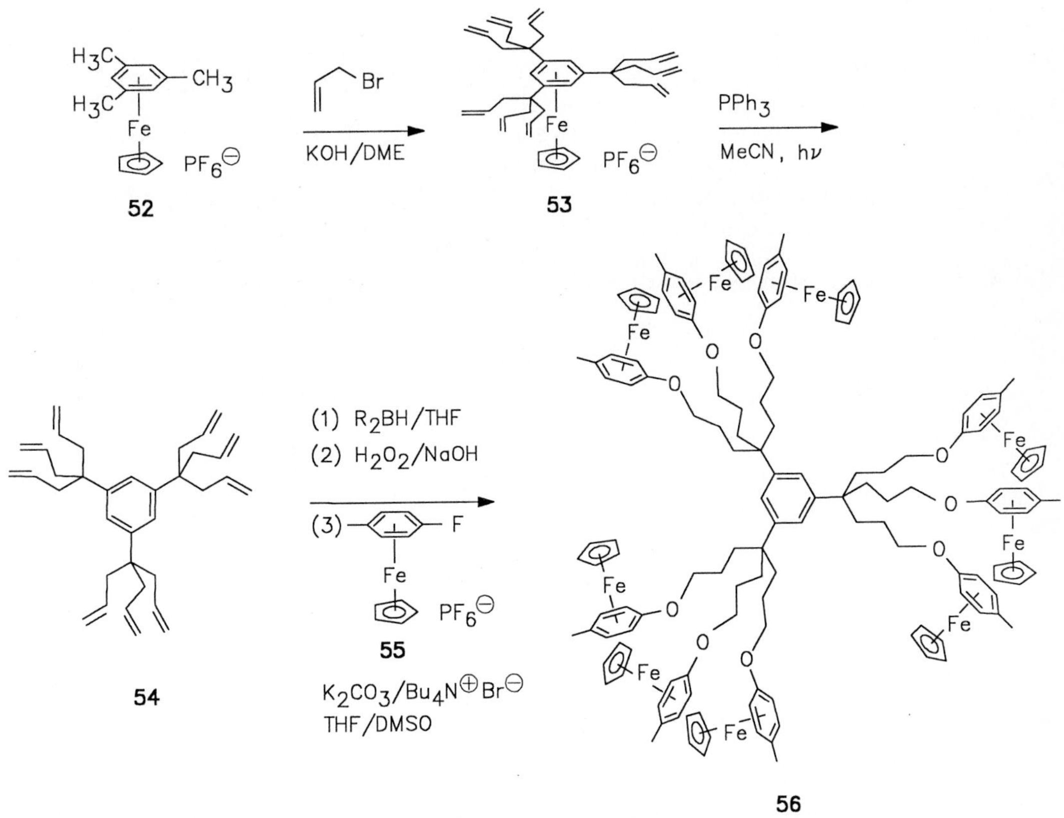

52

53

54

(1) R$_2$BH/THF
(2) H$_2$O$_2$/NaOH

(3) ⬡—F
Fe
PF$_6^\ominus$

55

K$_2$CO$_3$/Bu$_4$N$^\oplus$Br$^\ominus$
THF/DMSO

56

Scheme 8.13. Astruc et al.[46–51] have investigated iterative strategies focused on Fe-activated arene chemistry.

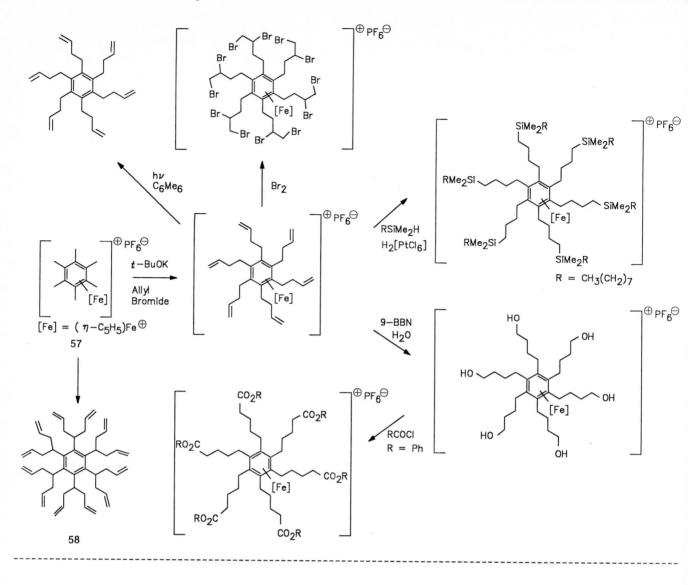

Scheme 8.14. Diverse branched structures are accessible via metal arene-based technologies.

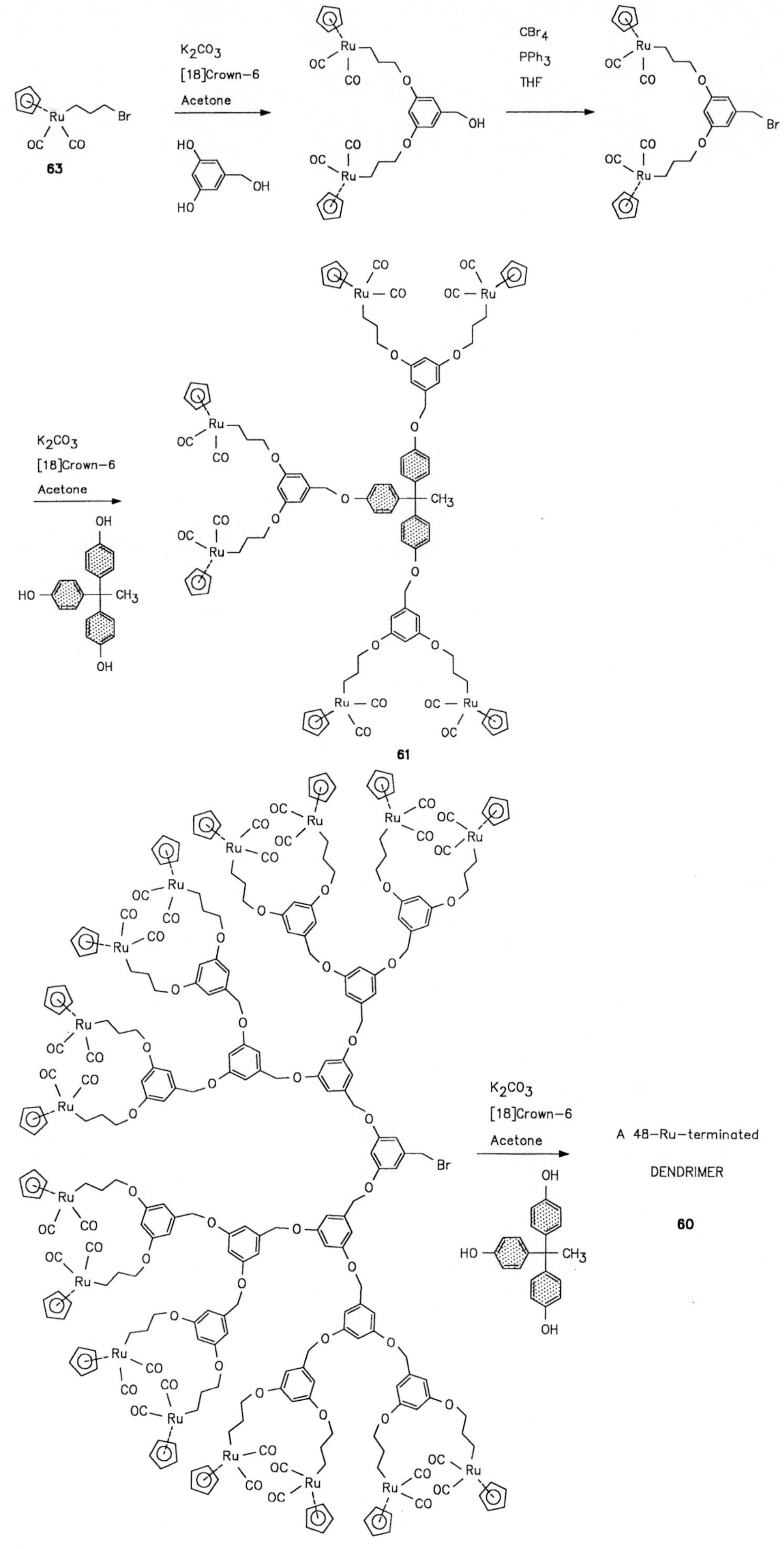

Scheme 8.15. Liao and Moss[53,54] employed Fréchet et al.'s[55,56] convergent benzyl ether chemistry for the construction of a series of $(\eta^5\text{-}C_5H_5)Ru(CO)_2$ terminated dendrimers.

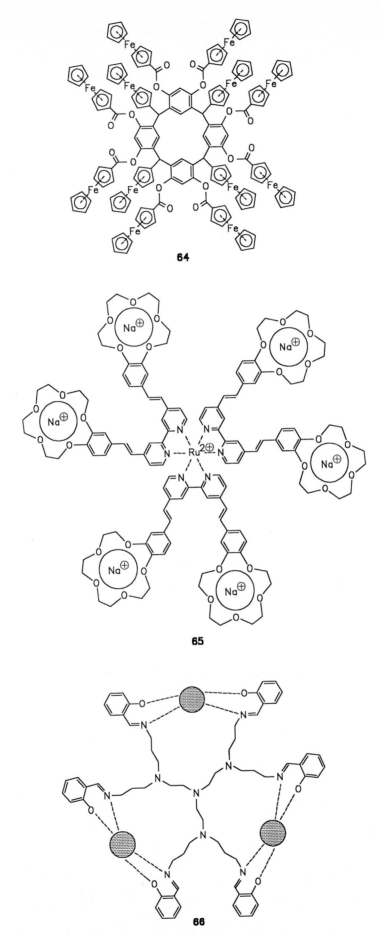

Figure 8.3. Beer et al.[57,58] have investigated branched structures incorporating multiple Fe (**64**) and Na+ (**65**); whereas Vögtle and Moors[60a] have studied a polycobalt assembly (**66**).

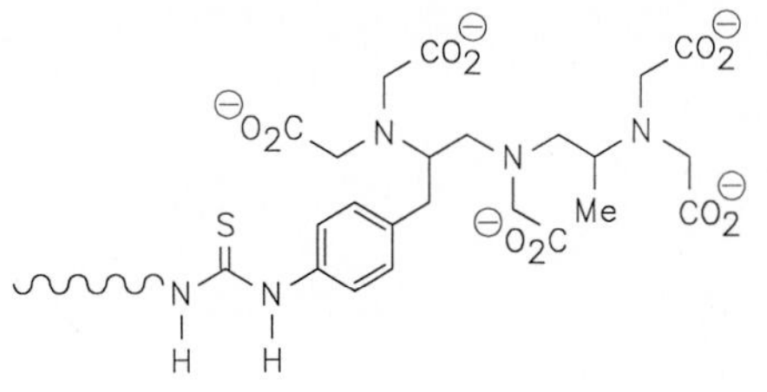

67 tier 2: 92% substituted; 50% yield
11 of 12 sites reacted

68 tier 6: 88% substituted; 41% yield
170 of 192 sites reacted

Figure 8.4. Wiener et al.[63] prepared surface modified PAMAM dendrimers for the coordination of Gd(III) for use as MRI contrast agents.

Wiener et al.[63] employed the PAMAM dendrimers for the construction of terminally functionalized Gd(III) dendrimers for use as *Magnetic Resonance Imaging* contrast agents. The second and sixth generation dendrimers were terminally functionalized with diethylenetriaminepentaacetic acid (Figure 8.4) to give (50% and 41%) the peripherally substituted macromolecules **67** and **68**, respectively. These agents were observed to significantly enhance conventional, as well as 3D time-of-flight, MR images and angiograms.

8.4 Metals as Transformation Auxiliaries

8.4.1 Site-Specific Metal Inclusion

Investigation into the reactivity of functional groups located within the dendritic framework and hence reactant accessibility to the interior 'void volumes' was recently reported by Newkome et al.[25,64] Treatment of the polyalkyne precursor (**69**)[65] to the hydrocarbon-based Micellanoic Acid™ with decaborane [$B_{10}H_{14}$] afforded nearly quantitative yields of the poly(1,2-dicarba-*closo*-dodecaboranes), supporting the ease of accessibility to the inner regions of these and related dendritic systems. Thus, treatment of this polyalkyne (**69**) with dicobalt octacarbonyl gave rise to excellent yields of the corresponding poly-Co-cluster (**70**), termed a Cobaltomicellane™ (Scheme 8.16). The structure was supported, in part, by the loss of the alkyne absorption and the appearance (^{13}C NMR) of a peak at δ 99.6, corresponding to the new *C*−metal bond. This was the first example of a *site-specific* molecular attachment of metal centers within these unimolecular micelle precursors.[65]

8.4.2 Random Metal Inclusion

Ottaviani et al.[66] have studied the structure and coordination of Cu(II) complexes (e.g. **71**; Figure 8.5) formed with PAMAM dendrimers[67] in aqueous solution using electron paramagnetic resonance spectrometry. The results support previous observations indicating a change in dendritic structure as cascade growth progresses beyond the third generation. EPR spectra indicate three distinct types of copper coordination domains resulting in three types of signals (e.g., 2N2O, 4N, and 2O donor atoms). Aging the samples affords an increase in one EPR signal at the expense of another; thus, supporting the mobility of the copper ions within the amine superstructure. Bonding parameters were evaluated suggesting substantial metal−heteroatom covalent bond character, especially for the interior coordination sites.

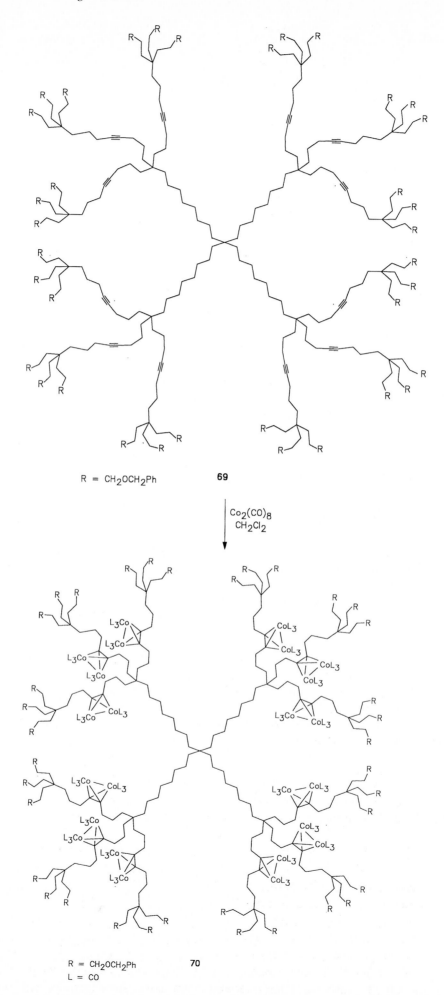

R = CH$_2$OCH$_2$Ph   **69**

Co$_2$(CO)$_8$
CH$_2$Cl$_2$

R = CH$_2$OCH$_2$Ph   **70**
L = CO

Scheme 8.16. Treatment of internal alkyne moieties with Co$_2$(CO)$_8$ yields a 24 cobalt complex.

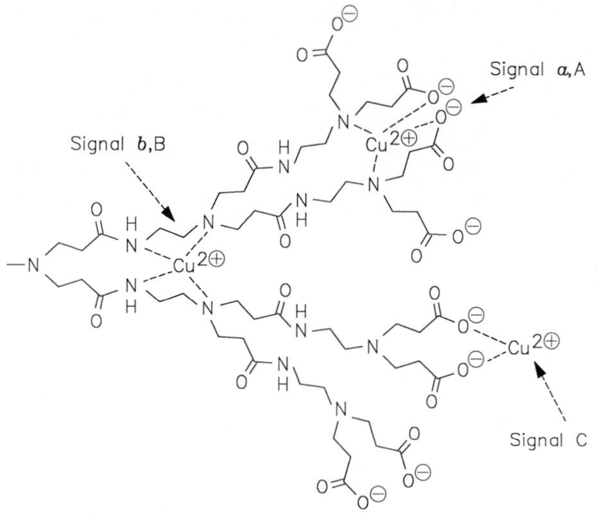

71

Figure 8.5. Coordination of Cu^{+2} at various sites on PAMAM dendrimers has been studied by Ottaviani et al.[66]

8.5 References

[1] G. Denti, S. Campagna, S. Serroni, M. Ciano, V. Balzani, *J. Am. Chem. Soc.* **1992**, *114*, 2944.

[2] G. Denti, S. Serroni, S. Campagna, V. Ricevuto, V. Balzani, *Inorg. Chim. Acta* **1991**, *182*, 127.

[3] S. Campagna, G. Denti, S. Serroni, M. Ciano, A. Juris, V. Balzani, *Inorg. Chem.* **1992**, *31*, 2982.

[4] C. N. Moorefield, G. R. Newkome, in *"Advances in Dendritic Macromolecules"* (Ed.: G. R. Newkome), JAI Press, Greenwich, CN, Vol. 1, Chapt. 1, **1994**.

[5] S. Campagna, G. Denti, S. Sabatino, S. Serroni, M. Ciano, V. Balzani, *J. Chem. Soc., Chem. Commun.* **1989**, 1500.

[6] S. Campagna, G. Denti, S. Serroni, M. Ciano, V. Balzani, *Inorg. Chem.* **1991**, *30*, 3728.

[7] S. Serroni, G. Denti, S. Campagna, A. Juris, M. Ciano, V. Balzani, *Angew. Chem.* **1992**, *109*, 1540; *Angew. Chem., Int. Ed. Engl.* **1992**, *31*, 1493.

[8] M. M. Richter, K. J. Brewer, *Inorg. Chem.* **1993**, *32*, 5762.

[9] A. Juris, V. Balzani, S. Campagna, G. Denti, S. Serroni, G. Frei, H. V. Güdel, *Inorg. Chem.* **1994**, *33*, 1491.

[10] G. Denti, S. Campagna, V. Balzani, *Mesomolecules: From Molecules to Materials*, (Eds.: G. D. Mendenhall, A. Greenberg, J. F. Liebman), Chapman & Hall, New York, **1995**, pp. 69–106.

[11] V. Balzani, S. Campagna, G. Denti, S. Serroni, M. Venturi, *Coord. Chem. Rev.* **1994**, *132*, 1.

[12] V. Balzani, A. Juris, M. Venturi, S. Serroni, S. Campagna, G. Denti, in *"Advances in Dendritic Macromolecules"*, (Ed.: G. R. Newkome), JAI Press, Greenwich, CN, Vol. 3, **1996**.

[13] S. Campagna, G. Denti, S. Serroni, A. Juris, M. Venturi, V. Ricevuto, V. Balzani, *Chem. Eur. J.* **1995**, *1*, 211.

[14] S. M. Molnar, G. Nallas, J. S. Bridgewater, K. J. Brewer, *J. Am. Chem. Soc.* **1994**, *116*, 5206.

[15] a) S. Rasmussen, E. Yi, M. M. Richter, H. Place, K. J. Brewer, *Inorg. Chem.* **1990**, *29*, 3926. b) S. Serroni, S. Campagna, A. Juris, M. Venturi, V. Balzani, G. Denti, *Gazz. Chim. Ital.* **1994**, *124*, 423.

[16] M. N. Bochkarev, V. B. Cilkin, L. P. Mayorova, G. A. Razuvaev, U. D. Cemchkov, V. E. Sherstyanux, *J. Organometal. Chem. U. S. S. R.* **1988**, *1*, 196.

[17] E. C. Constable, A. M. W. C. Thompson, *J. Chem. Soc., Chem. Commun.* **1992**, 617.

[18] J.-P. Collin, V. Heitz, J.-P. Sauvage, *Tetrahedron Lett.* **1991**, *32*, 5977.

[19] E. C. Constable, *Metal and Ligand Reactivity*, VCH, Weinheim, **1995**.

[20] G. R. Newkome, F. Cardullo, E. C. Constable, C. N. Moorefield, A. M. W. C. Thompson, *J. Chem. Soc., Chem. Commun.* **1993**, 925.

[21] Review: J.-P. Sauvage, J.-P. Collin, J.-C. Chambron, S. Guillerez, C. Coudret, V. Balzani, F. Barigelletti, L. De Cola, L. Flamigni, *Chem. Rev.* **1994**, *94*, 993.

[22] G. R. Newkome, C. N. Moorefield, G. R. Baker, *Aldrichim. Acta* **1992**, *25* (No. 2), 31.

[23] E. C. Constable, M. D. Ward, *J. Chem. Soc., Dalton Trans.* **1990**, 1405.

[24] a) G. R. Newkome, J. K. Young, G. R. Baker, R. L. Potter, L. Audoly, D. Cooper, C. D. Weis, K. F. Morris, C. S. Johnson, Jr., *Macromolecules* **1993**, *26*, 2394. b) E. C. Constable, P. Haverson, *J. Chem. Soc., Chem. Commun.* **1996**, 33.

[25] G. R. Newkome, C. N. Moorefield, *Macromol. Symp.* **1994**, *77*, 63.

[26] G. R. Newkome, C. N. Moorefield, R. Güther, G. R. Baker, *Polym. Preprints.* **1995,** *36*(1), 609.

[27] G. R. Newkome, R. Güther, C. N. Moorefield, F. Cardullo, L. Echegoyen, E. Pérez-Cordero, H. Luftmann, *Angew. Chem.* **1995,** *107*, 2159; *Angew. Chem., Int. Ed. Engl.* **1995,** *34*, 2023.

[28] G. R. Newkome, V. V. Narayanan, A. K. Patri, J. Groß, C. N. Moorefield, G. R. Baker, *Polym. Mater Sci. Eng.* **1995,** *73*, 222.

[29] E. C. Constable, A. M. W. C. Thompson, *J. Chem. Soc., Chem. Commun.* **1992,** 617.

[30] R. Breslow, G. A. Crispino, *Tetrahedron Lett.* **1991,** *32*, 601.

[31] A. J. Amoroso, A. M. W. C. Thompson, J. P. Maher, J. A. McCleverty, M. D. Ward, *Inorg. Chem.* **1995,** *34*, 4828.

[32] A. Achar, R. J. Puddephatt, *Angew. Chem.* **1994,** *106*, 895; *Angew. Chem. Int. Ed. Engl.* **1994,** *33*, 847.

[33] S. Achar, R. J. Puddephatt, *J. Chem. Soc., Chem. Commun.* **1994,** 1895. J. Achar, J. J. Vittal, R. J. Puddephatt, *Organometallics* **1996,** *15*, 43.

[34] B. Alonso, I. Cuadrado, M. Morán, J. Losada, *J. Chem. Soc., Chem. Commun.,* **1994,** 2575.

[35] A. W. van der Made, P. W. N. N. van Leeuwen, *J. Chem. Soc., Chem. Commun.* **1992,** 1400.

[36] L.-L. Zhou, J. Roovers, *Macromolecules* **1993,** *26*, 963.

[37] D. Seyferth, D. Y. Son, A. L. Rheingold, R. L. Ostrander, *Organometallics* **1994,** *13*, 2682.

[38] C. M. Casado, I. Cuadrado, M. Morán, B. Alonso, P. Lobete, J. Losada, *Organometallics* **1995,** *14*, 2618.

[39] C. M. Casado, M. Morán, J. Losada, I. Cuadrado, *Inorg. Chem.* **1995,** *34*, 1668.

[40] B. Alonso, N. Morán, C. M. Casado, P. Lobete, J. Losada, I. Cuadrado, *Chem. Mater.* **1995,** *7*, 1440.

[41] F. Lobete, I. Cuadrado, C. M. Casado, B. Alonso, C. Pascual, M. Morán, J. Losada, submitted.

[42] a) M. Morán, I. Cuadrado, C. Pascual, C. M. Casado, J. Lasada, *Organometallics* **1992,** *11*, 1210. b) D. Seyferth, T. Kugita, A. L. Rheingold, G. P. A. Yap, *Organometallics* **1995,** *14*, 5362.

[43] M. Morán, C. M. Casado, I. Cuadrado, J. Losada, *Organometallics* **1993,** *12*, 4237.

[44] J. W. J. Knappen, A. W. van der Made, J. C. de Wilde, P. W. N. M. van Leeuwen, P. Wijkens, D. M. Grove, G. van Koten, *Nature* **1994,** *372*, 659.

[45] A. Miedaner, C. J. Curtis, R. M. Barkley, D. L. DuBois, *Inorg. Chem.* **1994,** *33*, 5482.

[46] D. A. Astruc, *Top. Curr. Chem.* **1991,** *160*, 47.

[47] F. Moulines, B. Gloaguen, D. Astruc, *Angew. Chem.* **1992,** *104*, 452; *Angew. Chem., Int. Ed. Engl.* **1992,** *31*, 458.

[48] F. Moulines, L. Djakovitch, R. Boese, B. Gloaguen, W. Thiel, J.-L. Fillaut, M.-H. Delville, D. Astruc, *Angew. Chem.* **1993,** *105*, 1132; *Angew. Chem., Int. Ed. Engl.* **1993,** *32*, 1075.

[49] J.-L. Fillaut, D. Astruc, *J. Chem. Soc., Chem. Commun.* **1993,** 1320.

[50] E. Cloutet, J.-L. Fillaut, Y. Gnanou, D. Astruc, *J. Chem. Soc., Chem. Commun.* **1994,** 2433.

[51] J.-L. Fillaut, J. Linares, D. Astruc, *Angew. Chem.* **1994,** *106*, 2540; *Angew. Chem., Int. Ed. Engl.* **1994,** *33*, 2460.

[52] D. Astruc, M.-H. Desbois, B. Gloaguen, F. Moulines, J.-R. Hamon, in *Organic Synthesis via Organometallics,* (Eds.: K. H. Dötz, R. W. Hoffmann), Vieweg, Braunschweig, **1991,** 63.

[53] Y.-H. Liao, J. R. Moss, *Organometallics* **1995,** *14*, 2130.

[54] Y.-H. Liao, J. R. Moss, *J. Chem. Soc., Chem. Commun.* **1993,** 1774.

[55] C. J. Hawker, J. M. J. Fréchet, *J. Chem. Soc., Chem. Commun.* **1990,** 1010.

[56] C. J. Hawker, J. M. J. Fréchet, *J. Am. Chem. Soc.* **1990,** *112*, 7638.

[57] P. D. Beer, E. L. Tite, *Tetrahedron Lett.* **1988,** *29*, 2349.

[58] P. D. Beer, J. W. Wheeler, C. Moore, *Supramolecular Chemistry,* (Eds.: V. Balzani, L. DeCola), Kluwer, Dordrecht, **1992,** pp. 105–118.

[59] E. Buhleier, W. Wehner, F. Vögtle, *Synthesis* **1978,** 155.

[60] a) R. Moors, F. Vögtle, *Chem. Ber.* **1993,** *126*, 2133. b) P. Lhotak, S. Shinkai, *Tetrahedron* **1995,** *51*, 7681.

[61] R. Engel, K. Rengan, C.-S. Chan, *Heteroatom Chem.* **1993,** *4*, 181.

[62] M. Slany, M. Bardají, M.-J. Casanove, A.-M. Caminade, J. P. Majoral, B. Chaudret, *J. Am. Chem. Soc.* **1995,** *117*, 9764.

[63] E. C. Weiner, M. N. Brechbeil, H. Brothers, R. L. Magin, O. A. Gansow, D. A. Tomalia, P. C. Lauterbur, *Mag. Res. Med* **1994,** *31*, 1.

[64] G. R. Newkome, C. N. Moorefield, *Polym. Preprints* **1993,** *34*, 75.

[65] G. R. Newkome, C. N. Moorefield, G. R. Baker, A. L. Johnson, R. K. Behera, *Angew. Chem.* **1991,** *103*, 1205; *Angew. Chem., Int. Ed. Engl.* **1991,** *30*, 1176.

[66] M. F. Ottaviana, S. Bossmann, N. J. Turro, D. A. Tomalia, *J. Am. Chem. Soc.* **1994,** *116*, 661.

[67] D. A. Tomalia, H. Baker, J. Dewald, M. Hall, G. Kallos, S. Martin, J. Roeck, J. Ryder, P. Smith, *Polym. J.* **1985,** *17*, 117.

9 Dendritic Networks

9.1 Introduction: Dendritic Assemblies

A full appreciation of dendritic chemistry and hence the iterative method employed for generational construction would be deficient without consideration of 'higher order' macromolecular assemblies. To this end, in this Chapter we attempt to provide a suitable foundation for a preliminary debate concerning these higher order assemblies, or dendritic networks. For the purpose of our discussion, a dendritic network shall be defined as the deliberate connection, via covalent or non-covalent means, of multiple (usually pre-constructed) dendritic units resulting in an architecture with dimensions greater than would be obtained via the preparation of a standard dendrimer. This "deliberate connection" or "positioning" results in at least one degree of freedom less with respect to the relationship of individual dendrimers to other macromolecules in the network. Since there are but a few examples heralding this arena, this Chapter will be unique in the presentation in that it will include an examination of potential categories of networks as well as modes of formation.

Attention will focus in two primary modes of network assembly: (1) *random, uncontrolled connectivity* analogous to 'classical' polymer preparation, whereby monomer units or building blocks are essentially positioned relatively "unsystematically" via single-pot-type reactions; and (2) *ordered, controlled connectivity* analogous to tessellated dendritic polymer construction whereby elements, or building blocks, are "precisely juxtaposed" into a coherent motif.

Networks comprised of linear, classically synthesized polymers that are cross-linked have been reviewed[1–4] and will not be treated in this Chapter. Further, mathematical treatments of network properties will not be discussed herein.

As defined, dendritic networks are considered to result from the one-, two-, and three-dimensional orientation of dendrimers; thus "ordering" can be geometrically likened to rods; surfaces or sheets; and cubes, tetrahedrons, or spheres, respectively. Due to the broad scope and breadth of potential macromolecular architectures that can be obtained by application of different modes of connectivity, we will herein concentrate on networks constructed from the simplest dendritic structures, namely those that are pseudospherical or globular. The principles that are presented here pertaining to network formation should be easily adaptable to non-spheroidal dendritic structures as well as macromolecular assemblies possessing only limited dendritic character.

Construction of dendritic networks is a logical progression of the iterative synthetic method whereby the desired "positioning" of multiple nuclei components (such as in dendrimers) can be obtained. Realization of networks comprised of dendrimeric building blocks, as well as individual dendrimers, (Figure 9.1) thus has applications in such different areas of materials science as molecular electronics (largely pioneered by Forest L. Carter),[5–7] biomolecular engineering,[8] and (liquid) crystal engineering,[9,10] to mention but a few. In short, dendritic networks provide the material scientist with a potent mechanism for the construction of molecular devices that are capable of information processing.[11–19].

Furthermore, and perhaps more importantly, once the molecular weight ceiling was punctured by the advent of cascade chemistry, the construction of precise networks and assemblies have no limit. Current trends in chemistry and molecular design strongly suggest the potential for the creation of higher order architectures. This is clearly evident upon examination of current literature and the ubiquitous reports of complex structures contained therein. These include Lehn's helicates,[20–23] 3 X 3 Ag ion-based inorganic grid,[24] di- and tri-nuclear, rack-type Ru(II) complexes,[25a] and multi-porphyrin supramolecular macrocyles,[25b] Stoddart's proposed dendritic rotanes,[26] Swager's[27] and Gibson's[28,29] polyrotaxanes; Wuest's *H*-bonded, self-assembled networks;[30,31] Michel's "tin-

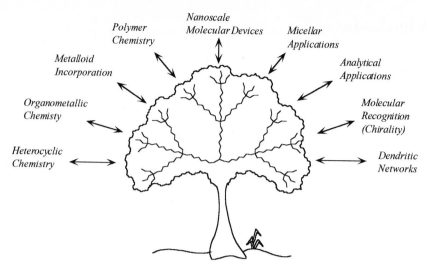

Figure 9.1. Areas of material science that have benefited from dendritic chemistry.

kertoy" assemblies,[32] Eaton's alkynylcubane structures;[33] Imanishi and Rings-dorf's,[34,35] Kimizuka's,[36] Hamilton's,[37] and Whitesides and MacDonald's[38] *H*-bonded nanostructures; Wegner's layered Langmuir–Blodgett (LB) assemblies,[39] Davis and Sterling's multilayered calix-4-resorcarenes;[40] Bunz's organometallic polyalkynes;[41,42] de Meijere's spirocyclopropanes;[43] Fujita and Oguara's interpenetrating molecular ladders and bricks;[44] Zimmermann's biomimetic self-assembling dendrimers;[45] Griffin and Pourcain's pyridine-based thermoreversible networks;[46] Diederich's alkynyl radial-enes,[47] acyclic tetraethynylethene molecular scaffolding,[48] and fullerene–acetylene hybrids;[49] McCord's hyper-cross-linked polyarylcarbinols;[50] Wagner's layered assemblies of hairy-rod macromolecules;[51] Wright's molecular chandeliers;[52] Poznoya and Buese's siloxane functionalized silica particle networks;[53] Schlüter's dendritic cylinders;[54] Hawker's architecturally controlled star and graft polymers;[55] Lindsey's,[56] Harriman and Sauvage's,[57] Higuchi and Ojima's,[58] Anderson, McPartlin and Sanders',[59] and Crossley's[60] linear and cyclic porphyrin arrays; Reetz's[61] nanostructured metal clusters; Subramanian and Zaworotko's[62] square channel, porous solids; Schreiber's[63] orthogonal receptor-ligand combinations; Reinhoudt's self-assembled, bifunctional receptor[64a], and self-assembled carceplex monolayers;[64b] Müller's[65] diamagnetic, mixed-valence, Mo(v)/Mo(vii) "big wheel" cluster; Nicolaou, Xio, and Nova's[66] semiconductor memory device for *R*adiofrequency *E*ncoded *C*ombinatorial *C*hemistry (REC^{TM}); Menger's giant synthetic vesicle,[67] octacationic cyclophanes[68a] and Prakash and Olah's tetrahedral tetracation[68b] as well as various specifically constructed oligomeric materials, such as Tour's thiophene–ethynylene molecular wires,[69] Moore's phenylacetylene ologimers[70] and "molecular turnstile",[71] Vögtle's nanometer scale molecular ribbons,[72a] Butler and Warrener's rigid "molecular racks",[72b] Sita's ferrocene-terminated polyphenylacetylenes,[73a] and Grime's penta- and hexadecker sandwiches.[73b]

It should be noted here that, currently, characterization, separation, and purification techniques of the larger cascades or dendrimers (previously discussed in earlier chapters) are straining the limits of present instrumentation. Thus, unequivocal characterization of dendritic networks is currently limited and will necessitate the development of new methods and instrumentation. However, dendritic network structural verification and elucidation should be facilitated by integration of known standard materials science methods, e.g., MS and EM.

9.2 Network Formation and Classification

9.2.1 Ordered versus Random

Dendritic networks, as defined in Section 9.1, can be considered to fall into two main classes: ordered and random assemblies (Figure 9.2). These two classifications are essentially ideal extremes of a continuum of possible geometries. Hence, many network structures will possess a higher or lower defined degree of "orderliness" (or increased "randomness") depending on the degrees of freedom inherent in the method of construction. A randomly prepared network thus lacks patterned connectivity due to the unrestricted manner in which the macrobuilding blocks (dendrimers) are positioned. Conversely, an ordered network results from specific restrictions (reduced degrees of freedom) that are set on the building block positioning process.

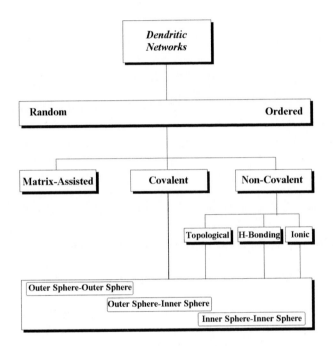

Figure 9.2. Dendritic network classification via dendrimer connectivity.

9.2.1.1 Methods of Formation

All ordered and random dendritic networks that are constructed via covalent or non-covalent means result from positioning of one dendrimer relative to another. Thus, macroassembly positioning can be effected via at least one of three different methods of connectivity. These methods are geometrically rooted in dendritic chemistry.

Since dendrimers are inherently globular or pseudospherical (particularly at higher generations) due to the branching patterns induced by the building blocks used for their construction, they may conveniently be considered geometrically as spheres. Spherical geometry dictates that the two major regions where chemical and physical transformations can occur are the outer region (surface area) and the inner region (internal superstructure area), as illustrated by the idealized dendritic regions in Figure 9.3.

Hence, dendritic connection can be brought about by a combination of these regions. "Regional" combinations can thus be classified into three distinct types: (1) outer sphere – outer sphere; (2) outer sphere – inner sphere; and (3) inner sphere – inner sphere. Each of these combinations can be employed either separately or in concert (Figure 9.4).

Outer sphere – outer sphere connectivity can be envisioned as the simple peripheral connection of dendrimers. Surface-to-surface connectivity can be facilitated by direct dendrimer – dendrimer attachment or by dendrimer – bridging unit – dendrimer attach-

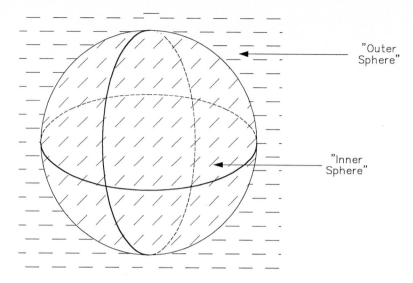

Figure 9.3. Idealized dendrimer regions available for attachment and manipulation.

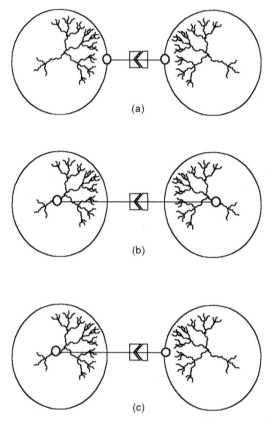

Figure 9.4. Idealized models of dendritic connectivity: a) outer sphere – outer sphere; b) inner sphere – inner sphere; and c) outer sphere – inner sphere.

ment(s). Inner sphere – inner sphere dendritic connection, without the use of a bridging unit, requires branched assemblies capable of interpenetration, which further dictates the avoidance of critical dense packing limits. The use of a sphere-to-sphere connector, or bridging unit, obviates the requirement of interdigitation. Whereas, the combination of these dendritic connections can be described as outer sphere – inner sphere and is self-evident.

9.2.1.2 Covalent and Non-Covalent Positioning

Covalent dendritic connection can result from any standard synthetic transformation capable of forming a covalent bond. These include nucleophilic, electrophilic, ionic, radical, and carbenoid reactions. Elements effecting bond formation include metals, non-metals, and metaloids.

Non-covalent means of dendritic connectivity can be further subdivided into three sub-categories comprised of topological-, hydrogen-, and ionic-bonding. Topological-bonding can be envisioned as any sterically induced association of two units each from two different macroassemblies. Topological, or mechanical bonding[74,75] can best be envisioned by considering interlocking rings (e.g., catenane- and rotaxane-type), physical entrapment of units within a designed cavity, or positioning based on other steric factors. Hydrogen- and ionic-bonding methods employed for dendrimer connectivity are un-ambiguous, as will be demonstrated.

9.3   Random Connectivity

9.3.1   Random, Covalently-Linked Dendrimer Networks

Consider the divergent synthesis of PAMAM dendrimers,[76] whereby undesired intra- or inter-molecular events lead to increased polydispersity and lost ideality. Although these events can be minimized or in selected cases eliminated, interestingly these processes can give rise to randomly positioned polydendritic systems. Thus, amidation of polyesters (e.g., **1**) with a diamine can lead to *intra*molecular bisamidation which can give rise to topological dendrimer connection (Scheme 9.1) as depicted by catenated bisdendri-

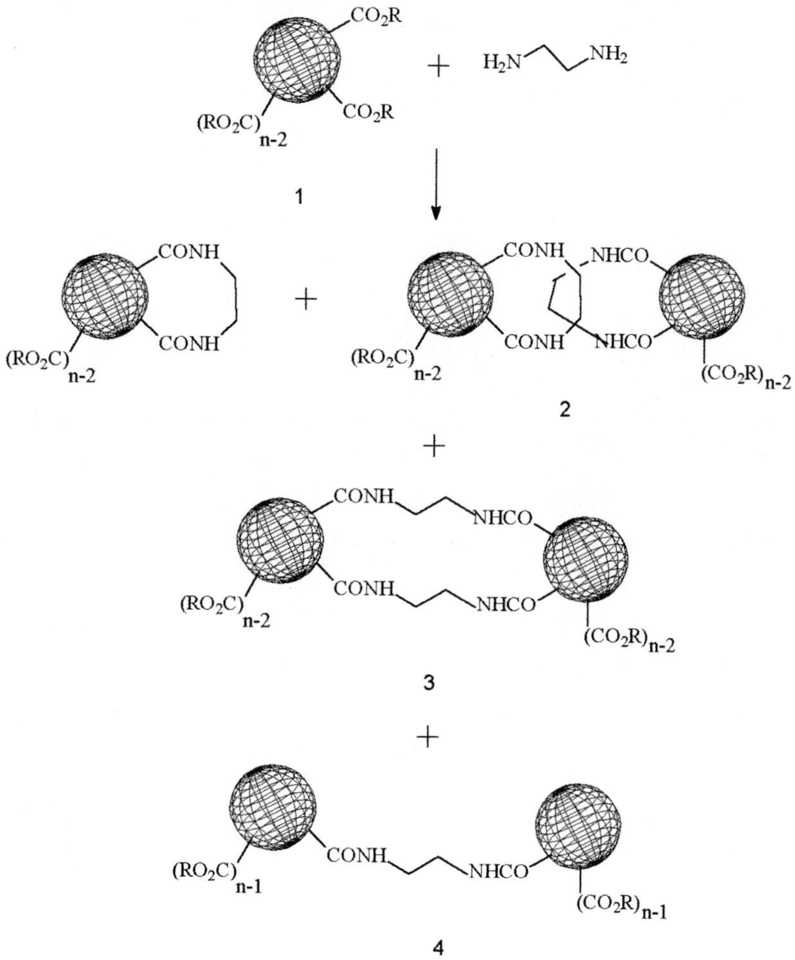

Scheme 9.1. Examples of possible *bis*dendrimer networks formed via the uncontrolled treatment of ester terminated dendrimers with an alkyl diamine.

mer **2**. Whereas, *inter*molecular bisamidation can lead to bridged dendrimer connection as illustrated by bisdendrimers **3** and **4**. Statistically-based, high-dilution reaction techniques can be employed to enhance the yields of these bridged dendrimers, yet a distribution of products will still be generated. Maximized production and isolation of individual components is thus generally more difficult using a statistical preparative method rather than a more directed approach. Random connectivity[76] of amine-terminated PAMAM dendrimers via treatment with di- or tri-halides later introduced the general concept of randomly constructed networks. No specific characterization of these bridged dendrimers was reported,[76] although electron micrographs have been shown[77] to support the linkage of two dendrimers possessing dissimilar surfaces (i.e., surface amines and surface carboxylic acids). To date, evidence for the presence of a bisdendrimer via this method is limited primarily to mass spectrometry.

9.3.2   Coupling via Surface-to-Surface Interaction

Treatment of a dendritic polyester **5** with a dendritic polyamine **6** is perhaps the simplest example of coupling between two dissimilar cascade macromolecules.[78–80] Scheme 9.2 illustrates some potential products (i.e., **7** and **8**) available by this general procedure, as well as a continuation of surface amidation to afford polymeric species such as random network **9**. As envisioned, it would be difficult to stop this procedure after the formation of a single amide bond, due to the close proximity of adjacent ester–amine groups.

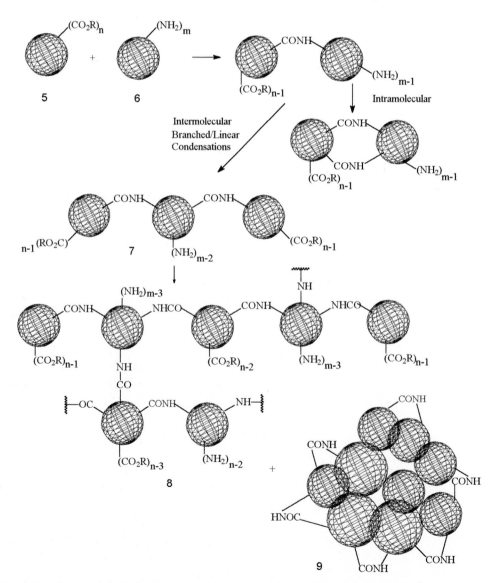

Scheme 9.2. Representations of potential randomly constructed networks via the reaction of ester terminated dendrimer and amine terminated dendrimer.

Potentially, a limited number of juxtaposed bridging amide bonds could be possible at the initial junction locus, depending on the contact surface area defined by the interaction of the initial reagents. Other methods of direct surface-to-surface connectivity that have been reported[79,80] include the reaction of olefinic terminated dendrimers with initiator-terminated dendrimers.

9.3.3 Coupling via Surface-to-Surface Bridging Units

Connection of dendrimers via treatment with multifunctional cross-linking-type reagents such as the addition of polyhalides to amine-terminated dendrimers[78] results in randomly positioned dendritic assemblies. Dendrimers have also been bridged by the incorporation of copolymerizable units into reactions with dendrimers possessing polymerizable, terminal olefins.[79,80]

9.4 Ordered Dendritic Networks

9.4.1 Multilayer Construction

Watanabe and Regen[81] reported the construction of ordered, dendritic multilayers (**10**) via a bridged, outer sphere − outer sphere mode of assembly (Figure 9.5) whereby the transition metal Pt was used as a connector moiety. Although amine-terminated PAMAM-type dendrimers[76] were employed for this particular example, this process could easily be extended to other types of macromolecules.

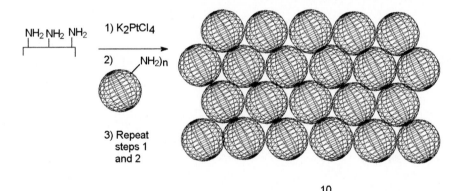

10

Figure 9.5. An idealized representation of an ordered dendritic network constructed by the procedure of Watanabe and Regen.[81]

The ordering procedure does not rely on site-specific reactions but rather the packing efficiency of the dendritic species, or building blocks. This example possesses characteristics of both random and ordered networks and exemplifies the broad spectrum of the random-ordered network continuum.

9.4.2 Directed Network Construction

9.4.2.1 Covalent, Metal-Based Assembly

Ideally, from a macromolecular subunit position control perspective, directed approaches toward network construction are desirable. Indeed, the construction of dendrimers possessing metal-ligating, 4-substituted, 2,2′ : 6′,2″-terpyridine moieties have been reported.[82,83]

Ligand incorporation was achieved by facile alkoxylation of 4-chloro-substituted terpyridine by a hydroxy-terminated carboxylic acid followed by divergent dendrimer con-

struction employing amide-based connectivity (see Section 4.12.1.4) and a 1 → 3 branching multiplicity. Formation of the bisdendrimer assembly **11** (Figure 9.6) was achieved
by treatment of a larger (third tier), cascade "compartmentalized", terpyridine ligand
with a smaller, lower generation Ru(III), dendritic complex. While complex **11** is not a
multidentate network, it constitutes the first example of an inner sphere − outer sphere,
specifically positioned dendritic assembly.

Metal connectivity is advantageous. Transition metal characterization such as with
electrochemical and microscopic analysis compliments standard "organic"-type methods
(e.g., ^{13}C and ^{1}H NMR) and further enhances characterization of the large multidendri-
tic structure(s). Cyclic voltammetry data strongly suggest the connection of branched
structures.[82]

Metal connectivity is adaptable to higher order and more complex architectures such
as the pseudotetrahedral, pentadendrimer **12** depicted in Figure 9.7. Structure **12** is readily accessible from integration of current technology[82,83] with novel unsymmetrical
dendritic building blocks possessing ligating sites and the capacity to continue the branching process.

9.4.2.2 Hyperbranched-Type, Metal-Based Assembly

Extending the concept of metal-based, ordered, dendritic network construction,
hyperbranched networks can be envisioned. Employing known synthetic methods, complex networks should be accessible via the propensity of deliberately constructed bridges
and dendrimers to self-assemble into intrinsically stable arrays based on isotopically
extended connectivity in three dimensions.[84] Thus, treatment of dendrimers[85a] possessing four tetrahedrally arranged terpyridine moieties (i.e., **13**) with rigid, bisruthenium
connector rods (**14**) of an appropriate length should afford "diamondoid" architectures,
such as **15** depicted in Figure 9.8.

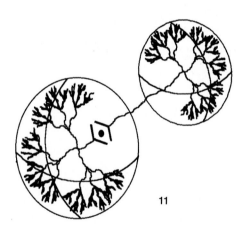

11

Figure 9.6. Specifically connected *bis*dendrimer assemblies attached via metal ligation.

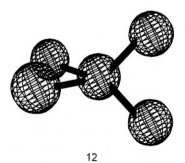

12

Figure 9.7. A representative "supermethane"-based motif obtained via the tetrahedral ordering of four "terminal" dendrimers about a single "core" dendrimer by the use of rigid rod connectors.

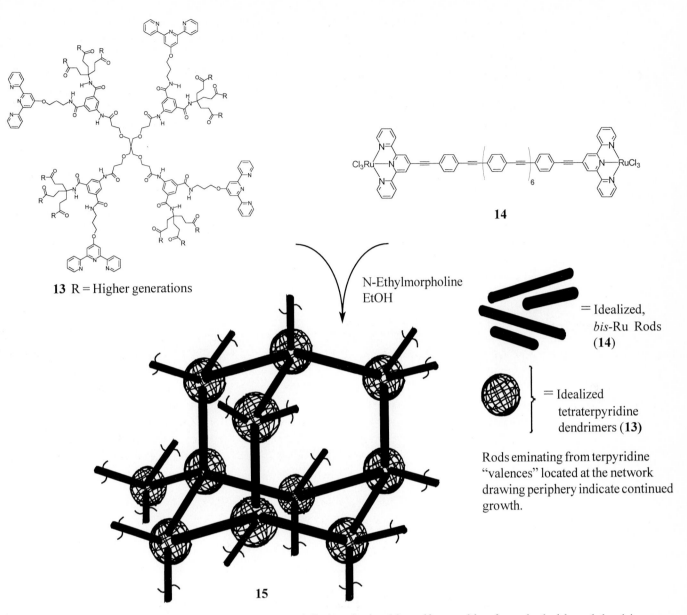

13 R = Higher generations

14

N-Ethylmorpholine
EtOH

= Idealized,
bis-Ru Rods
(14)

= Idealized
tetraterpyridine
dendrimers (**13**)

Rods eminating from terpyridine
"valences" located at the network
drawing periphery indicate continued
growth.

15

Figure 9.8. Diamondoid architectures that can potentially be obtained by self-assembly of tetrahedral-based dendrimers.

Reinhoudt et al.[85b] have reported the preparation of metal-based dendritic assemblies employing a novel self-assembly process. Complimentary building block positioning was predicated on labile acetonitrile substitution at tetravalent, square planar Pd(II) complexes with a kinetically inert arylcyanomethyl monomer moiety. Average aggregate diameters, as determined by transmission electron and atomic force microscopies, were found to be 205 nm with a narrow distribution. Energy dispersive X-ray spectroscopy (EDX) confirmed elemental *S* and *Pd* existence in the aggregates. Triblock and multiblock copolymers possessing polydisperse segments of well defined architecture capable of forming chain-folded crystallites or metal complex-based helices have been reported by Eisenach et al.[85c]

9.4.2.3 Hydrogen-Bonding Assembly

Incorporation of *H*-bonding moieties capable of self-assembly into or onto a dendritic superstructure can lead to ordered networks. One of the first dendrimers to be developed[86] for the purpose of exploring potential network formation is depicted in Scheme 9.3. Aminopyridine triester **16** was reacted with a tetraacyl halide core to afford the first generation tetrapyridine **17**, which was subsequently deprotected with HCO_2H and treated with amino triester **18** to give the second tier, 36-ester **19**. As envisioned, treatment

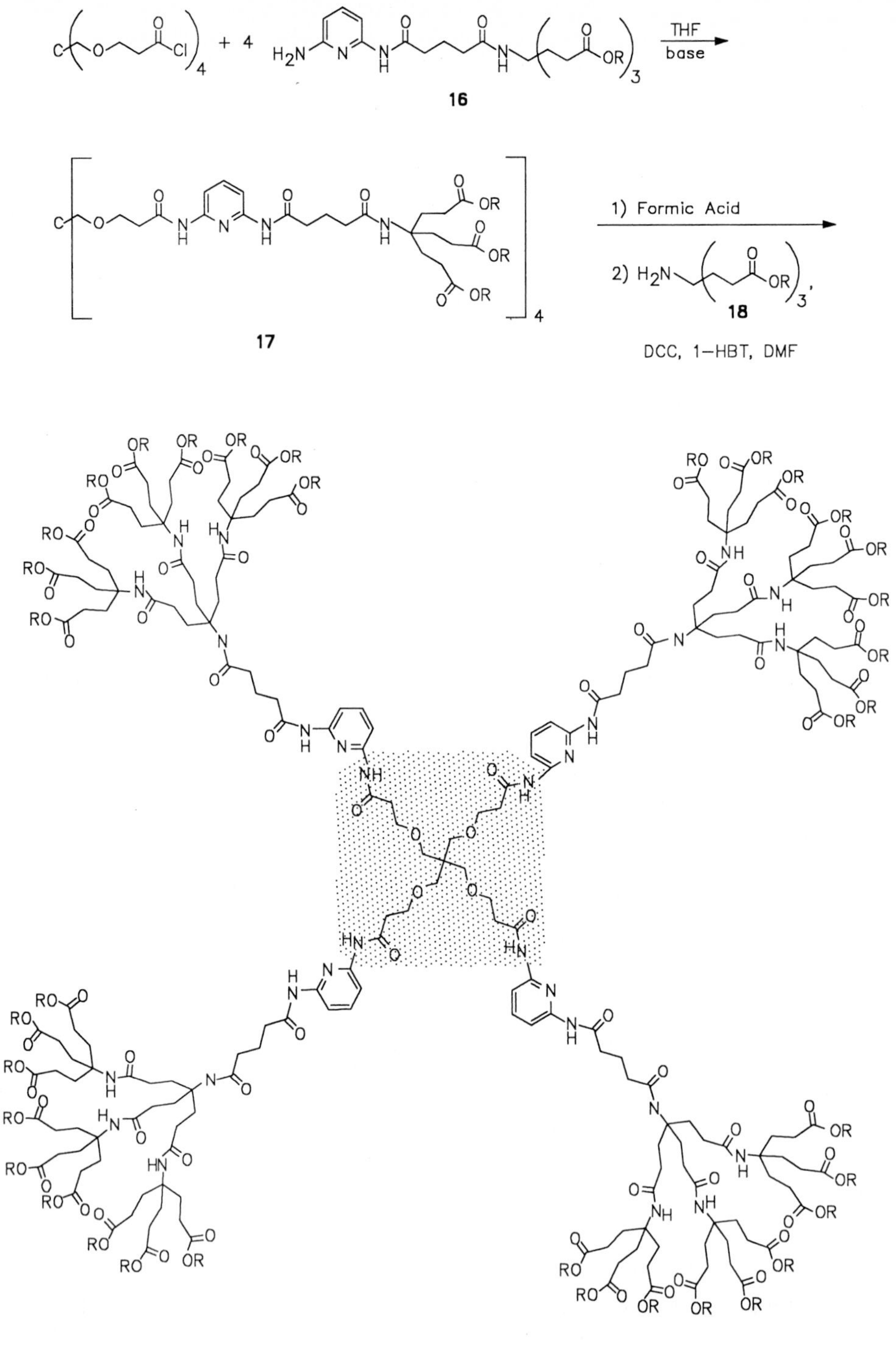

Scheme 9.3. Construction of a dendrimer with four sites that are capable of molecular recognition via donor-acceptor-donor hydrogen bonding interactions.

of the tetrakis(diaminopyridine) polyester (**19**) with either dendrimers connected to single imide moieties or α,ω-bisimide rods should lead to tetrahedral dendritic arrays and "adamantanoid" hyperbranched-like networks (i.e., **20**; Figure 9.9).

As early as 1986 dumb-bell-shaped dendrimers were reported to form linear networks.[87,88] These surfactant-like dendrimers possessed external, branched, ball-shaped architecture connected to either sides of a linear alkyl chain interior (Figure 9.10); these dendrimers were termed "arborols" and were discussed earlier in detail (Section 4.12.1.1).

When arborol **21** is added in low concentrations to an aqueous environment gelation occurs. The propensity of minimize lipophilic-hydrophilic interactions as well as maximize *H*-bonding and packing effects is proposed as the reason for the formation of the

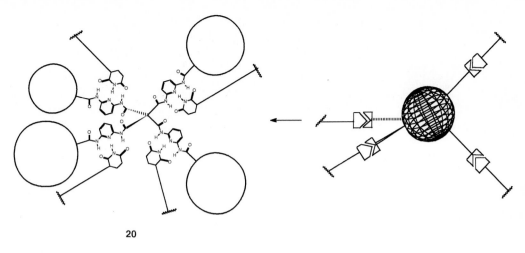

20

Figure 9.9. Idealized depiction of a network assembled by hydrogen-bonding interactions.

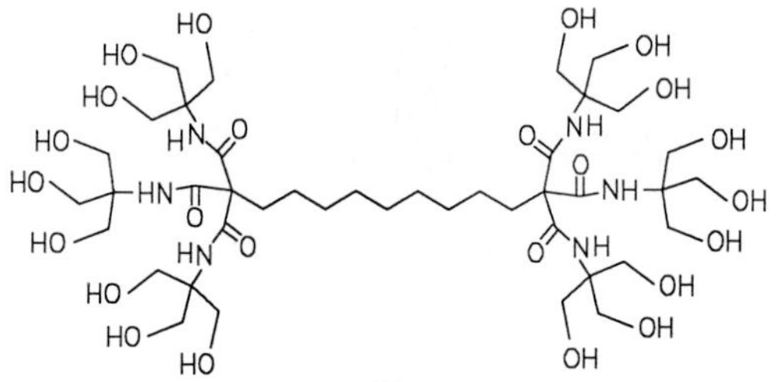

21 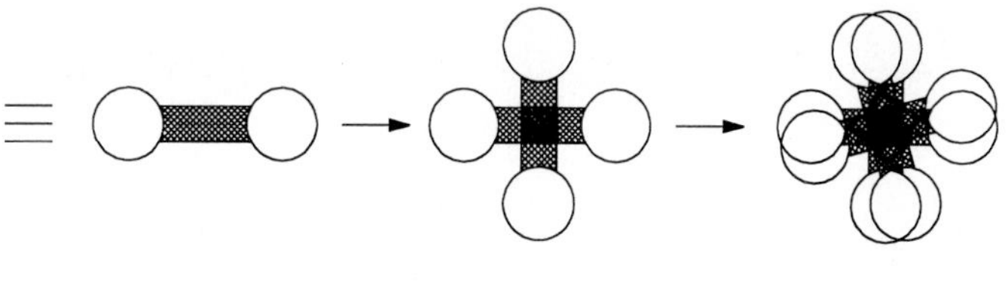   **22**

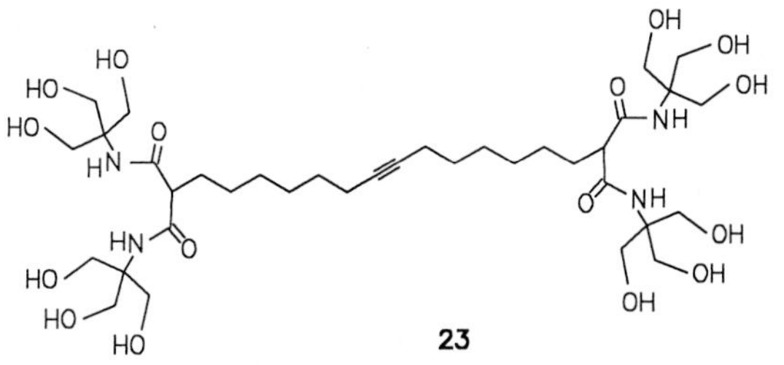

23 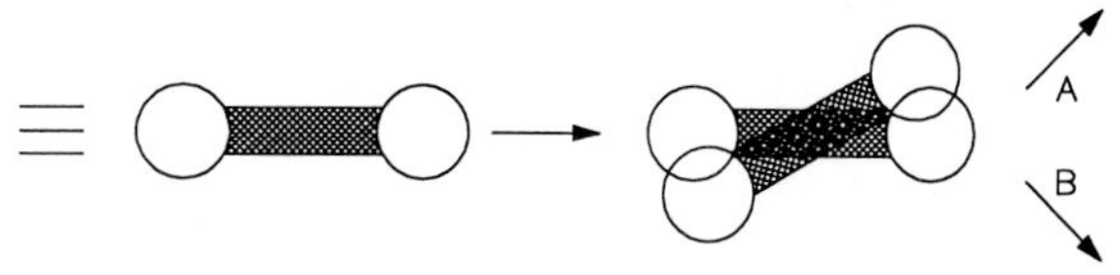 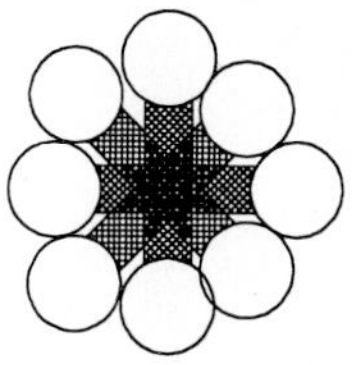   **24**

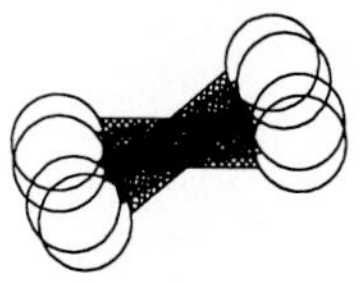

                                                             25

Figure 9.10. Network formation via the stacked aggregation of dumbbell shaped dendritic molecules.

ordered, linear hyperbranched-like networks (**22**) that are clearly visible via electron microscopy of the dried gel.

Incorporation of an alkyne moiety within the lipophilic core (e.g., **23**) imparts helical and scissors-like morphology to the stacked array (e.g., **24** and **25**, respectively). Disc-like, benzene-based arborols have also been reported[89] to form *H*-bonded aggregates, or networks, comprised of approximately 50–60 dendrimers. The approximation is based on network diameters determined from electron microscopy.

Zimmermann et al.[45,90] have disclosed a pioneering effort based on the convergent preparation of dendritic wedges possessing tetraacid moieties (i.e., **26**) that self-assemble into a hexameric, disk-like network **27** (Fig. 9.11). The tetraacid unit **28** is known to form cyclic as well as linear structures in solution via carboxylic acid dimerization. However, when the tetraacid unit is incorporated onto large dendritic wedges, the hexamer form is preferred. It is postulated that the cyclic form is favored due to the less sterically demanding environment than would be the case for linear aggregates. The argument is supported by SEC experiments. Thus, employing the fourth generation wedges, prepared by the method of Fréchet,[91a] disk-shaped ordered networks 90 Å in diameter and

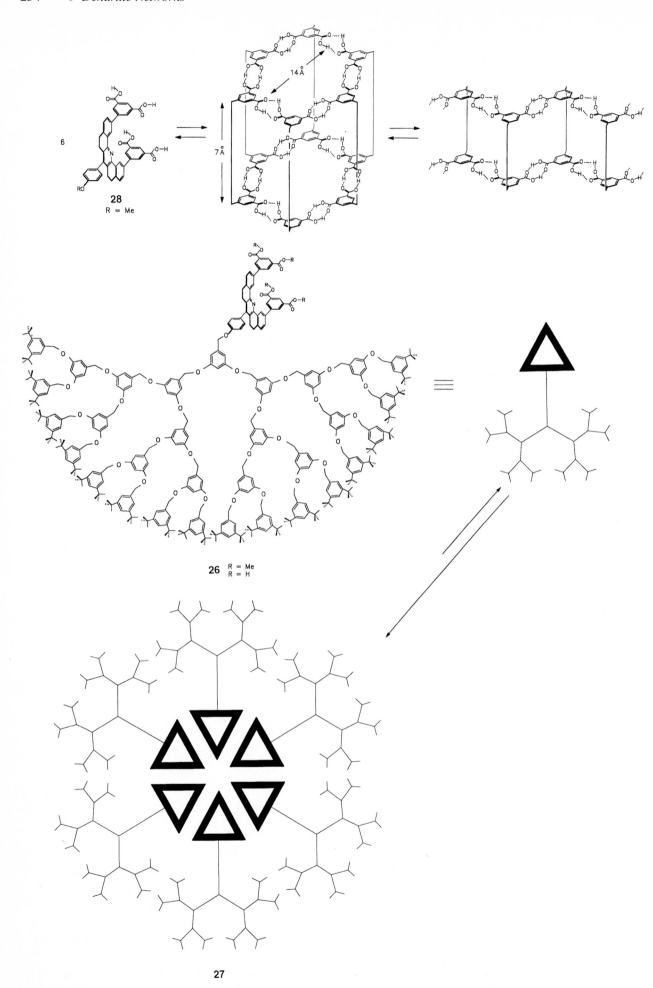

Figure 9.11. Zimmermann et al.'s[45,90] dendritic ordering based on well-known carboxylic acid dimer formation.

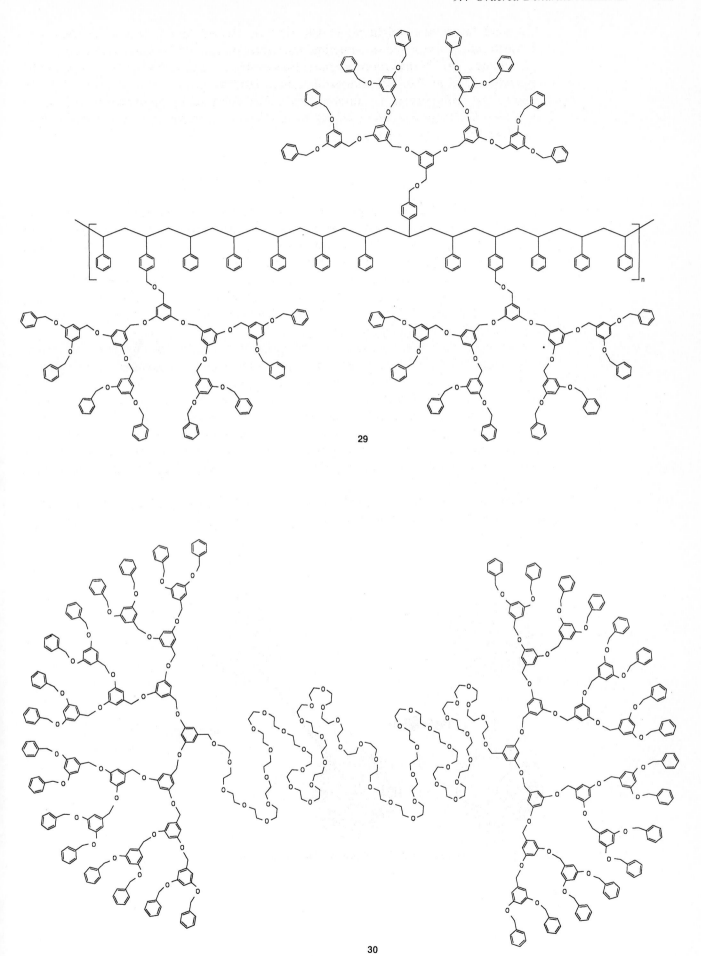

29

30

Figure 9.12. Fréchet et al.'s[92–94] dendritic assemblies ordered via a polystyrene backbone and polyethylene glycol chains.

20 Å thick (molecular weight ca. 34,000 amu) are formed via a dual, non-covalent mode of construction comprised of complimentary topological and *H*-bonding aspects.

Stoddart et al.[91b] have reported the self-assembly of branched [*n*]rotaxanes in an investigation aimed at the preparation of larger, dendritic rotaxanes. Kato et al.[91c] have reported the preparation of supramolecular liquid crystalline networks based on self-assembly of carboxylic acid-based, trigonally branched, *H*-bonding donors and bipyridine-type *H*-bonding acceptors.

9.4.2.4   Covalent Assembly

Fréchet and coworkers[92] have significantly contributed to the area of dendritic chemistry with the introduction of the convergent method whereby dendritic wedges are connected to a core producing a final dendrimer. This method is tantamount to covalent, dendritic positioning. Architectures constructed via the integration of this method with traditional chemistry include the hybrid linear-dendritic block copolymer[93] **29** (Figure 9.12), prepared by the free radical copolymerization of styrene-functionalized dendritic wedges with styrene and the bisdendrimer[94] **30** synthesized by reaction of brominated dendritic wedges with poly(ethylene glycol).

As envisioned, covalent dendritic ordering can be realized by using secondary (embedded or latent) protection–deprotection schemes in concert with those already developed for dendrimer construction. For example, consider the preparation of a tetrahedrally-based dendrimer **31** possessing internal sites of attachment whereby three sites are protected and one is available for connection; additional peripheral and internal functionalities are inert to the chosen attachment and deprotection conditions. Furthermore, consider the connection of two of these dendrimers to afford a bisdendrimer **32**. Deprotection of the internal moieties (in no case NO_2 reduction) allows further dendrimer attachment, etc. (Scheme 9.4).

Use of this concept for the preparation of networks leads to unlimited architectures. It should facilitate and expand ordered (as well as random) network construction in the future.

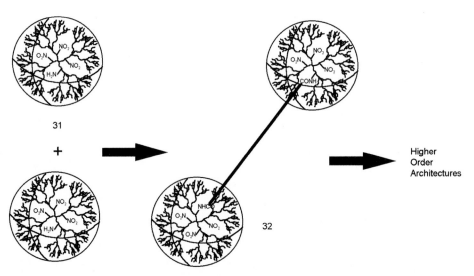

Scheme 9.4. Use of iterative protection–deprotection technology for the construction of network architectures.

9.5 References

[1] P. J. Flory, *Principles of Polymer Chemistry*, Cornell University Press, Ithaca, New York, **1953**.

[2] G. Odian, *Principles of Polymerization*, 3rd ed., Wiley, New York, **1991**.

[3] M. Adam, J. Bastide, S. Candau, A. Coniglio, M. Delsanti, J. E. Mark, D. Stauffer, A. J. Staverman, in *Polymer Networks*, (Ed.: K. Dušek), Springer, Berlin, **1982**.

[4] S. M. Aharoni, S. F. Edwards, *"Rigid Polymer Networks"* in *Advances in Polymer Science*, Springer, Berlin, **1994**. N. K. Müller-Nedebock, S. M. Aharoni, S. F. Edwards, *Macromol. Symp.* **1995**, *98*, 701.

[5] *Molecular Electronics: Biosensors and Biocomputers*, (Ed.: F. T. Hong) Plenum, New York, **1989**.

[6] *Molecular Eletronic Devices*, (Ed.: F. L. Carter), Dekker, New York, **1982**.

[7] *Molecular Electronic Devices II*, (Ed.: F. L. Carter), Dekker, New York, **1987**.

[8] J. G. Tirrell, M. J. Fournier, T. L. Mason, D. A. Tirrell, *Chem. Eng. News*, **1994**, (12/19), 40.

[9] J. Zhang, J. S. Moore, *J. Am. Chem. Soc.* **1994**, *116*, 2655.

[10] G. R. Desirajer, *Angew. Chem.* **1995**, *107*, 2541; *Angew. Chem. Int. Ed. Engl.* **1995**, *34*, 2311.

[11] J.-M. Lehn, *Angew. Chem., Int. Ed. Engl.* **1988**, *27*, 89; *Angew. Chem.* **1988**, *100*, 91.

[12] J.-M. Lehn, *Angew. Chem., Int. Ed. Engl.* **1990**, *29*, 1304; *Angew. Chem.* **1990**, *102*, 1347.

[13] H. Ringsdorf, B. Schlarb, J. Venzmer, *Angew. Chem., Int. Ed. Engl.* **1988**, *27*, 113; *Angew. Chem.* **1988**, *100*, 117.

[14] D. Philp, J. F. Stoddart, *Synlett* **1991**, *7*, 445.

[15] J. S. Lindsey, *New J. Chem.* **1991**, *15*, 153.

[16] H.-J. Schneider, *Angew. Chem., Int. Ed. Engl.* **1991**, *30*, 1417; *Angew. Chem.* **1991**, *103*, 1419.

[17] L. Yu, M. Chen, L. R. Dalton, *Chem. Mater.* **1990**, *2*, 649.

[18] R. J. Pieters, I. Huc, J. Rebek, Jr., *Chem. Eur. J.* **1995**, *1*, 183.

[19] M. D. Ward, *Chem. Soc. Rev.* **1995**, *24*, 121.

[20] J.-M. Lehn, A. Rigault, J. Siegel, J. Harrowfield, B. Chevrier, D. Moras, *Proc. Natl. Acad. Sci., U. S. A.* **1987**, *84*, 2565.

[21] J.-M. Lehn, A. Rigault, *Angew. Chem., Int. Ed. Engl.* **1988**, *27*, 1095; *Angew. Chem.* **1988**, *100*, 1121.

[22] A. Pfeil, J.-M. Lehn, *J. Chem. Soc., Chem. Commun.* **1992**, 838.

[23] G. S. Hanan, J.-M. Lehn, N. Kyritsakas, J. Fischer, *J. Chem. Soc., Chem. Commun.* **1995**, 765.

[24] P. N. W. Baxter, J.-M. Lehn, J. Fischer, M.-T. Youinow, *Angew. Chem., Int. Ed. Engl.* **1994**, *33*, 2284; *Angew. Chem.* **1994**, *106*, 2432.

[25] a) G. S. Hanan, C. R. Arana, J.-M. Lehn, D. Fenske, *Angew. Chem., Int. Ed. Engl.* **1995**, *34*, 1122; *Angew. Chem.* **1995**, *107*, 1191. b) C. M. Drain, K. C. Russell, J.-M. Lehn, *J. Chem. Soc., Chem. Commun.* **1996**, 337.

[26] D. B. Amabilino, P. R. Ashton, M. Belohradský, F. M. Raymo, J. F. Stoddart, *J. Chem. Soc., Chem. Commun.* **1995**, 751.

[27] M. J. Marsella, P. J. Carroll, T. M. Swager, *J. Am. Chem. Soc.* **1995**, *117*, 9832. Q. Zhou, T. M. Swager, *J. Am. Chem. Soc.* **1995**, *117*, 12593.

[28] H. W. Gibson, S. Liu, S.-h. Lee, H. Marand, A. Prasad, F. Wang, M. Bheda, D. Nagvekar, *Macromol. Symp.* **1995**, *98*, 501.

[29] H. W. Gibson, H. Marand, *Adv. Mater.* **1993**, *5*, 11.

[30] M. Simard, D. Su, J. D. Wuest, *J. Am. Chem. Soc.* **1991**, *113*, 4696.

[31] J. D. Wuest, *"Molecular Tectonics"* in *Mesomolecules: From Molecules to Materials*, (Eds.: G. D. Mendenhall, A. Greenberg, J. F. Liebman), Chapman & Hall, SEARCH Series, *Vol. 1*, Chapt. 4, **1995**, pp. 107–131.

[32] J. Michl, *"Supramolecular Assemblies from 'Tinkertoy' Rigid-Rod Molecules"* in *Mesomolecules: From Molecules to Materials*, G. D. Mendenhall, A. Greenberg, J. F. Liebman, Chapman & Hall, SEARCH Series, Vol. 1, Chapt. 5, **1995**, pp. 132–160.

[33] P. E. Eaton, E. Galoppini, R. Gilardi, *J. Am. Chem. Soc.* **1994**, *116*, 7588.

[34] K. Fijita, S. Kimura, Y. Imanishi, E. Rump, J. van Esch, H. Ringsdorf, *J. Am. Chem. Soc.* **1994**, *116*, 5479.

[35] T. M. Bohanon, S. Denzinger, R. Fink, W. Paulus, H. Ringdorf, M. Weck, *Angew. Chem., Int. Ed. Engl.* **1995**, *34*, 58; *Angew. Chem.* **1995**, *107*, 102.

[36] N. Kimizuka, T. Kawasaki, K. Hirata, T. Kunitake, *J. Am. Chem. Soc.* **1995**, *117*, 6360.

[37] E. Fan, J. Yang, S. J. Geib, T. C. Stoner, M. D. Hopkins, A. D. Hamilton, *J. Chem. Soc., Chem. Commun.* **1995**, 1251.

[38] J. C. MacDonald, G. M. Whitesides, *Chem. Rev.* **1994**, *94*, 2383.

[39] M. Seufert, M. Schaub, G. Wenz, G. Wegner, *Angew. Chem., Int. Ed. Engl.* **1995**, *34*, 340; *Angew. Chem.* **1995**, *107*, 363.

[40] F. Davis, C. J. M. Stirling, *J. Am. Chem. Soc.* **1995**, *117*, 10385.

[41] J. E. C. Wiegelmann-Kreiter, U. H. F. Bunz, *Organometallics* **1995**, *14*, 4449.

[42] U. H. F. Bunz, *Angew. Chem.* **1994**, *106*, 1127; *Angew. Chem., Int. Ed. Engl.* **1994**, *33*, 1073. M. Altmann, V. Enkelmann, F. Beer, U. H. F. Bunz, *Organometallics* **1996**, *15*, 394. J. E. C. Wiegelmann-Kreiter, U. H. F. Bunz, *Organometallics* **1995**, *14*, 4449.

[43] S. I. Kozhuskov, T. Haumann, R. Boese, A. de Meijere, *Angew. Chem.* **1993**, *105*, 426; *Angew. Chem., Int. Ed. Engl.* **1993**, *32*, 401.

[44] M. Fujita, Y. J. Kwon, O. Sasaki, K. Yamaguchi, K. Ogura, *J. Am. Chem. Soc.* **1995**, *117*, 7287.

[45] S. C. Zimmerman, F. Zeng, D. E. C. Reichert, S. V. Kolotuchin, 210th National ACS Meeting, Chicago, IL, Aug. 20–24, **1995**, Org. Chem. Abstr. No. 146.

[46] C. B. St. Pourcain, A. C. Griffin, *Macromolecules* **1995**, *28*, 4116.

[47] T. Lange, V. Gramlich, W. Amrein, F. Diederich, M. Gross, C. Boudon, J.-P. Gisselbrecht, *Angew. Chem., Int. Ed. Engl.* **1995**, *34*, 805; *Angew. Chem.* **1995**, *107*, 898.

[48] A. M. Boldi, J. Anthony, V. Gramlich, C. B. Knobler, C. Boudon, J.-P. Gisselbrecht, M. Gross, F. Diederich, *Helv. Chim. Acta*, **1995**, *78*, 779.

[49] L. Isaacs, P. Seiler, F. Diederich, *Angew. Chem., Int. Ed. Engl.* **1995**, *34*, 1466.

[50] C. Urban, E. F. McCord, O. W. Webster, L. Abrams, H. W. Long, H. Gaede, P. Tang, A. Pines, *Chem. Mater.* **1995**, *7*, 1325.

[51] G. Wagner, M. Schulze, M. Seufert, M. Schaub, T. Vahlenkamp, *Polym. Preprints* **1995**, *36*(1), 598.

[52] M. E. Wright, S. Mullick, J. M. Hoover, M. D. Seltzer, B. C. Lee, *Polym. Preprints* **1995**, *36*(1), 596.

[53] D. Poznoya, M. A. Buese, *Polym. Preprints* **1995**, *36*(1), 221.

[54] A.-D. Schlüter, W. Claussen, R. Freudenberger, *Macromol. Symp.* **1995**, *98*, 475. B. Karakaija, W. Claussen, A.-D. Schülter, *Polym. Preprints* **1996**, *37*(1), 216.

[55] C. J. Hawker, *Angew. Chem., Int. Ed. Engl.* **1995**, *34*, 1456. C. J. Hawker, J. M. J. Fréchet, R. B. Grubbs, J. Dao, *J. Am. Chem. Soc.* **1995**, *117*, 10763.

[56] R. W. Wagner, J. S. Lindsey, *J. Am. Chem. Soc.* **1994**, *116*, 9750.

[57] A. Harriman, F. Odobel, J.-P. Sauvage, *J. Am. Chem. Soc.* **1995**, *117*, 9461.

[58] H. Higuchi, K. Shimizu, J. Ojimak, *Tetrahedron Lett* **1995**, *36*, 5359.

[59] S. Anderson, H. L. Anderson, A. Bashall, M. McPartlin, J. K. M. Sanders, *Angew. Chem., Int. Ed. Engl.* **1995**, *34*, 1096; *Angew. Chem.* **1995**, *107*, 1196.

[60] M. J. Crossley, P. L. Burn, S. J. Langford, J. K. Prashar, *J. Chem. Soc., Chem. Commun.* **1995**, 1921.

[61] M. T. Reetz, S. A. Quaiser, *Angew. Chem., Int. Ed. Engl.* **1995**, *34*, 2240; *Angew. Chem.* **1995**, *107*, 2461.

[62] S. Subramanian, M. J. Zaworotko, *Angew. Chem., Int. Ed. Engl.* **1995**, *34*, 2127; *Angew. Chem.* **1995**, *107*, 2295.

[63] P. J. Belshaw, J. G. Schoepfer, K.-Q. Liu, K. L. Morrison, S. L. Schreiber, *Angew. Chem., Int. Ed. Engl.* **1995**, *34*, 2129; *Angew. Chem.* **1995**, *107*, 2313.

[64] a) D. M. Rudkevich, A. N. Shivanyuk, Z. Brzozka, W. Verboom, D. N. Reinhoudt, *Angew. Chem., Int. Ed. Engl.* **1995**, *34*, 2124; *Angew. Chem.* **1995**, *107*, 2300. b) B.-H. Huisman, D. M. Rudkevich, F. C. J. M. van Veggal, D. N. Reinhoudt, *J. Am. Chem. Soc.* **1996**, *118*, 3523.

[65] A. Müller, E. Krickemeyer, J. Meyer, H. Bögge, F. Peters, W. Plass, E. Diemann, S. Dillinger, F. Nonnenbruch, M. Randerath, C. Menke, *Angew. Chem., Int. Ed. Engl.* **1995**, *34*, 2122; *Angew. Chem.* **1995**, *107*, 2293.

[66] K. C. Nicolaou, X.-Y. Xiao, Z. Parandoosh, A. Senyei, M. P. Nova, *Angew. Chem., Int. Ed. Engl.* **1995**, *34*, 2289; *Angew. Chem.* **1995**, *107*, 2476.

[67] F. M. Menger, S. J. Lee, *Langmuir* **1995**, *11*, 3685.

[68] a) F. M. Menger, K. K. Catlin, *Angew. Chem., Int. Ed. Engl.95*, *34*, 2147; *Angew. Chem.* **1995**, *107*, 2330. b) N. J. Head, G. K. S. Prakash, A. Bashir-Hashemi, G. A. Olah, *J. Am. Chem. Soc.* **1995**, *117*, 12005.

[69] a) D. L. Pearson, J. M. Tour, *Polym. Preprints* **1995**, *36*(2), 344. L. Jones, II, D. L. Pearson, J. S. Schumm, J. M. Tour, *Pure & Appl. Chem.* **1996**, *68*, 145. b) X. Wang, M. Sabat, R. N. Grimes, *J. Am. Chem. Soc.* **1995**, *117*, 12227.

[70] J. Zhang, Z. S. Moore, Z. Xu, R. A. Aguirra, *J. Am. Chem. Soc.* **1992**, *114*, 2273.

[71] T. C. Bedard, J. S. Moore, *J. Am. Chem. Soc.* **1995**, *117*, 10662.

[72] a) S. Breidenbach, S. Ohren, M. Nieger, F. Vögtle, *J. Chem. Soc., Chem. Commun.* **1995**, 1237. b) D. N. Butler, P. M. Tepperman, R. A. Gaw, R. N. Warrener, *Tetrahedron Lett.* **1996**, *37*, 2825.

[73] a) R. P. Hsung, C. E. D. Chidsey, L. R. Sita, *Organometallics* **1995**, *14*, 4808. b) X. Wang, M. Sabat, R. N. Grimes, *J. Am. Chem. Soc.* **1995**, *117*, 12227.

[74] H. L. Frisch, E. Wasserman, *J. Am. Chem. Soc.* **1991**, *83*, 3789.

[75] E. Wasserman, *J. Am. Chem. Soc.* **1960**, *82*, 4434.

[76] D. A. Tomalia, H. Baker, J. Dewald, M. Hall, G. Kallas, S. Martin, J. Roeck, J. Ryder, P. Smith, *Polymer J.* **1985**, *17*, 117.

[77] D. A. Tomalia NATO Workshop, Estes Park, Colorado, U. S. A., **1995**.

[78] D. A. Tomalia, J. R. Dewald, U. S. Pat. 4,568,737 (Feb. 4, **1988**).

[79] D. A. Tomalia, U. S. Pat. 4,737,550 (Apr. 12, **1988**).

[80] D. A. Tomalia, L. R. Wilson, *Eur. Pat. Appl.* (87107833.3, **1987**). US Pat. 4,713,975 (**1987**).

[81] S. Watanabe, S. Regen, *J. Am. Chem. Soc.* **1994**, *116*, 8855.

[82] G. R. Newkome, R. Güther, C. N. Moorefield, F. Cardullo, L. Echegotyen, E. Pérez-Corero, H. Luftmann, *Angew. Chem.* **1995**, *107*, 2159; *Angew. Chem. Int. Ed. Engl.* **1995**, *34*, 2023.

[83] E. C. Constable, P. Harverson, *J. Chem. Soc., Chem. Commun.* **1996**, 33.

[84] M. J. Zaworotko, *Chem. Soc. Rev.* **1994**, *23*, 284.

[85] a) C. N. Moorefield, G. R. Newkome, **1996**, unpublished results. b) W. T. S. Huck, F. C. J. M. van Veggel, B. L. Kropman, D. H. A. Blank, E. G. Keim, M. M. A. Smithers, D. N. Reinhoudt, *J. Am. Chem. Soc.* **1996**, *117*, 8293. c) C. D. Eisenbach, W. Degelmann, A. Göldel, J. Heinlein, M. Terskan-Reinhold, U. S. Schubert, *Macromol. Symp.* **1995**, *98*, 565.

[86] G. R. Newkome, C. N. Moorefield, R. Güther, G. R. Baker, *Polym. Preprints* **1995**, *36*(1), 609.

[87] G. R. Newkome, G. R. Baker, M. J. Saunders, P. S. Russo, V. K. Gupta, Z.-q. Yao, J. E. Miller, K. Bouillon, *J. Chem. Soc., Chem. Commun.* **1986**, 752.

[88] G. R. Newkome, G. R. Baker, S. Arai, M. J. Saunders, P. S. Russo, K. J. Theriot, C. N. Moorefield, L. E. Rogers, J. E. Miller, T. R. Lieux, M. E. Murray, B. Phillips, L. Pascal, *J. Am. Chem. Soc.* **1990**, *112*, 8458.

[89] G. R. Newkome, Z.-q. Yao, G. R. Baker, V. K. Gupta, P. S. Russo, M. J. Saunders, *J. Am. Chem. Soc.* **1986**, *108*, 849.

[90] S. C. Zimmermann, F. Zeng, D. E. C. Reichert, S. V. Kolotuchin, *Science* **1996**, *271*, 1095.

[91] a) C. J. Hawker, J. M. J. Fréchet, *J. Chem. Soc., Chem. Commun.* **1990**, 1010. b) D. B. Amabilino, P. R. Ashton, M. Belohrasky, F. M. Raymo, J. F. Stoddart, *J. Chem. Soc., Chem. Commun.* **1995**, 751. c) H. Kihara, T. Kato, T. Uryu, J. M. J. Fréchet, *Chem. Mater.* **1996**, *8*, 961.

[92] J. M. J. Fréchet, C. J. Hawker, K. L. Wooley, *J. Macromol. Sci. − Pure Appl. Chem.* **1994**, *A31*(11), 1627.

[93] C. J. Hawker, J. M. J. Fréchet, *Polymer,* **1992**, *33*, 1507.

[94] I. Gitsov, K. L. Wooley, C. J. Hawker, P. T. Ivanova, J. M. J. Fréchet, *Macromolecules* **1993**, *26*, 5621.

10 Utilitarian Aspects

Applications* for dendrimers were hypothesized in the early years. Recently dendritic exploitation is indicative of future discoveries and uses. The following references and patents have been included as an initial guide to researchers interested in work related to this polymeric area. Utilitarian aspects are thus broken down into broad general topics; these are not intended to be inclusive but rather illustrative in nature.

10.1 General Applications

'Arboreal polymers. Report, DRET-87/1410DS/DR [Order No. PB89-163752' *Gov. Rep. Announce, Index* **1988**, *89*(14), 938,733] [#543].*

'Dendrimers: dream molecules approach real applications.' [R. F. Service, *Science* **1995**, *267*, 458–459] [#1717].

'Synthesis of polymers for new materials.' [G. Smets, *Chem. Mag.* **1989** (Sept.), 481–483; 485] [#552].

'Application areas for dendritic macromolecules.' [Anonymous, *Res. Discl.* **1994**, *365*, 492] [#1818].

'Dendritic polymers: from aesthetic macromolecules to commercially interesting materials.' [B. I. Voit, *Acta Polym.* **1995**, *46*(2), 87–99] [#1728].

'Polyester-type dendritic macromolecules and their manufacture and use.' [A Hult, E. Malmström, M. Johansson, K. Sörensen, Swed. Pat. 468,771 **1993**; *Chem. Abstr.* **119**: 181 545] [#418].

'Functionalized comb-like polymers: synthesis, modification and application.' [H. Ritter, *Angew. Makromol. Chem.* **1994**, *223*, 165–175] [#1777].

10.2 Chromatography

10.2.1 Ion Exchange Processes

'Ion exchange/chelation resins containing dense star polymers having ion exchange or chelate capabilities.' [G. R. Killat, D. A. Tomalia, U. S. Pat. 4,871,779 **1989**] [#541].

'Dendrimeric ion-exchange materials.' [A. Cherestes, R. Engel, *Polymer* **1994**, *35*(15), 3343–3344] [#1790].

'Ion-exchange electrokinetic capillay chromatography with Starburst (PAMAM) dendrimers: a route towards high-performance electrokinetic capillary chromatography.' [M. Castagnola, L. Cassiano, A. Lupi, I. Messana, M. Patamia, R. Rabino, D. V. Rossetti, B. Giardina, *J. Chromatogr., A* **1995**, *694*(2), 463–469] [#1820].

10.2.2 Electrokinetic Capillary Chromatography

'Starburst dendrimers as carriers in electrokinetic chromatography.' [N. Tanaka, T. Tanigawa, K. Hosoya, K. Kimata, K. Araki, S. Terabe, *Chem. Lett.* **1992**, 959–962] [#617].

---

* [#] denotes reference to a dendrimer and supportive materials data base: G. R. Baker and G. R. Newkome, "Cascade Macromolecules: A Compendium of Dendritic Macromolecules, Building Blocks and Supportive Materials", **1996**; see: http://dendrimers.US.Fl.com, http://www.dendrimers.com.

'Electrokinetic Chromatography.' [K. Hibi, M. Senda, N. Tanaka, Jpn. Kokai Tokkyo Koho 05322849, **1993**] [#343].

'Dendrimer electrokinetic capillary chromatography: unimolecular micellar behavior of carboxylic acid terminated cascade macromolecules.' [S. A. Kuzadal, C. A. Monnig, G. R. Newkome, C. N Moorefield, *J. Chem. Soc., Chem. Commun.* **1994**, 2139–2140] [#638].

'Ion-exchange electrokinetic capillary chromatography with Starburst (PAMAM) dendrimers: a route towards high-performance electrokinetic capillary chromatography.' [M. Castagnola, L. Cassiano, A. Lupi, I. Messana, M. Patamia, R. Rabino, D. V. Rossetti, B. Giardina, *J. Chromatogr., A* **1995**, *692*(2), 463–469] [#1820].

'Dendrimers as pseudo-stationary phases in electrokinetic chromatography.' [P. G. H. M. Muijselaar, H. A. Claessens, C. A. Cramers, J. F. G. A. Jansen, E. W. Meijer, E. M. M. de Brabander-van den Berg, S. van der Wal, *J. High Resol. Chromatogr.* **1995**, *18*(2), 121–123] [#1739].

10.2.3   Size Exclusion Chromatography

'Carboxylated starburst dendrimers as calibration standards for aqueous size exclusion chromatography.' [P. L. Dubin, S. L. Edwards, J. I. Kaplan, M. S. Mehta, D. Tomalia, J. Xia, *Anal. Chem.* **1992**, *64*(20), 2344–2347] [#198].

'Quantitation of non-ideal behavior in protein size-exclusion chromatography.' [P. L. Dubin, S. L. Edwards, M. S. Mehta, D. Tomalia, *J. Chromatogr.* **1993**, *635*(1), 51–60] [#432].

10.2.4   Gas Separation Membranes

'Asymmetric poly(vinyltrimethylsilane) gas separation membrane.' [L. F. Borisova, B. I. Nakhmanovich, A. V. Pakhomov, N. V. Platonova, G. I. Fajdel, V. K. Shatalov, USSR Pat. 1,808,362, **1993**; *Chem. Abstr. 122*: 241 808] [#412].

10.3   Electronic Properties

10.3.1   Semiconduction

'"Starburst" polyacrylamines and their semiconducting complexes as potentially electroactive materials.' [H. K. Hall, Jr., D. W. Polis, *Polym. Bull.* **1987**, *17*(5), 409–416] [#604].

10.3.2   Photochemical Molecular Devices

'Towards a supramolecular photochemistry: assembly of molecular components to obtain photochemical molecular devices.' [V. Balzani, L. Moggi, F. Scandola, in *Supramolecular Photochemistry;* V. Balzani; Reidel: Dordrecht, The Netherlands, **1987**] [#819].

'Luminescent and redox-reactive building blocks for the design of photochemical molecular devices: mono-, di-,tri-, and tetranuclear ruthenium(II) polypyridine complexes.' [G. Denti, S. Campagna, L. Sabatino, S. Serroni, M. Ciano, V. Balzani, *Inorg. Chem.* **1990**, *29*(23), 4750–4758] [#830].

'Towards an artificial photosynthesis. Di-, tri-, tetra,- and heptanuclear luminescent and redox-active metal complexes.' [G. Denti, S. Campagna, L. Sabatino, S. Serroni, M. Ciano, V. Balzani, in *Photochem. Convers. Storage Sol. Energy;* E. Pelizzetti; M. Schiavello; Kluwer, Dordrecht, The Netherlands, **1991**] [#762].

'Photoelectron transfer between molecules adsorbed in restricted spaces.' [N. J. Turro, J. K. Barton, D. Tomalia, In *Photochem. Convers. Storage Sol. Energy*, E. Pelizzetti; M. Schiavello; Kluwer, Dordrecht, The Netherlands, **1991**] [#469].

'Photo-induced electron transfer at polyelectrolyt-water interface.' [G. Caminati, D. A. Tomalia, N. J. Turro, *Prog. Colloid Polym. Sci.* **1991**, *84*(5), 219–222] [#480].

'Decanuclear homo- and heterometallic polypyridine complexes: syntheses, absorption spectra, luminescence, electrochemical oxidation, and intercomponent energy transfer.' [G. Denti, S. Campagna, S. Serroni, M. Ciano, V. Balzani, *J. Am. Chem. Soc.* **1992**, *114*(8), 2944–2950] [#109].

'Organometallic molecular trees as multielectron and multiproton reservoirs: CpFe$^+$-induced non-allylation of mesitylene and phase-transfer-catalyzed synthesis of a redox-active nonairon complex.' [F. Moulines, L. Djakovitch, R. Boese, B. Gloaguen, W. Theil, J. L. Fillaut, M. H. Delville, D. Astruc, *Angew. Chem., Int. Ed. Engl.* **1993**, *32*(7), 1075–1077] [#126].

'Light-harvesting polymer systems.' [M. A. Fox, W. E. Jones, Jr., D. M. Watkins, *Chem. Eng. News* **1993**, *71*, Mar. 15, 38–48] [#1287].

'Supramolecular Ru and/or Os complexes of tris(bipyridine) bridging ligands. Syntheses, absorption spectra, luminescence properties, electrochemical behavior, intercomponent energy, and electron transfer.' [P. Belser, A. von Zelewsky, M. Frank, C. Seel, F. Vögtle, L. De Cola, F. Barigelletti, V. Balzani, *J. Am. Chem. Soc.* **1993**, *115*(10), 4076–4086] [#23].

'Design and synthesis of a convergent and directional molecular antenna.' [Z. Xu, J. S. Moore, *Acta Polym.* **1994**, *45*(2), 83–87] [#223].

'New materials based on highly-functionalised tetrathiafulvalene derivatives.' [M. R. Bryce, A. S. Batsanov, W. Devonport, J. N. Heaton, J. A. K. Howard, G. J. Marshallsay, A. J. Moore, P. J. Skabara, S. Wegener, in *Molecular Engineering for Advanced Materials*, (Eds.: J. Becher, K. Schaumburg), Kluwer, The Netherlands, **1995**] [#877].

10.4 Physical Properties

10.4.1 Solubilization/Drug Delivery

'Unimolecular micelles and method of making the same.' [G. R. Newkome, C. N. Moorefield, U.S. Pat. 5,154,853 **1992**] [#778].

'Unimolecular micelles and globular amphiphiles: dendritic macromolecules as novel recyclable solubilization agents.' [C. J. Hawker, K. L. Wooley, J. M. J. Fréchet, *J. Chem. Soc., Perkin Trans. 1* **1993**, 1287–1297] [#129].

'Starburst dendrimers as solubilizers for hardly soluble drugs.' [H. Ibuki, M. Fujiwara, O. Kono, H. Nakajima, Jpn. Kokai Tokkyo Koho 06228071, **1994**; *Chem. Abstr. 122:* 64314] [#1832].

'Synthesis, metal-binding properties and polypeptide solubilization of 'Crowned' arborols.' [T. Nagasaki, O. Kimura, M. Ukon, S. Arimori, I. Hamachi, S. Shinkai, *J. Chem. Soc., Perkin Trans. 1* **1994**, 75–81] [#315].

'Cascade polymers as drug carriers.' [K. Akioyshi, *Kagaku (Kyoto)* **1994**, *49*(6), 422] [#385].

'Prospects of using star polymers and dendrimers in drug delivery and other pharmaceutical application.' [N. A. Peppas, T. Nagai, M. Miyajima, *Pharm. Techn. Jpn.* **1994**, *10*(6), 611–617] [#1862].

'Encapsulation of guest molecules into a dendritic box.' [J. F. G. A. Jansen, E. M. M. de Brabander-van den Berg, E. W. Meijer, *Science* **1994**, *266*, 1226–1229] [#1559].

'The dendritic box: shape-selective liberation of encapsulated guests.' [J. F. G. A. Jansen, E. W. Meijer, E. M. M. de Brabander-van den Berg, *J. Am. Chem. Soc.* **1995**, *117*(15), 4417–4418] [#1703].

10.4.2   Size Calibration/Pore Creation

'Dense star polymers having two dimensional molecular diameter.' [D. A: Tomalia, J. R. Dewald, U.S. Pat. 4,587,329 **1986**] [#164].

'Dense star polymers for calibrating/characterizing sub-micron apertures.' [D. A. Tomalia, L. R. Wilson, Eur. Pat. 247,629 **1987**; *Chem. Abstr. 109:*145845] [#167].

'Carboxylated starburst dendrimers as calibration standards for aqueous size exclusion chromatography.' [P. L. Dubin, S. L. Edwards, J. I. Kaplan, M. S. Mehta, D. Tomalia, J. Xia, *Anal. Chem.* **1992**, *64*(20), 2344–2347] [#198].

'Manufacture of porous silica having controlled uniform pore size.' [T. Saegusa, Y. Nakajo, S. Kure, Jpn. Kokai Tokkyo Koho 04285081, **1992**; *Chem. Abstr. 118:*127748] [#443].

'Cascade polymers: pH dependence of hydrodynamic radii of acid terminated dendrimers.' [G. R. Newkome, J. K. Young, G. R. Baker, R. L. Potter, L. Audoly, D. Cooper, C. D. Weis, K. F. Morris, C. S. Johnson, Jr., *Macromolecules* **1993**, *26*(9), 2394–2396] [#217].

'Quantitation of non-ideal behavior in protein size-exclusion chromatography.' [P. L. Dubin, S. L. Edwards, M. S. Mehta, D. Tomalia, *J. Chromatogr.* **1993**, *635*(1), 51–60] [#432].

''Smart' Cascade Polymers. Modular Syntheses of Four-directional Dendritic Macromoleculers with Acidic, Neutral, or Basic Terminal Groups and the Effect of pH Changes on their Hydrodynamic Radii.' [J. K. Young, G. R. Baker, G. R. Newkome, K. F. Morris, C. S. Johnson, Jr., *Macromolecules* **1994**, *27*, 3464–3471] [#233].

'Control of pore size of porous silica by means of pyrolysis of an organic−inorganic polymer hybrid.' [Y. Chujo, H. Matsuki, S. Kure, T. Saegusa, T. Yazawa, *J. Chem. Soc., Chem. Commun.* **1994**, 635–636] [#1761].

'Synthetic uniform polymers and their use in polymer science.' [K. Hatada, K. Ute, N. Miyatake, *Prog. Polym. Sci.* **1994**, *19*(6), 1067–1082] [#1829].

10.4.3   Antioxidants

'Antioxidant dispersant polymer dendrimer.' [C. A. Migdal, U.S. Pat. 4,938,885 **1990**] [#536].

'Preparation of phenolic tertiary amide antioxidants and compositions containing them.' [V. J. Gatto, PCT Int. Appl. 9216495 **1992**] [#724].

10.4.4   Inks and Toners

'Inks with dendrimer colorants.' [F. M. Winnik, A. R. Davidson, M. P. Breton, U.S. Pat. 5,098,475, **1992**] [#459].

'Dendrimer-containing yet-printing inks.' [F. M. Winnik, A. R. Davidson, M. P. Breton, U.S. Pat. 5,120,361, **1992**] [#461].

'Ink compositions.' [K. B. Gundlach, G. A. R. Nobes, M. P. Breton, R. L. Colt, U.S. Pat. 5,254,159 **1993**] [#334].

'Porphyrin chromophore and dendrimer ink compositions.' [F. M. Winnik, A. R. Davidson, M. P. Breton, U.S. Pat. 5,256,193 **1993**] [#356].

'Toner compositions with dendrimer charge enhancing additives.' [F. M. Winnik, J. M. Duff, G. G. Sacripante, A. R. Davidson, U.S. Pat. 5,256,516 **1993**] [#362].

'Ink compositions with dendrimer grafts.' [F. M. Winnik, M. P. Breton, U.S. Pat. 5,266,106 **1993**] [#350].

10.4.5　Liquid Crystals

'A non-aqueous lyotropic liquid crystal with a Starburst dendrimer as a solvent.' [S. E. Friberg, M. Podzimek, D. A. Tomalia, D. M. Hedstrand, *Mol. Cryst. Liq. Cryst.* **1988**, *164*, 157–165] [#796].

'Synthesis and characterization of a thermotropic nematic liquid crystalline dendrimeric polymer.' [V. Percec, M. Kawasumi, *Macromolecules* **1992**, *25*(15), 3843–3850] [#466].

'Synthesis and characterization of a thermotropic liquid crystalline dendrimer.' [V. Percec, M. Kawasumi, *Polym. Preprints* **1992**, *33*(1), 221–222] [#194].

'Synthesis and characterization of thermotropic liquid crystalline dendrimers.' [V. Percec, M. Kawasumi, *Polym. Preprints* **1993**, *34*(1), 158–15] [#119].

'Toward "Willow-like" thermotropic dendrimers.' [V. Percec, P. Chu, M. Kawasumi, *Macromolecules* **1994**, *27*(16), 4441–4453] [#316].

10.4.6　Chemical Sensitivity

"Interaction between Organized, Surface-confined Monolayers and Vapor-Phase Probe Molecules. 10. Preparation and Properties of Chemically Sensitive Dendrimer Surfaces" [M. Wells, R. M. Crooks, *J. Am. Chem. Soc.* **1996**, *118*, 3988].

10.5　Catalysis

'Synthesis and use as a catalyst support of porous polystyrene with bis(phosphonic acid)-functionalized surfaces.' [M. J. Sundell, E. O. Pajunen, O. E. O. Hormi, J. H. Näsman, *Chem. Mater.* **1993**, *5*(3), 372–376] [#21].

'Aminolysis of phenyl esters by microgel and dendrimer molecules possessing primary amines.' [D. J. Evans, A. Kanagasooriam, A. Williams, R. J. Pryce, *J. Mol. Catal.* **1993**, *85*(1), 21–32] [#359].

'Electrochemical reduction of CO_2 catalyzed by small organophosphine dendrimers containing palladium.' [A. Miedaner, C. J. Curtis, R. M. Barkley, D. L. DuBois, *Inorg. Chem.* **1994**, *33*(24), 5482–5490] [#1543].

'Reactivity of organic promoted by a quaternary ammonium ion dendrimer.' [J. J. Lee, W. T. Ford, J. A. Moore, Y. Li, *Macromolecules* **1994**, *27*(16), 4632–4634] [#236].

'Catalysis, what promise for dendrimers?' [D. A. Tomalia, P. R. Dvornic, *Nature* **1994**, *372*, 617–618] [#1768].

'Homogeneous catalysts based on silane dendrimers functionalized with arylnickel(II) complexes.' [J. W. J. Knapen, A. W. van der Made, J. C. de Wilde, P. W. N. M. van Leeuwen, P. Wijkens, D. M. Grove, G. van Koten, *Nature* **1994**, *372*, 659–663] [#1769].

'Dendrimers engineered to provide new catalysts.' [J. Haggin, *Chem. Eng. News* **1995**, 2/6/95, 26–27] [#1694].

'Dendrimers as macroinitiators for anionic ring-opening polymerization. Polmyerization of ε-caprolactone.' [I. Gitsov, P. T. Ivanova, J. M. J. Fréchet, *Macromol. Rapid Commun.* **1994**, *15*(5), 387–393] [#391].

10.6　Biochemical/Pharmaceutical

10.6.1　Sucrose Mimetic

'New non-ionic polyol derivatives with sucrose mimetic properties.' [G. E. DuBois, B. Zhi, G. M. Roy, S. Y. Stevens, M. Yalpani, *J. Chem. Soc., Chem. Commun.* **1992**, 1604–1605] [#1277].

10.6.2   MAP/Antigen

'Synthetic peptide vaccine design: synthesis and properties of a high-density multiple antigenic peptide system.' [J. P. Tam, *Proc. Natl. Acad. Sci. USA* **1988**, *85*(9), 5409–5413] [#626].

'Vaccine engineering: enhancement of immunogenicity of synthetic peptide vaccines related to hepatitis in chemically defined models consisting of T- and B-cell epitopes.' [J. P. Tam, Y. A. Lu, *Proc. Natl. Acad. Sci. USA* **1989**, *86*(12), 9084–9088] [#627].

'Long-term high-titer neutralizing activity induced by octameric synthetic HIV-1 antigen.' [C. Y. Wang, D. J. Looney, M. L. Li, A. M. Walfield, J. Ye, B. Hosein, J. P. Tam, F. Wong-Staal, *Science* **1991**, *254*, 285–288] [#628].

'Macromolecular assemblage in the design of a synthetic AIDS vaccine.' [J. P. Defoort, B. Nardelli, W. Huang, D. D. Ho, J. P. Tam, *Proc. Natl. Acad. Sci. USA* **1992**, *89*(5), 3879–3883] [#889].

'Preparation of multiple antigen peptide systems for vaccines and diagnostics.' [J. P. Tam, U.S. Pat. 5,229,490 **1993**] [#358].

'Synthesis and antigenic properties of sialic acid based dendrimers.' [R. Roy, D. Zanini, S. J. Meunier, A. Romanowska, in *Synthetic Oligosaccharides*, (Ed.: P. Kovác), American Chemical Society, Washington, DC, **1994**] [#1837].

10.6.3   Antibody Conjugates/Labeling

'Using starburst dendrimers as linker molecules to radiolabel antibodies.' [J. C. Roberts, Y. E. Adams, D. Tomalia, J. A. Mercer-Smith, D. K. Lavallee, *Bioconj. Chem.* **1990**, *1*(5), 305–308] [#179].

'Delivery of boron-10 for neutron capture therapy by means of monoclonal antibody – starburst dendrimer immunoconjugates.' [R. F. Barth, A. H. Soloway, D. M. Adams, F. Alam, In *Prog. Neutron Capture Ther. Cancer*, (Eds.: B. J. Allen, D. E. Moore, B. V. Harrington, Plenum, New York, **1992**] [#1871].

'In vivo distribution of boronated monoclonal antibodies and starburst dendrimers.' [R. F. Barth, D. M. Adams, A. H. Soloway, M. V. Darby, In *Adv. Neutron Capture Ther.*, (Eds.: A. H. Soloway, R. F. Barth, D. E. Carpenter), Plenum, New York, **1993**] [#1867].

'Epidermal growth factor as a potential targeting agent for delivery of 10B to malignant gliomas.' [J. Capala, R. F. Barth, D. M. Adams, A. H. Soloway, J. Carlsson, in *Adv. Neutron Capture Ther.*, (Eds.: A. H. Soloway, R. F. Barth, D. E. Carpenter), Plenum, New York, **1993**] [#1870].

'Starburst polymer conjugates with agrochemicals.' [D. A. Tomalia, L. R. Wilson, Can. Pat. 1,316,364 **1993**] [#373].

'Starburst dendrimers: characterization, derivatization and bioconjugation.' [P. Singh, S. Lin, F. Moll, III, K. Yu, C. Ferzli, R. Saul, S. Diamond, *Polym. Mater. Sci. Eng.* **1993**, *70*, 237–238] [#1853].

'Boronated starburst dendrimer-monoclonal antibody immunoconjugates: Evaluation as a potential delivery system for neutron capture therapy.' [R. F. Barth, D. M. Adams, A. H. Soloway, F. Alam, M. V. Darby, *Bioconj. Chem.* **1994**, *5*(1), 58–66] [#364].

'Dynamic contrast-enhanced MR imaging of the upper abdomen: enhancement properties of gadobutrol, gadolinium – DTPA – polylysine, and gadolinium – DTPA – cascade-polymer.' [G. Adam, J. Neuerburg, E. Spuentrup, A. Muehler, K. Scherer, R. W. Guenther, *Magn. Reson. Med.* **1994**, *32*(5), 622–628] [#1846].

'Starburst conjugates.' [D. A. Toamalia, D. A. Kaplan, W. J. Kruper, Jr., R. C. Cheng, I. A. Tomlinson, M. J. Fazio, D. M. Hedstrand, L. R. Wilson, U.S. Pat. 5,338,532 **1994**] [#1384].

10.6.4   Nucleic Acids

'Nucleic acid dendrimers: novel biopolymer structures.' [R. H. E. Hudson, M. J. Damha, *J. Am. Chem. Soc.* **1993**, *115*(6), 2119–2124] [#216].

10.6.5   Gene Therapy

'Polyamidoamine cascade polymers mediate efficient transfection of cells in culture.' [J. Haensler, F. C. Szoka, Jr., *Bioconj. Chem.* **1993**, *4*(5), 372–379] [#420].

'Self-assembling polynucleotide delivery system for genetic transformation and gene therapy.' [F. C. Szoka, Jr., J. Haensler, PCT Int. Appl. 9319768 **1993**; *Chem. Abstr. 120:* 184 644] [#399].

10.6.6   Pharmaceutical Complexes/Boron Neutron Capture Therapy

'Complexes of complexing agents bonded to cascade polymers for use in pharmaceuticals.' [J. Platzek, H. Schmitt-Willich, H. Gries, G. Schuhmann-Giampieri, H. Vogler, H. J. Weinmann, H. Bauer, Ger. Pat. Offen. 3,938,992 **1991**; *Chem. Abstr. 115:* 280 870] [#495].

'Novel tetrahydroxamate chelators for Actinide complexation: synthesis and binding studies.' [A. S. Gopalan, V. J. Huber, O. Zincircioglu, P. H. Smith, *J. Chem. Soc., Chem. Commun.* **1992**, 1266–1268] [#13].

'Solid-phase syntheses of dendritic sialoside inhibitors of Influenza A virus haemagglutinin.' [R. Roy, D. Zanini, S. J. Meunier, A. Romanowska, *J. Chem. Soc., Chem. Commun.* **1993**, 1869–1872] [#215].

'Dendrimeric polychelates as imaging agents.' [A. D. Watson, PCT Int. Appl. 9306868 **1993**; *Chem. Abstr. 119:* 220702] [#414].

'Chemistry within a unimolecular micelle precursor: boron superclusters by site- and depth-specific transformations of dendrimers.' [G. R. Newkome, C. N. Moorefield, J.M. Keith, G. R. Baker, G. H. Escamilla, *Angew. Chem., Int. Ed. Engl.* **1994**, *33*(6), 666–668] [#149].

'Reagents for detecting and assaying nucleic acid sequences.' [T. W. Nilsen, W. Prensky, U.S. Pat. 5,175,270 **1992**] [#411].

'Starburst™ dendrimers: enhanced performance and flexibility for immunoassays.' [P. Singh, F. Moll, III, S. H. Lin, C. Ferzli, K. S. Yu, R. K. Koski, R. G. Saul, P. Cronin, *Clin. Chem.* **1994**, *40*(9), 1845–1849] [#905].

'Dendrimer-based metal chelates: a new class of magnetic resonance imaging contrast agents.' [E. C. Wiener, M. W. Brechbiel, H. Brothers, R. L. Magin, O. A. Gansow, D. A. Tomalia, P. C. Lauterbur, *Magn. Reson. Med.* **1994**, *31*(1), 1–8] [#394].

'Dynamic contrast-enhanced MR imaging of the upper abdomen: enhancement properties of gadobutrol, gadolinium−DTPA−polylysine, and gadolinium−DTPA−cascade-polymer.' [G. Adam, J. Neuerburg, E. Spuentrup, A. Muehler, K. Scherer, R. W. Guenther, *Magn. Reson. Med* **1994**, *32*(5), 622–628] [#1846].

'Immunobilization of specific binding assay reagents.' [S. H. Lin, K. S. Yu, P. Singh, S. E. Diamond, PCT Int. Appl. 9419693 **1994**; *Chem. Abstr. 121:* 225 877] [#1859].

'Polychelants containing macrocyclic chelant moieties.' [P. F. Sieving, A. D. Watson, S. C. Quay, S. M. Rocklage, U.S. Pat. 5,364,613 **1994**] [#1831].

'Polyols of a cascade type as a water-solubilizing element of carborane derivatives for boron neutron capture therapy.' [H. Nemoto, J. G. Wilson, H. Nakamura, Y. Yamamoto, *J. Org. Chem.* **1992**, *57*, 435–435] [#811].

'Delivery of boron-10 for neutron capture therapy by means of monoclonal antibody−starburst dendrimer immunoconjugates.' [R. F. Barth, A. H. Soloway, D. M. Adams, F. Alam, in *Prog. Neutron Capture Ther. Cancer,* B. J. Allen, D. E. Moore, B. V. Harrington, Plenum, New York, **1992**] [#1871].

'In vivo distribution of boronated monoclonal antibodies and starburst dendrimers.' [R. F. Barth, D. M. Adams, A. H. Soloway, M. V. Darby, in *Adv. Neutron Capture Ther.,* (Eds.: A. H. Soloway, R. F. Barth, D. E. Carpenter), Plenum, New York, **1993**] [#1867].

'Epidermal growth factor as a potential targeting agent for delivery of 10B to malignant gliomas.' [J. Capala, R. F. Barth, D. M. Adams, A. H. Soloway, J. Carlsson, In *Adv. Neutron Capture Ther.,* (Eds.: A. H. Soloway, R. F. Barth, D. E. Carpenter), Plenum, New York, **1993**] [#1870].

'Dendrimers and star polymers for pharamceutical and medical applications.' [N. A. Peppas, A. B. Argade, In *Proc. Int. Symp. Controlled Release Bioact. Mater.*, (Eds.: T. J. Roseman N. A. Peppas; H. L. Gabelnick), Controlled Release Society: Deerfield, Ill., **1993**] [#413].

'A combined adjuvant and carrier system for enhancing synthetic peptides immunogenicity utilizing lipidic amino acids.' [I. Toth, M. Danton, N. Flinn, W. A. Gibbons, *Tetrahedron Lett.* **1993**, *34*(24), 3925–3928] [#135].

'Metallo- and Metalloido-Micellane™ derivatives: incorporation of metals and non-metals within unimolecular superstructures.' [G. R. Newkome, C. N. Moorefield, *Makromol. Chem.* **1994**, *77*, 63] [#631].

'Metal-chelate-dendrimer-antibody constructs for use in radioimmunotherapy and imaging.' [C. Wu, M. W. Brechbiel, R. W. Kozak, O. A. Gansow, *Bioorg. Med. Chem. Lett.* **1994**, *4*(3), 449–454] [#1763].

'Iron transport-mediated drug delivery: synthesis and biological evaluation of cyanuric acid-based siderophore analogs and β-lactam conjugates.' [M. Ghosh, M. J. Miller, *J. Org. Chem.* **1994**, *59*(5), 1020–1026] [#46].

'Boron neutron capture therapy of primary and metastatic brain tumors.' [R. F. Barth, A. H. Soloway, *Mol. Chem. Neuropathol.* **1994**, *21*(2–3), 139–154] [#1863].

'Boronated starburst dendrimer-monoclonal antibody immunoconjugates: Evaluation as a potential delivery system for neutron capture therapy.' [R. F. Barth, D. M. Adams, A. H. Soloway, F. Alam, M. V. Darby, *Bioconj. Chem.* **1994**, *5*(1), 58–66] [#364].

'Chemistry within a unimolecular micelle precursor: boron superclusters by site- and depth-specific transformations of dendrimers.' [G. R. Newkome, C. N. Moorefield, J. M. Keith, G. R. Baker, G. H. Escamilla, *Angew. Chem., Int. Ed. Engl.* **1994**, *33*(6), 666–668] [#149].

'Perfluoro-1H,1H-neopentyl-containing aryl contrast agents and method to use same for imaging.' [R. Lohrmann, A. Krishnan, PCT Int. Appl. 9422368 **1994**; *Chem. Abstr. 122:* 50279] [#1839].

'Detection and therapy of lesions with biotin/avidin polymer conjugates for amplification of targeting.' [G. L. Griffiths, PCT Int. Appl. 9423759 **1994**; *Chem. Abstr. 122:* 38823] [#1841].

'Starburst dendrimers as solubilizers for hardly soluble drugs.' [H. Ibuki, M. Fujiwara, O. Kono, H. Nakajima, Jpn. Kokai Tokkyo Koho 06228071, **1994**; Chem. Abstr. 122: 64314] [#1832].

10.6.7 Specific Receptors

'Preparation of cluster glycosides of *N*-acetylgalactosamine that have subnanomolar binding constants towards the mammalian hepatic Gal/GalNAc-specific receptor.' [R. T. Lee, Y. C. Lee, *Glycoconjugate J.* **1987**, *4*, 317–328] [#38].

10.7 Synthetic Aspects

10.7.1 Building Block Preparation (Patents)

'Multifunctional synthons as used in the preparation of cascade polymers or unimolecular micelles.' [G. R. Newkome, C. N. Moorefield, U.S. Pat. 5,136,096 **1992**; [#455]; U.S. Pat. 5,206,410 **1993** [#2261]; U.S. Pat. 5,210,309 **1993** [#2262].

'Tertiary-butyl esters as monomers for cascade polymers.' [G. R. Newkome, C. N. Moorefield, R. K. Behera, PCT Int. Appl. 9321144 **1993**; *Chem. Abstr. 120:* 299577] [#392].

10.7.2 Dendrimer Preparation (Patents)

'Macromolecular highly branched α,ω-diaminocarboxylic acids.' [R. G. Denkewalter, J. F. Kolc, W. J. Lukasavage, U.S. Pat. 4,410,688 **1979**] [#1597].

'Macromolecular highly branched homogeneous compound based on lysine units.' [R. G. Denkewalter, J. Kolc, W. J. Lukasavage, U.S. Pat. 4,289,872 **1981**] [#317].

'Preparation of lysine based macromolecular highly branched homogeneous compound.' [R. G. Denkewalter, J. Kolc, W. J. Lukasavage, U.S. Pat. 4,360,646 **1982**] [#1527].

'Macromolecular highly branched homogeneous compound.' [R. J. Denkewalter, J. Kolc, W. J. Lukasavage, U.S. Pat. 4,410,688 **1983**] [#318].

'Dense star polymers and a process for producing dense star polymers.' [D. A. Tomalia, J. R. Dewald, PCT Int. Appl. 8402705 **1984**] [#589].

'Branched polyamidoamines.' [D. A. Tomalia, L. R. Wilson, J. R. Conklin, US Pat. 4,435,548 **1984**] [#161].

'Dense star polymers having core, core branches, terminal groups.' [D. A. Tomalia, J. R. Dewald, U.S. Pat. 4,507,466 **1985**] [#162].

'Dense star polymer.' [D. A. Tomalia, J. Dewald, U.S. Pat. 4,558,120 **1985**] [#165].

'Dense star polymers and dendrimers.' [D. A. Tomalia, J. R. Dewald, U.S. Pat. 4,568,737 **1986**] [#163].

'Dense star polymers having two dimensional molecular diameter.' [D. A. Tomalia, J. R. Dewald, U.S. Pat. 4,587,329 **1986**] [#164].

'Hydrolytically-stable dense star polyamine.' [D. A. Tomalia, J. R. Dewald, U.S. Pat. 4,631,337 **1986**] [#166].

'Rod-shaped dendrimers.' [D. A. Tomalia, P. M. Kirchhoff, Eur. Pat. 234,408 **1987** [#571]; [U. S. Pat. 4,694,064 **1987**] [#168].

'Dense star polymers for calibrating/characterizing sub-micron apertures.' [D. A. Tomalia, L. R. Wilson, Eur. Pat. 247,629 **1987**] [#167].

'Bridged dense star polymers.' [D. A. Tomalia, U.S. Pat. 4,737,550 **1988**] [#169].

'Manufacture of dense amine dendrimer star polymers.' [#. A. Tomalia, J. R. Stahlbush, U.S. Pat. 4,857,599 **1989**] [#550].

'Hyperbranched polyarylene.' [Y. H. Kim, U.S. Pat. 4,857,630 **1989**] [#733].

'Preparation of siloxane dendrimers.' [Jpn. Kokai Tokkyo Koho 03263431, **1991**; *Chem. Abstr.* **116:** 236 379] [#1875].

'Dendritic molecules and method of production.' [J. M. J. Fréchet, C. J. Hawker, A. E. Philippides, U.S. Pat. 5,041,516 **1991**] [#185].

'Noncrosslinked polybranched polymers.' [D. M. Hedstrand, D. A. Tomalia, Eur. Pat. 473,088 **1992**] [#471].

'Unimolecular micelles and their preparation.' [G. R. Newkome, C. N. Moorefield, PCT Int. Appl. 9214543 **1992**] [#471].

'Unimolecular micelles and method of making the same.' [G. R. Newkome, C. N. Moorefield, U.S. Pat. 5,154,853 **1992**] [#778].

'Preparation of phenolic tertiary amide antioxidants and compositions containing them.' [V. J. Gatto, PCT Int. Appl. 9216495 **1992**] [#724].

'Star or dendrimer block copolymers based on acrylate and methacrylate units.' [N. Niessner, F. Seitz, Eur. Pat. 545,184, **1993**] [#376].

'Dendritic aromatic polyamides.' [Y. H. Kim, PCT Int. Appl. 9309162 **1993**] [#402].

'Dendritic macromolecule and the preparation thereof.' [E. M. M. de Brabander-van den Berg, E. W. Meijer, F. H. A. M. J. Vandenbooren, H. J. M. Bosman, PCT Int. Appl. 9314147 **1993** [#396].

'Hyperbranched polyester and a process for its manufacture.' [G. Hardeman, T. A. Misev, A. Heyenk, PCT Int. Appl. 9318079 **1993**] [#398].

'Dendritic based macromolecules and method of production.' [J. M. J. Fréchet, C. J. Hawker, K. L. Wooley, PCT Int. Appl. 9321259 **1993**] [#393].

'Polyester-type dendritic macromolecules and their manufacture and use.' [A. Hult, E. Malmström, M. Johansson, K. Sörensen, Swed. Pat. 468,771 **1993**] [#418].

'Preparation of multipule antigen peptide systems for vaccines and diagnostics.' [J. P. Tam, U.S. Pat. 5,229,490 **1993**] [#358].

'Polyesters having predeterminated monomeric sequence.' [M. E. Hermes, R. R. Muth, B. Huang, Eur. Pat. 579,173 **1994**] [#333].

'Metallospheres and Superclusters.' [G. R. Newkome, C. N. Moorefield, U. S. Pat 5,376,690 **1994**] [#2259].

'Starburst conjugates.' [D. A. Tomalia, D. A. Kaplan, W. Kruper, Jr., R. C. Cheng, I. A. Tomlinson, M. J. Fazio, D. M. Hedstrand, L. R. Wilson, U.S. Pat. 5,338,532 **1994**] [#1384].

'Metallospheres and Superclusters.' [G. R. Newkome, C. N. Moorefield, U.S. Pat. 5,422,379 **1995**] [#2260].

'Metallospheres and Superclusters.' [G. R. Newkome, C. N. Moorefield, U.S. Pat. 5,516,810 **1996**] [#2491].

11 Appendices

11.1 Key Reviews and Highlights

11.1.1 1996 Reviews

"Dendrimere – von der Ästhetik zur Anwendung?" [H. Frey, K. Lorenz, C. Lach, *Chem. Unserer Zeit* **1996**, *30*, 75].

"Conjugated Macromolecules of Precise Length and Constitution. Organic Synthesis for the Construction of Nanoarchitectures" [J. M. Tour, *Chem. Rev.* **1996**, *96*, 537–553].

11.1.2 1995 Reviews

"Dendritic polymers obtained by a single-stage synthesis" [M. N. Bochkarev, M. A. Katkova, *Russ. Chem. Rev.* **1995**, *64*, 1035–1048].

"Interlocking and Intertwined Structures and Superstructures" [D. B. Amabilino, J. F. Stoddart, *Chem. Rev.* **1995**, *95*, 2725–2828].

"Self-Assembling Supramolecular Complexes" [D. S. Lawrence, T. Jiang, M. Levett, *Chem. Rev.* **1995**, *95*, 2229–2260].

'The Saga of Spherical Dendrimers: On the Scent of Cyclophosphazenic Dandelions' [J.-F. Labarre, F. Sournies, F. Crasnier, M.-C. Labarre, C. Videl, J.-P. Faucher, M. Graffeuil, *Main Group Chem. News* **1995**, *3*, 4–8].

'Main Group Element-Based Dendrimers' [A.-M. Caminade, J.-P. Majoral, *Main Group Chem. News* **1995**, *3*, 14–24].

'Dendrimers: Dream Molecules Approach Real Applications' [R. F. Service, *Science* **1995**, *267*, 458–459].

'Dendrimers Engineered to Provide New Catalysts' [J. Haggin, Chem. Eng. News **1995**, (Feb. 6), 26].

'Dendritic polymers: from aesthetic macromolecules to commercially interesting materials' [B. I. Voit, *Acta Polymer.* **1995**, *46*(2), 87–99].

'Chemistry of dendrimer dendron' [M. Kakimoto, *Kagaku (Kyoto)* **1995**, *50*(3), 192–193].

'New materials based on highly-functionalised tetrathiafulvalene derivatives' [M. R. Bryce, A. S. Batsanov, W. Devonport, J. N. Heaton, J. A. K. Howard, G. J. Marshallsay, A. J. Moore, P. J. Skabara, S. Wegener, in *Molecular Engineering for Advanced Materials*, (Eds.: J. Becher, K. Schaumburg) Kluwer. The Netherlands, **1995**, pp. 235–250].

'"Smart" cascade macromolecules. From arborols to unimolecular micelles and beyond' [G. R. Newkome, G. R. Baker, in *Molecular Engineering for Advanced Materials*, Eds.: J. Becher, K. Schaumburg, Kluwer, The Netherlands, **1995**, pp. 59–75].

'Supramolecular organic and inorganic photochemistry: radical pair recombination in micelles, electron transfer on starburst dendrimers, and the use of DNA as a molecular wire' [N. J. Turro, *Pure Appl. Chem.* **1995**, *67*(1), 199–208].

'Dendrimer molecules' [D. A. Tomalia, *Sci. Am.* **1995**, *272*(5), 62–66; *Spekt. Wiss.* **1995** (Sept.), 42–46].

'Noncovalent Synthesis: Using Physical Organic Chemistry to Make Aggregates' [G. M. Whitesides, E. E. Simanek, J. P. Mathias, C. T. Seto, D. N. Chin, M. Mammen, D. M. Gordon, *Acc. Chem. Res.* **1995**, *28*, 37.

11.1.3 1994 Reviews

'Dendrimers: From Generations and Functional Groups to Functions' [J. Issberner, R. Moors, F. Vögtle, *Angew. Chem.* **1994**, *106*, 2507; *Angew. Chem. Int. Ed. Engl.* **1994**, *33*, 2413].

'Functional Polymers and Dendrimers: Reactivity, Molecular Architecture, and Interfacial Energy' [J. M. J. Fréchet, *Science* **1994**, *263*, 1710–1715].

'A Family Tree for Polymers' [P. R. Dvornic, D. A. Tomalia, *Chem. Brit.* **1994** (Aug.), 641].

'Exploitation of the Hydrogen Bond: Recent Developments in the Context of Crystal Engineering' [S. Subramanian, M. J. Zaworotko, *Coord. Chem. Rev.* **1994**, *137*, 57].

'The Convergent Route to Globular Dendritic Macromolecules: A Versatile Approach to Precisely Functionalized Three-Dimensional Polymers and Novel Block Copolymers' [J. M. J. Fréchet, C. J. Hawker, K. L. Wooley, *J. M. S.-Pure Appl. Chem.* **1994**, *A31*, 1627].

'Organic Diradicals and Polyradicals: From Spin Coupling to Magnetism?' [A. Rajca, *Chem. Rev.* **1994**, *94*, 871].

'Solid-State Structures of Hydrogen-Bond Tapes Based on Cyclic Secondary Diamides' [J. C. MacDonald, G. M. Whitesides, *Chem. Rev.* **1994**, *94*, 2383].

'Metallo- and Metalloido-Micellane Derivatives: Incorporation of Metals and Nonmetals within Unimolecular Superstructure' [G. R. Newkome, C. N. Moorefield, *Macromol. Symp.* **1994**, *77*, 63–72].

'Synthesis and properties of aramid dedrimers' [S. C. E. Backson, P. M. Bayliff, W. J. Feast, A. M. Kenwright, D. Parker, R. W. Richards, *Macromol. Symp.* **1994**, *77*, 1–10].

'Novel macromolecular architectures: Globular block copolymers containing dendritic components' [C. J. Hawker, K. L. Wooley, J. M. J. Fréchet, *Macromol. Symp.* **1994**, *77*, 11–20].

'Highly branched aromatic polymers prepared by single step syntheses' [Y. H. Kim, *Macromol. Symp.* **1994**, *77*, 21–33].

'Dendritic analogues of engineering plastics – A general one-step synthesis of dendritic polyaryl ethers' [T. M. Miller, T. X. Neenan, E. W. Kwock, S. M. Stein, *Macromol. Symp.* **1994**, *77*, 35–49].

'Large-scale production of polypropylenimine dendrimers' [E. M. M. de Brabander-van den Berg, A. Nijenhuis, M. Mure, J. Keulen, R. Reintjens, F. Vandenbooren, B. Bosman, R. de Raat, T. Frijns, S. v. d. Wal, M. Castelijns, J. Put, E. W. Meijer, *Macromol. Symp.* **1994**, *77*, 51–62].

'Polyynes – Fascinating Monomers for the Construction of Carbon Networks' [U. H. F. Bunz, *Angew. Chem.* **1994**, *106*, 1127; *Angew. Chem., Int. Ed. Engl.* **1994**, *33*, 1073].

'Bottom-Up Strategy to Obtain Luminescent and Redox-Active Metal Complexes of Nanometric Dimensions' [V. Balzani, S. Campagna, G. Denti, A. Juris, S. Serroni, M. Venturi, *Coord. Chem. Rev.* **1994**, *132*, 1].

'Biomolecular Materials' [J. G. Tirrell, M. J. Fournier, T. L. Mason, D. A. Tirrell, *Chem. Eng. News* **1994** (Dec. 19), 40].

'Heavy-Metal Organic Chemistry: Building with Tin' [L. R. Sita, *Acc. Chem. Res.* **1994**, *27*, 191].

'Starburst cascade dendrimers: fundamental building-blocks for a new nanoscopic chemistry set' [D. A. Tomalia, *Adv. Mat. (Weinheim, Ger.)* **1994**, *6*(7–8), 529–539].

'Designer magnets' [J. S. Miller, A. J. Epstein, *Angew. Chem.* **1994**, *106*, 399].

'Sticky solutions' [Y. Ichikawa, R. L. Halcomb, C. H. Wong, *Chem. Brit.* **1994**, *30*(2), 117–121].

'Dendrimers' [I. Stibor, V. Lellek, *Chem. Listy* **1994**, *88*(7), 423–450].

'Functionalized polymers with comb-like, rotaxanic-, and dendrimeric structures' [H. Ritter, *GIT Fachz. Lab.* **1994**, *38*(6), 615–619].

'Molecular architecture of polymers' [O. Vogl, J. Bartus, M. F. Qin, P. Zarras, *J. Macromol. Sci., Pure & Appl. Chem.* **1994**, *A31*(10), 1329–1353].

'Molecular recognition directed self-assembly of supramolecular architectures' [V. Percec, J. Heck, G. Johansson, D. Tomazos, M. Kawasumi, P. Chu, G. Ungar, *J. Macromol. Sci., Pure Appl. Chem.* **1994**, *A31*(11), 1719–1758].

'Cascade polymers as drug carriers' [K. Akiyoshi, *Kagaku (Kyoto)* **1994**, *49*(6), 422].

'Chemistry of unimolecular micelle "micellane"' [T. Seki, *Kagaku (Kyoto)* **1994**, *49*(8), 586].

'Starburst dendrimers: a conceptual approach to nanoscopic chemistry and architecture' [P. R. Dvornic, D. A. Tomalia, *Macromol. Symp.* **1994**, *88*, 123–148].

'Boron neutron capture therapy of primary and metastatic brain tumors' [R. F. Barth, A. H. Soloway, *Mol. Chem. Neuropathol* **1994**, *21*(2–3), 139–154].

'Prospects of using star polymers and dendrimers in drug delivery and other pharmaceutical application' [N. A. Peppas, T. Nagai, M. Miyajima, *Pharm. Tech. Jpn.* **1994**, *10*(6), 611–617].

'Synthetic uniform polymers and their use in polymer science' [K. Hatada, K. Ute, N. Miyatake, *Progr. Polym. Sci* **1994**, *19*(6), 1067–1082].

'Application areas for dendritic macromolecules' [Anonymous, *Res. Discl.* **1994**, *365*, 492].

'Starburst polymers' [M. Kakimoto, A. Morikawa, *Shinsozai* **1994**, *5*(8), 75–81].

11.1.4 1993 Reviews

'Chemistry of Dendritic Molecules Holds Growing Allure for Researchers' [R. Dagani, *Chem. Eng. News* **1993** (Feb. 1), 28].

'Synthesis of Structure-Controlled Macromolecule' [J. S. Moore, *Polym. News* **1993**, *18*, 5].

'Starburst™/Cascade Dendrimers: Fundamental Building Blocks for a New Nanoscopic Chemistry Set' [D. A. Tomalia, *Aldrichim. Chem.* **1993**, *26*(4), 91].

'Genealogically Directed Synthesis: Starburst/Cascade Dendrimers and Hyperbranched Structures' [D. A. Tomalia, H. D. Durst, in *Top. Curr. Chem.*, Vol. 165, Springer, Berlin, **1993**, p. 193].

'Directionally Aligned Helical Peptides on Surfaces' [J. K. Whitesell, H. K. Chang, *Science*, **1993**, *261*, 73].

'Dendrimeric silanes' [A. W. van der Made, P. W. N. M. van Leeuwen, J. C. de Wilde, R. A. C. Brandes, *Adv. Mat. (Weinheim, Ger.)* **1993**, *5*(6), 466–468].

'A new category of dendritic synthetic polymers' [Y. Li, *Gaofenzi Tongbao* **1993**, (3), 155–164].

'Dendrimeric hyperbranched alkylaromatic polyradicals with mesoscopic dimensions and high-spin ground states' [N. Ventosa, D. Ruiz, C. Rovira, J. Veciana, *Mol. Cryst. Liq. Cryst. Sci. Technol., Sect. A* **1993**, *232*, 333–342].

'Silicon and starburst dendrimers' [Y. Yamamoto, *Organometal. News* **1993**, (2), 40–42].

'Dendrimers and star polymers for pharmaceutical and medical application' [N. A. Peppas, A. B. Argade, in *Proceedings of the International Symposium on Controlled Release of Bioactive Materials*' [T. J. Roseman, N. A. Peppas, H. L. Gabelnick, Controlled Release Society, Deerfield, Fl., USA, **1993**].

'Iptycenes, cuppedophanes and cappedophanes' [H. Hart, *Pure, Appl. Chem.* **1993**, *65*, 27–34].

'Photochemical conversion in organized media' [D. Yogev, D. Meisel, in *Photochemistry, Conversion and Storage of Solar Energy*, **1993**].

11.1.5 1992 Reviews

'Building Blocks for Dendritic Macromolecules' [G. R. Newkome, C. N. Moorefield, G. R. Baker, *Aldrichim. Acta* **1992**, *25*(2), 31–38].

'Dendrimers, Arborols, and Cascade Molecules: Breakthrough into Generations of New Molecules' [H.-B. Mekelburger, W. Jaworek, F. Vögtle, *Angew. Chem.* **1992**, *104*, 1609; *Angew. Chem., Int. Ed. Engl.* **1992**, *31*, 1571].

'Cascade Polymers' [R. Engel, *Polym. News* **1992**, *17*, 301].

'Starburst dendrimers: control of size, shape, surface chemistry, topology and flexibility from atoms to macroscopic matter' [D. A. Tomalia, D. M. Hedstrand, *Actualite Chim.* **1992**(5), 347–349].

'Highly branched polymer' [Y. H. Kim, *Adv. Mat. (Weinheim, Ger.)* **1992**, *4*(11), 764–766].

'Dendritic macromolecules' [C. J. Hawker, K. L. Wooley, J. M. J. Fréchet, *Chem. J. Austr.* **1992**, *59*(12), 620–622].

'Highlights. A selection of recent topics from the chemical literature' [G. R. Stephenson, A. B. Holmes, K. R. Burgess, *Chem. Ind. (London)* **1992**, 698–700].

'The Ins and Outs of Macromolecules' [G. R. Newkome, in *Crown Compounds: Toward Future Applications;* S. R. Cooper; VCH, New York, **1992**, Chapt. 3].

'Starburst polymers (dendrimers')' [A. Morikawa, M. Kakimoto, Y. Imai, *Nippon Gomu Kyokaishi* **1992**, *65*(4), 205–212].

11.1.6   1991 Reviews

'Starburst Dendrimers and Arborols' [K. Krohn, *Org. Syn. Highlights*; (Ed.: J. Mulzer), VCH, New York, **1991**, 378].

'Starburst Polymers' [*Mechanical Engineering,* **1991** (Aug.), p.60].

'Self-Assembly in Organic Synthesis' [D. Philp, J. F. Stoddart, *Synlett* **1991**, 445].

'The Use of π-Organoiron Sandwiches in Aromatic Chemistry' [D. Astruc, *Top. Curr. Chem.* **1991**, *160*, 47].

'Molecular Recognition and Chemistry in Restricted Reaction Spaces. Photophysics and Photoinduced Electron Transfer on the Surfaces of Micelles, Dendrimers and DNA' [N. J. Turro, J. K. Barton, D. A. Tomalia, *Acc. Chem. Res.* **1991**, *24*, 332].

'Self-Assembly in Synthetic Routes to Molecular Devices. Biological Principles and Chemical Perspectives: A Review' [J. S. Lindsey, *New J. Chem.* **1991**, *15*, 153].

'Amorphous molecular materials: synthesis and properties of a novel starburst molecule, 4,4′,4″-tri(N-phenothiazinyl)triphenylamine' [A. Higuchi, H. Inada, T. Kobata, Y. Shirota, *Adv. Mat. (Weinheim, Ger.)* **1991**, *3*(11), 549–550].

'Synthesis of new functional silicon-based condensation polymers' [Y. Imai, *J. Macromol. Sci., Chem.* **1991**, *A28*(11,12), 1115–1135].

'Synthesis of new silicon-based condensation polymers' [Y. Imai, *Kagaku (Kyoto)* **1991**, *46*(4), 280–281].

'Synthesis of branched and network state polymers. End-reactive polymer' [Y. Tezuka, K. Imai, *Kobunshi* **1991**, *40*(5), 314–317].

'Stable polyradicals with high spin ground states' [J. Veciana, C. Rovira, in *Magnetic Molecular Materials*, D. Gatteschi et al., Kluwer, The Netherlands **1991**.

'Meet the molecular superstars' [D. Tomalia, *New Scientist* **1991**, *132*, 30–34].

'"Starburst dendrimers" and "arborols"' [K. Krohn, in *Organic Synthesis Highlights*, (Eds.: H. J. Altenbach, M. Braun, K. Krohn, H.-U. Reissig, VCH, Weinheim, **1991**].

'Organo-iron complexes in aromatic synthesis' [D. Astruc, M. H. Desbois, B. Gloaguen, F. Moulines, J. R. Hamon, in *Organic Synthesis via Organometallics,* (Eds.: K. H. Dötz, R. W. Hoffmann), Vieweg, Braunschweig, **1991**].

'Towards an artificial photosynthesis. Di-, tri-, tetra-, and hepta-nuclear luminescent and redox-active metal complexes' [G. Denti, S. Campagna, L. Sabatino, S. Serroni, M. Ciano, V. Balzani, in *Photochemistry, Conversion and Storage of Solar Energy*, (Eds.: E. Pelizzetti; M. Schiavello), Kluwer, Dordrecht, The Netherlands, **1991**].

'Rising chemical "Stars" could play many roles' [J. Alper, *Science* **1991**, *251*, 1562–1564].

'Unimolecular micelles' [G. R. Newkome, In *Supramolecular Chemistry*, Eds.: V. Balzani, L. De Cola), Kluwer, Dordrecht, The Netherlands, **1991**].

'Cascade Polymers. Soft soap' [Anonymous, *The Economist* **1991**, *321*, 93].

'Spatially growing polyorganosiloxanes. Possibilities of molecular construction in highly functional systems' [A. M. Muzafarov, E. A. Rebrov, V. S. Papkov, *Uspekhii Khim.* **1991**, *60*(7), 1596–1612].

11.1.7   1990 Reviews

'Trekking in the Molecular Forest' [I. Amoto, *Science News* **1990**, *138*, 298].

'Starburst Dendrimers: Molecular-Level Control of Size, Shape, Surface Chemistry, Topology, and Flexibility from Atoms to Macromolecules' [D. A. Tomalia, A. M. Naylor, W. A. Goddard III, *Angew. Chem., Int. Ed. Engl.* **1990**, *29*, 138; *Angew. Chem.* **1990**, *102*, 119].

'Cascade–A New Family of Multibranched Macromolecules' [Y.-X. Chen, *Youji Huaxue* **1990**, *10*, 289].

'Perspectives in Supramolecular Chemistry–From Molecular Recognition towards Molecular Information Processing and Self-Organization' [J.-M. Lehn, *Angew. Chem.* **1990**, *102*, 147; *Angew. Chem., Int. Ed. Engl.* **1990**, *29*, 1304].

'Dendritic Polymers' [D. A. Tomalia, D. M. Hedstrand, L. R. Wilson, *Encyclopedia of Polymer Science and Engineering*, Index Volume, 2nd Ed., Wiley, New York, **1990**, p. 46].

'Designer Solids and Surfaces' [T. E. Mallouk, H. Lee, *Symp. on Molecular Architecture* **1990**, *67*(10), 829].

'Design and Application of Lipid Microstructures' [B. P. Gaber, *Navel Research News* **1990**, *42*(1), 2].

'Recent advances in topologically well-defined polymers' [M. Sawamoto, *Kagaku (Kyoto)* **1990**, *45*(8), 537–539].

11.1.8   1989 Reviews

'Structure-Directed Synthesis of New Organic Materials' [F. H. Kohnke, J. P. Mattias, J. F. Stoddart, *Angew. Chem., Int. Ed. Engl. Adv. Mater.* **1989**, *28*, 1103].

'Application of simulation and theory to biocatalysis and biomimetics' [A. M. Naylor, W. A. Goddard III, in *Biocatalysis and Biomimetics*, (Eds.: J. D. Burrington, D. S. Clark), American Chemical Society, Washington, D.C., **1989**].

'Synthesis of polymers for new materials' [G. Smets, *Chemie Magazine (Belgium)* **1989**, (Sept.), 481–483, 485].

'The ability of macromonomers to copolymerize: a critical review with new developments' [Y. Gnanou, P. Lutz, *Makromol. Chem.* **1989,** *190*, 577–588].

'Syntheses of cascade molecules' [S. Arai, *J. Syn. Org. Chem., Jpn.* **1989**, *47*(1), 62–68].

11.1.9   1988 Reviews

'New Families of Multibranched Macromolecules Synthesized' [W. Worthy, *Chem. Eng. News* **1988** (Feb. 22), 19].

'Molecular Architecture and Function of Polymeric Oriented Systems: Models for the Study of Organization, Surface Recognition, and Dynamics of Biomembranes' [H. Ringsdorf, B. Schlarb, J. Venzmer, *Angew. Chem.* **1988**, *100*, 117; *Angew. Chem., Int. Ed. Engl.* **1988**, *27*, 113].

'Supramolecular Chemistry–Scope and Perspectives; Molecules supermolecules, and Molecular devices' [J.-M. Lehn, *Angew. Chem.* **1988**, *100*, 91; *Angew. Chem., Int. Ed. Engl.* **1988**, *27*, 90].

'Surfactants, Micelles, and Fascinating Phenomena' [H. Hoffmann, G. Ebert, *Angew. Chem.* **1988**, *100*, 933; *Angew. Chem., Int. Ed. Engl.* **1988**, *27*, 902].

11.1.10   1987 Review

'Cascade Syntheses' [S. Shinkai, *Kagaku (Kyoto)* **1987**, *42*, 74–75].

'Bürgenstock **1987**' [D. Spitzner, *Nachr. Chem. Tech. Lab.* **1987**, *35*(7), 693–696].

'Starburst dendrimers and arborols' [K. Krohn, *Nachr. Chem. Tech. Lab.* **1987**, *35*(12), 1252–1255].

'Towards a supramolecular photochemistry: assembly of molecular components to obtain photochemical molecular devices' [V. Balzani, L. Moggi, F. Scandola, in *Supramolecular Photochemistry*, V. Balzani, Reidel, Dordrecht, The Netherlands, **1987**].

11.1.11   1986 and Earlier Reviews

'C. Tanford, *The Hydrophobic Effect: Formation of Micelles and Biological Membranes*, 2nd ed., Wiley, New York, **1980**].

'Micellization, Solubilization, and Microemulsions' [Proceedings from the International Symposium on Micellization, Solubilization and Microemulsions; 7th Northeast Regional Meeting of the American Chemical Society, Albany, N.Y., Aug. 8–11, **1976**); K. L. Mittal, P. Mukerjee, Vol. 1, (Ed.: K. L. Mittal), Plenum, New York, **1977**].

'New type of oligomers–starburst and calixarene oligomers' [T. Otsu, T. Matsunaga, *Kagaku (Kyoto)* **1986**, *41*(3), 206–207].

'Chemistry of multi-armed organic compounds' [F. M. Menger, *Top. Curr. Chem.* **1986**, *36*, 1–15].

11.2   Advances Series

"Advances in Dendritic Macromolecules" Series (Ed.: G. R. Newkome), JAI Press, Greenwich, Connecticut (USA).

Volume 1 (1994)

"A Review of Dendritic Macromolecules' [C. N. Moorefield, G. R. Newkome, Chapt. 1, pp. 1–67].

'Stiff Dendritic Macromolecules Based on Phenylacetylene' [Z. Xu, B. Kyan, J. S. Moore, Chapt. 2, pp. 69–104].

'Preparation and Properties of Monodisperse Aromatic Dendritic Macromolecules' [T. X. Neenan, T. M. Miller, E. W. Kwock, H. E. Blair, Chapt. 3, pp. 105–132].

'High-spin Polyarylmethyl Polyradicals' [A. Rajca, Chapt. 4, pp. 133–168].

'A Systematic Nomenclature for Cascade Dendritic Polymers' [G. R. Baker, J. K. Young, Chapt. 5, pp. 169–186].

Volume 2 (1995)

'The Convergent Growth Approach to Dendritic Macromolecules' [C. J. Hawker, K. L. Wooley, Chapt. 1, pp. 1–39].

'Cascade Molecules: Building Blocks, Multiple Functionalization, Complexing Units, Photoswitching' [R. Moors, F. Vögtle, Chapt. 2, pp.41–71].

'Ionic Dendrimers and Related Materials' [R. Engel, Chapt. 3, pp. 73–99].

'Silicon-based Stars, Dendrimers, and Hyperbranched Polymers' [L. J. Mathias, T. W. Carothers, Chapt. 4, pp. 101–121].

'Highly Branched Stable Aromatic Polymers: Synthesis and Applications' [Y. Kim, Chapt. 5, pp. 123–156].

'Dendritic Bolaamphiphiles and Related Molecules' [G. H. Escamilla, Chapt. 6, pp. 157–190].

Volume 3 (1997)

"Dendrimers Based on Metal Comlexes" [S. Serroni, S. Campagna, G. Denti, A. Juris, M. Venturi, V. Balzani, Chapt. 1].

"Redox-Active Dendrimers, Related Building Blocks, and Oligomers" [M. R. Bryce, W. Davonport, Chapt. 2].

"Dendrimers and Hyperbranched Aliphatic Polyesters Based on 2,2-Bis(hydroxyme-thyl)propionic Acid (bis-MPA)" [H. Ihre, M. Johansson, E. Malmström, A. Hult, Chapt. 3].

"Organometallic Dendritic Macromolecules: Organosilicon and Organometallic Entities as Cores or Building Blocks" [I. Cuadrado, M. Morán, J. Losada, C. M. Casado, C. Pascual, B. Alonso, F. Lobete, Chapt. 4].

"Consequences of the Fractal Character of Dendritic High-Spin Molecules on Some of their Physicochemical Properties" [N. Ventosa, D. Ruiz, C. Rovira, J. Veciana, Chapt. 5].

"Poly(propyleneimine) Dendrimers" [E. W. Meijer et al., Chapt. 6].

SEARCH Series, Volume 1. "Mesomolecules, From Molecules to Materials", (Eds.: G. D. Mendenhall, A. Greenberg, J. F. Liebman), Chapman & Hall, New York, **1995**.

'From Molecules to Materials' [G. D. Mendenhall, Chapt. 1, pp. 1–26].

'Cascade Molecules' [G. R. Newkome, C. N. Moorefield, Chapt. 2, pp. 27–68].

'Dendritic Polynuclear Metal Complexes with Made-to-Order Luminescent and Redox Properties' [G. Denti, S. Campagna, V. Balzani, Chapt. 3, pp. 69–106].

'Molecular Tectonics' [J. D. Wuest, Chapt. 4, pp. 107–131].

'Supramolecular Assemblies from "Tinkertoy" Rigid-Rod Molecules' [J. Michl, Chapt. 5, pp. 132–160].

'Graphite: Flat, Fibrous, and Spherical' [J. A. Jaszczak, Chapt. 6, pp. 161–180].

'Fractal Index and Fractal Nomenclature' [G. D. Mendenhall, pp. 181–194].

11.3   Glossary of Terms

amu: atomic mass unit
arbor [(arbour Brit.)]: (Latin) tree
bishomotris: 4-amino-4-[1-(3-hydroxypropyl)]-1,7-heptanediol
BOC: *tert*-butoxycarbonyl
bpy: 2,2'-bipyridine
cascade synthesis: reaction sequences conducted in a repetitive manner
convergent synthesis: construction of the (macro)molecule from the outside in
CV: cyclic voltammetry
dba: dibenzylideneacetone [e.g., Pd(dba)]
DCC: dicyclohexylcarbodiimide
DDQ: dichlorodicyanoquinone
DEKC: dendrimer electrokinetic chromatography
dendro-: (Greek) tree-like
dendron: dendritic building block
Divergent synthesis: construction of the (macro)molecule from the inside out.
DMAc: *N,N*-dimethylacetamide
DMF: *N,N*-dimethylformamide
DMAP: *N,N*-dimethylaminopyridine
DMSO: dimethylsulfoxide
DOSY NMR: 2-dimensional diffusion ordered spectroscopy nuclear magnetic resonance (spectroscopy)
dpb: bis(dibenzylideneacetone)
dppe: 1,2-bis(diphenylphosphino)ethane
DSC: differential scanning calorimetry
DV: differential viscometry

ES-MS: electrospray ionization mass spectroscopy

EPR: electron paramagnetic resonance

FAB MS: fast atom bombardment mass spectrometry focal point: site of connectivity on convergently prepared building block or dendron

GALA: an amphipathic peptide

GPC: gas phase chromatography

1-HBT: 1-hydroxybenzotriazole

HPLC: high performance liquid chromatography

hypercores: large dendrons or wedges

Iteration (or iterative procedure): repeating a synthetic step with the product of the previous step

IUPAC: International Union of Pure & Applied Chemistry

LALLS: low-angle laser light scattering

LCAA-CPG: long chain alkylamine-controlled pore glass

MALDI-MS: matrix-assisted laser desorption ionization mass spectrometry

MALDI-TOF: matrix-assisted laser desorption ionization time-of-flight

MAP: multiple antigen peptide

MECC: micellar electrokinetic capillary chromatography

MEM: 2-methoxyethoxymethyl (group)

Micellane™ series: registered trademark (University of South Florida) descriptor of the all-hydrocarbon based dendrimers

MRI: magnetic resonance imaging

NBS: *N*-bromosuccinimide

network: any collection of dendrimers whereby at least one degree of freedom with respect fo macromolecular juxtaposition has been removed

NMP: *N*-methylpyrrolidone

ORD/CD: optical rotatory dispersion/circular dichroism ordered network: controlled, multiple, dendritic positioning

PAMs: phenylacetylene macrocycles

PAMAM: *polyamidoamine*

PCS: photo correlation spectroscopy

PEG: poly(ethylene glycol)

PEO: poly(ethylene oxide)

PE-TMAI: pentaerythritol trimethylammonium iodide

phen: 1,10-phenanthroline

PMMA: poly(methyl methacrylate)

PPI: *polypropylenimine*

random network: uncontrolled, multiple dendritic positioning

SDS: sodium dodecyl sulfate

SEC: size exclusion chromatography

SPEI: "*starburst*" *polyethylenimine*

Starburst® dendrimers: registered trademark (The Dow) descriptor of PAMAM dendrimers

SVPDE/AP: snake venom phosphodiesterase and alkaline phosphatase

t-BOC: *tert*-butoxycarbonyl (group)

TBDMS: *tert*-butyldimethylsilyl (group)

TFA: trifluoroacetic acid

TGA: thermal gravimetric analysis

THF: tetrahydrofuran

TMEDA: Tetramethylethylenediamine

TMS: trimethylsilyl (group)

topological dendritic connectivity: non-covalent, sterically or mechanically induced dendritic attachments such as exhibited by interlocking arms

tpy: 2,2':6',2"-terpyridine

tris: 1,1,1-tris(hydroxymethyl)aminomethane

TTF: tetrathiafulvalene

Index